National Key Book Publishing Planning Project of the 13th Five-Year Plan

"十三五"国家重点图书出版规划项目

International Clinical Medicine Series Based on the Be

"一带一路"背景下国际化临

Learning Guidance of Biochemistry

生物化学学习指导

Chief Editor Li Ling Fei Xiaowen

主编 李 凌 费小雯

图书在版编目(CIP)数据

生物化学学习指导 = Learning Guide of Biochemistry：英文／李凌，费小雯主编. —2版. —郑州：郑州大学出版社，2020. 12
（“一带一路”背景下国际化临床医学丛书）
ISBN 978-7-5645-7304-1

Ⅰ. ①生… Ⅱ. ①李…②费… Ⅲ. ①生物化学-医学院校-教学参考资料-英文 Ⅳ. ①Q5

中国版本图书馆 CIP 数据核字(2020)第182369号

生物化学学习指导 = Learning Guide of Biochemistry：**英文**

项目负责人	孙保营　杨秦予	策划编辑	李龙传
责任编辑	李龙传	装帧设计	苏永生
责任校对	薛　晗	责任监制	凌　青　李瑞卿

出版发行	郑州大学出版社有限公司	地　址	郑州市大学路40号(450052)
出 版 人	孙保营	网　址	http://www.zzup.cn
经　销	全国新华书店	发行电话	0371-66966070
印　刷	新乡市豫北印务有限公司		
开　本	850 mm×1 168 mm　1/16		
印　张	20	字　数	772千字
版　次	2020年12月第1版	印　次	2020年12月第1次印刷
书　号	ISBN 978-7-5645-7304-1	定　价	79.00元

本书如有印装质量问题，请与本社联系调换。

Staff of Expert Steering Committee

Chairmen

Zhong Shizhen Li Sijin Lü Chuanzhu

Vice Chairmen

Bai Yuting Chen Xu Cui Wen Huang Gang Huang Yuanhua
Jiang Zhisheng Li Yumin Liu zhangsuo Luo Baojun Lü Yi
Tang Yingjie Tang Shiying

Committee Member

An Dongping Bai Xiaochun Cao Shanying Chen jun Chen Yijiu
Chen Zhesheng Chen Zhihong Chen Zhiqiao Ding Yueming Du hua
Duan Zhongping Guan Chengnong Huang Xufeng Jian Jie Jiang Yaochuan
Jiao xiaomin Ji Ling Li Cairui Li Guoxin Li Guoming
Li Jiabing Li Zhijie Liu Hongmin Liu Huifan Liu Kangdong
Song Weiqun Tang Chunzhi Wang Yuan Wang Huamin Wang Huixin
Wang Jiahong Wang Jiangang Wang Wenjun Wei Jia Wen Xiaojun
Wu Jun Wu Weidong Wu Xuedong Xie Xieju Xue Qing
Yan Wenhai Yan Xinming Yang Donghua Yu Feng Yu Xiyong
Zhang Mao Zhang Ming Zhang Lirong Zhang Yu'an Zhang Junjian
Zhao Song Zhao Yumin Zheng Weiyang Zhu Lin

专家指导委员会

主任委员

钟世镇　李思进　吕传柱

副主任委员（以姓氏汉语拼音排序）

白育庭　陈　旭　崔　文　黄　钢　黄元华　姜志胜

李玉民　刘章锁　雒保军　吕　毅　唐世英

委　　员（以姓氏汉语拼音排序）

安东平　白晓春　曹山鹰　陈　君　陈忆九　陈哲生

陈志宏　陈志桥　丁跃明　杜　华　段钟平　官成浓

黄旭枫　简　洁　蒋尧传　焦小民　李　玲　李才锐

李国新　李果明　李家斌　李志杰　刘宏民　刘会范

刘康栋　宋为群　唐纯志　王　渊　王华民　王慧欣

王家宏　王建刚　王文军　韦　嘉　温小军　吴　军

吴卫东　吴学东　谢协驹　薛　青　鄢文海　闫新明

杨冬华　余　峰　余细勇　张　茂　张　明　张莉蓉

张玉安　章军建　赵　松　赵玉敏　郑维扬　朱　林

Staff of Editor Steering Committee

Chairmen

Cao Xuetao　Liang Guiyou　Wu Jiliang

Vice Chairmen

Chen Pingyan　Chen Yuguo　Huang Wenhua　Li Yaming　Wang Heng
Xu Zuojun　Yao Ke　Yao Libo　Yu Xuezhong　Zhao Xiaodong

Committee Member

Cao Hong　Zeng Qinbing　Chen Guangjie　Chen Kuisheng　Chen Xiaolan
Dong Hongmei　DuJian　Du Ying　Fei Xiaowen　Gao Ning
Gao Jianbo　Guan Ging　Guo Xiuhua　Han Liping　Han Xingmin
He Wei　He fanggang　Huang Yan　Huang Yong　Jiang Haishan
Jin Qing　Jin Chengyun　Li Lin　Li Ling　Li Wei
Li Mincai　Li Youchang　Li Qiuming　Li Xiaodan　Li Youhui
Liang Li　Lin Jun　Liu Fen　Liu Hong　Liu Hui
Lu Jin　Lü Bing　Lü Quanjun　Ma Wang　Ma Qingyong
Mei Wuxuan　Nie Dongfeng　Peng Biwen　Peng Hongjuan　Qiu Xinguang
Song Chuanjun　Tan Dongfeng　Tu Jiancheng　Wang Ling　Wang Peng
Wang Huijun　Wang Rongfu　Wang Shusen　Wang Chongjian　Xia Chaoming
Xiao Zheman　Xie Xiaodong　Xu Xia　Xu Falin　Xu Jitian
Yang Aimin　Xue Fuzhong　Yang Xuesong　Yi Lan　Yi Kai
Yu Zujiang　Yu Hong　Yue Baohong　Zhang Hui　Zhang Ling
Zhang Lu　Zhang Yanru　Zhao Dong　Zhao Hongshan　Zhao Wen
Zheng Yanfang　Zhou Huaiyu　Zhu Changju　Zhu Lifang

编审委员会

主任委员

曹雪涛　梁贵友　吴基良

副主任委员（以姓氏汉语拼音排序）

陈平雁　陈玉国　黄文华　李亚明　王　恒

徐作军　姚　克　药立波　于学忠　赵晓东

委　　员（以姓氏汉语拼音排序）

曹　虹　曾庆冰　陈广洁　陈奎生　陈晓岚　董红梅

都　建　杜　英　费晓雯　高　宇　高剑波　关　颖

郭秀花　韩丽萍　韩星敏　何　巍　何方刚　黄　艳

黄　泳　蒋海山　金　清　金成允　金润铭　李　琳

李　凌　李　薇　李敏才　李迺昶　李秋明　李晓丹

李幼辉　梁　莉　林　军　刘　芬　刘　红　刘　晖

路　静　吕　滨　吕全军　马　望　马清涌　梅武轩

聂东风　彭碧文　彭鸿娟　邱新光　宋传君　谈东风

涂建成　汪　琳　王　鹏　王慧君　王荣福　王树森

王重建　夏超明　肖哲曼　谢小冬　徐　霞　徐发林

许继田　薛付忠　杨爱民　杨雪松　易　岚　尹　凯

余祖江　喻　红　岳保红　张　慧　张　琳　张　璐

张雁儒　赵　东　赵红珊　赵　文　郑燕芳　周怀瑜

朱长举　朱荔芳

Editorial Staff

Chief Editors

Li Ling	Southern Medical University
Fei Xiaowen	Hainan Medical College

Vice Chief Editors

Liu Xiaoyu	Naval Medical University
Sun Wei	College of Basic Medical Sciences, Jilin University
Wang Xuejun	Nanjing Medical University
Li Jianning	Ningxia Medical University

Editorial Staffs

Guo Rui	Shanxi Medical University
Lai Mingming	School of Basic Medical Sciences, Dali University
Li Chongqi	Hainan Medical University
Li Dongmin	Xi 'an Jiaotong University Health Science Center
Li Jiao	Tongji University School of Medicine
Li Meining	Shanxi Medical University
Li Zhihong	Medical college of China Three Gorges University
Lin Li	School of Basic Medical Sciences, Lanzhou University
Liu Baoqin	China Medical University
Lu Xiaoling	Naval Medical University
Lü Lixia	Tongji University School of Medicine
Meng Liesu	Xi 'an Jiaotong University Health Science Center
Peng Fan	Medical college of China Three Gorges University
Ren Fangfang	Suzhou University
Su Xiong	Suzhou University
Wang Huaqin	China Medical University
Wang Kai	Hainan Medical College
Wang Lianghua	Naval Medical University
Wang Yanfei	Guangzhou Medical University
Xie Jianjun	Shantou University Medical college
Yang Guang	College of Basic Medical Sciences, Jilin University

Yu Hong	Wuhan University School of Basic Medical Sciences
Yuan Ping	Tongji Medical College, Huazhong University of Science and Technology
Zang Mingxi	School of Basic Medical Sciences, Zhengzhou University
Zhang Baifang	Wuhan University School of Basic Medical Sciences
Zhang Haifeng	School of Basic Medical Sciences, Zhengzhou University
Zhang Weijuan	School of Basic Medical Sciences, Henan University
Zhang Yanling	Suzhou University
Zhang Yuzhe	School of Basic Medical Sciences, Dali University
Zhu Lina	Southern Medical University

Secretary

Zhu Lina	Southern Medical University

作者名单

主　编

　　李　凌　南方医科大学

　　费小雯　海南医学院

副主编

　　刘小宇　海军军医大学

　　孙　巍　吉林大学基础医学院

　　王学军　南京医科大学

　　李建宁　宁夏医科大学

编　委（以姓氏汉语拼音排序）

　　郭　睿　山西医科大学

　　来明名　大理大学基础医学院

　　李　姣　同济大学医学院

　　李崇奇　海南医学院

　　李冬民　西安交通大学医学部

　　李美宁　山西医科大学

　　李志红　三峡大学医学院

　　林　利　兰州大学基础医学院

　　刘宝琴　中国医科大学

　　刘小宇　海军军医大学

　　卢小玲　海军军医大学

　　吕立夏　同济大学医学院

　　孟列素　西安交通大学医学部

　　彭　帆　三峡大学医学院

　　任芳芳　苏州大学

　　苏　雄　苏州大学

　　王　凯　海南医学院

　　王华芹　中国医科大学

　　王梁华　海军军医大学

　　王燕菲　广州医科大学

　　谢剑君　汕头大学医学院

杨　光　吉林大学基础医学院
喻　红　武汉大学基础医学院
袁　萍　华中科技大学同济医学院
臧明玺　郑州大学基础医学院
张百芳　武汉大学基础医学院
张海风　郑州大学基础医学院
张维娟　河南大学基础医学院
张艳岭　苏州大学
张钰哲　大理大学基础医学院
朱利娜　南方医科大学

秘　书

朱利娜

Preface

At the Second Belt and Road Summit Forum on International Cooperation in 2019 and the Seventy-third World Health Assembly in 2020, General Secretary Xi Jinping stated the importancefor promoting the construction of the "Belt and Road" and jointly build a community for human health. Countries and regions alongthe "Belt and Road" have a large number of overseas Chinese communities, and shared close geographic proximity, similarities in culture, disease profiles and medical habits. They also shared a profound mass base with ample space for cooperation and exchange in Clinical Medicine. The publication of the International Clinical Medicine series for clinical researchers, medical teachers and students in countries along the "Belt and Road" is a concrete measure to promote the exchange of Chinese and foreign medical science and technology with mutual appreciation and reciprocity.

Zhengzhou University Press coordinated more than 600 medical experts from over 160 renowned medical research institutes, medical schools and clinical hospitalsacross China. It produced this set of medical tools in English to serve the needs for the construction of the "Belt and Road". It comprehensively coversaspects in the theoretical framework and clinical practicesinClinical Medicine, including basic science, multiple clinical specialities and social medicine. It reflects the latest academic and technological developments, and the international frontiers of academic advancements in Clinical Medicine. It shared with the world China's latest diagnosis and therapeuticapproaches, clinical techniques, and experiences in prescription and medication. It has an important role in disseminating contemporary Chinese medical science and technology innovations, demonstrating the achievements of modern China's economic and social development, and promoting the unique charm of Chinese culture to the world.

The series is the first set of medical tools written in English by Chinese medical experts to serve the needs of the "Belt and Road" construction. It systematically and comprehensivelyreflects the Chinese characteristics in Clinical Medicine. Also, it presents a landmark achievement in the implementation of the "Belt and Road" initiative in promoting exchanges in medical science and technology. This series is theoretical in nature, with each volume built on the mainlines in traditional disciplines but at the same time introducing contemporary theories that guide clinical practices, diagnosis and treatment methods, echoing the latest research findings in Clinical Medicine.

As the disciplines in Clinical Medicine rapidly advances, different views on knowledge, inclusiveness, and medical ethics may arise. We hope thiswork will facilitate the exchange of ideas, build common ground while allowing differences, and contribute to the building of a community for human health in a broad spectrum of disciplines and research focuses.

Nick Lemoine

Foreign Academician of the Chinese Academy of Engineering
Dean, Academy of Medical Sciences of Zhengzhou University Director,
Barts Cancer Institute, London, UK
6th August, 2020

Foreword

Biochemistry is an important basic course for a major in medicine. Moreover, it is an indispensable aspect of the Medical Licensing Examination (MLE) and the postgraduate entrance examination. Biochemistry is regarded as one of the most difficult courses for medical students due to its abstract theory and the complexity of its knowledge systems.

We wrote this learning guide to help medical students more effectively study biochemistry and successfully pass their examinations. We will highlight the following features in the organization of the content.

1. Construct a framework of basic biochemistry knowledge.

In each chapter, we first provided the*Examination syllabus* for MLE, after which we made mind maps as a road map for each section, connected the *Major points* of knowledge in a conceptual framework by using words, figures and tables refined according to the authors' teaching experience. Its objective is to strengthen the internal connection of knowledge, strengthen the key points and simplify the difficult points, while also enabling students to quickly and systematically understand the basic concepts and theories of biochemistry.

2. Strengthen learning effects with practical examples

Based on analyzing the knowledge points in each chapter, we supplied specific multiple-choice questions (A1, A2, B1, X-type) and some clinical case problems for intensive *Practice*, and provided the reference answers for each question and brief explanations for practice.

This book is suitable for undergraduate students in medical schools (especially Bachelor of Medicine & Bachelor of Surgery, MBBS) to review major concepts and information while also preparing for course exams, MLE, and the postgraduate entrance exam. This book can also be used as a teaching reference book for biochemistry teachers as well as a reference book of continuing education for relevant professionals.

The participating universities have greatly supported the compilation of this book. We are very grateful to all experts and all universities.

The book inevitably has omissions and inaccuracies, and we greatly appreciate your feedback so we can improve this guide.

Authors

Brief Introduction

This guide includes 25 chapters divided into four parts: Structure and function of biomolecule (Part Ⅰ), Metabolism and regulation (Part Ⅱ), Information pathways (Part Ⅲ) and Molecular medicine (Part Ⅳ).

There are three portions in each chapter: *Examination syllabus*, *Major points* and *Practices*. Firstly the examination syllabus for MLE (Medical Licensing Examination) is provided. Then the major points are connected in a conceptual framework by making mind maps, and using words, figures and tables refined. Finally, multiple choice questions (A1, A2, B1, X-type) and some clinical case problems are supplied for practice and testing the learning effect.

This book is suitable for undergraduate students in medical schools (especially Bachelor of Medicine & Bachelor of Surgery, MBBS) to review major concepts and information while also preparing for course exams, MLE, and the postgraduate entrance exam. This book can also be used as a teaching reference book for biochemistry teachers as well as a reference book of continuing education for relevant professionals.

CONTENT

Part Ⅰ STRUCTURE AND FUNCTION OF BIOMOLECULE

Chapter 1 Structure and Function of Proteins ········ 1

Ⅰ. Amino acids and Peptides ········ 3

Ⅱ. Protein Structure ········ 6

Ⅲ. Protein Function ········ 8

Ⅳ. Working with proteins ········ 10

Chapter 2 Structure and Function of Nucleic Acids ········ 17

Ⅰ. Nucleic acid ········ 17

Ⅱ. Structure and function of DNA ········ 19

Ⅲ. Structure and function of RNA ········ 22

Ⅳ. Physical and chemical properties of nucleic acids and their applications ········ 25

Chapter 3 Structure and Function of Glycoconjugates ········ 32

Ⅰ. Glycans in glycoproteins ········ 32

Ⅱ. Glycosaminoglycans in proteoglycan molecules ········ 33

Ⅲ. Glycolipids ········ 34

Ⅳ. Functions of Glycan structure ········ 35

Chapter 4 Enzyme ········ 37

Ⅰ. General Properties of Enzymes ········ 37

Ⅱ. Mechanism of Enzyme Action ········ 40

Ⅲ. Enzyme kinetics ········ 41

Ⅳ. Regulation of enzyme ········ 45

Chapter 5 Vitamins ········ 51

Ⅰ. Lipid-soluble vitamins ········ 51

Ⅱ. Water-soluble vitamins ········ 52

Part Ⅱ METABOLISM AND REGULATION

Chapter 6 Carbohydrate metabolism ········ 59

Ⅰ. The anaerobic oxidation of glucose ········ 59

Ⅱ. Aerobic oxidation ········ 62

Ⅲ. Pentose phosphate pathway ········ 65
Ⅳ. Glycogen metabolism ········ 67
Ⅴ. Gluconeogenesis ········ 69
Ⅵ. Blood glucose and the regulation ········ 71
Chapter 7 Biological Oxidation ········ 78
Ⅰ. Electron transport chain ········ 78
Ⅱ. Oxidative phosphorylation and ATP synthesis ········ 80
Ⅲ. Regulation of oxidative phosphorylation ········ 82
Ⅳ. Selective transport across the inner mitochondrial membrane ········ 83
Ⅴ. Other biological oxidations without ATP producing ········ 84
Chapter 8 Metabolism of Lipids ········ 95
Ⅰ. The physiological function of lipids ········ 95
Ⅱ. Digestion and absorption of fats ········ 96
Ⅲ. Anabolism of triacylglycerol ········ 98
Ⅳ. Synthesis of fatty acids ········ 99
Ⅴ. Fat catabolism ········ 100
Ⅵ. Metabolism of glycerophospholipids ········ 102
Ⅶ. Cholesterol metabolism ········ 103
Ⅷ. Plasma lipoprotein metabolism ········ 105
Chapter 9 Metabolism of Amino acid ········ 116
Ⅰ. The physiological function and nutritive value of protein ········ 116
Ⅱ. Digestion, absorption and putrefaction of the protein in the intestinal tract. ········ 117
Ⅲ. General metabolism of amino acids ········ 118
Ⅳ. Metabolism of ammonia ········ 120
Ⅴ. Metabolisms of individual amino acid ········ 123
Chapter 10 Nucleotide Metabolism ········ 130
Ⅰ. Functions of nucleotides ········ 130
Ⅱ. Synthesis and degradation of purine nucleotides ········ 131
Ⅲ. Synthesis and degradation of pyrimidine nucleotides ········ 132
Chapter 11 Blood Biochemistry ········ 139
Ⅰ. Plasma proteins ········ 140
Ⅱ. Synthesis of hemoglobin ········ 141
Ⅲ. Metabolism of red blood cell and leukocyte ········ 142
Chapter 12 Liver Biochemistry ········ 146
Ⅰ. The Role of Liver in Metabolism ········ 146
Ⅱ. Hepatic Biotransformation ········ 148
Ⅲ. Bile and the metabolism of bile acid ········ 150
Ⅳ. Metabolism of bile pigment ········ 152
Chapter 13 Integration and Regulation of Metabolism ········ 158
Ⅰ. The relationralation and integration of metabolism ········ 158
Ⅱ. The major manner of regulation of metabolism ········ 159
Ⅲ. Characteristic of metabolism in different organs ········ 161
Ⅳ. The abnormal metabolism in disease ········ 162

Part Ⅲ INFORMATION PATHWAYS

Chapter 14 DNA Synthesis ······ 169

Ⅰ. Three basic rules of DNA replication ······ 169

Ⅱ. Various enzymes and proteins involved in DNA replication ······ 170

Ⅲ. Prokaryotic DNA replication ······ 172

Ⅳ. Eukaryotic DNA replication ······ 172

Ⅴ. Reverse Transcription and Other Models of DNA Replication ······ 174

Chapter 15 DNA Damage and Repair ······ 181

Ⅰ. DNA damage ······ 181

Ⅱ. DNA repair ······ 183

Ⅲ. Significance of DNA damage and repair ······ 184

Chapter 16 RNA Synthesis ······ 187

Ⅰ. Introductiont of RNA biosynthesis ······ 187

Ⅱ. The process of RNA synthesis process ······ 189

Ⅲ. Post-transcriptional processing ······ 195

Chapter 17 Protein Synthesis ······ 203

Ⅰ. The concept of protein biosynthesis ······ 203

Ⅱ. Protein biosynthesis system and genetic code ······ 203

Ⅲ. The basic process of protein biosynthesis ······ 206

Ⅳ. The relationship between protein biosynthesis and medicine ······ 208

Chapter 18 Regulation of Gene Expression ······ 215

Ⅰ. Gene expression ······ 215

Ⅱ. Mechanism of gene regulation ······ 216

Chapter 19 Signal Transduction ······ 224

Ⅰ. Extracellular signal molecules ······ 226

Ⅱ. Receptors and signal transduction molecules ······ 227

Ⅲ. Membrane receptor-mediated intracellular signal transduction ······ 232

Ⅳ. Intracellular receptor-mediated signaling pathway ······ 236

Ⅴ. Abnormal intracellular signal Transduction and Medicine ······ 237

Part Ⅳ MOLECULAR MEDICINE

Chapter 20 The Basics of Omics ······ 245

Ⅰ. Genomics ······ 245

Ⅱ. Proteomics ······ 247

Ⅲ. Other kinds of Omics ······ 247

Ⅳ. Applications of omics in medicine and health ······ 248

Chapter 21 Gene Manipulation and Omics-based Technologies ······ 252

Ⅰ. Extraction and purification of nucleic acids ······ 252

Ⅱ. PCR technology ······ 256

Ⅲ. Nucleic acid hybridization ······ 258

Ⅳ. DNA sequencing …… 260
Ⅴ. Gene function analysis techniques …… 261
Ⅵ. Functional genomics …… 262
Ⅶ. Techniques to investigate macromolecular interaction …… 263
Ⅷ. OmicsTechniques …… 263
Chapter 22 Recombinant DNA Technology …… 266
Ⅰ. The necessary elements for recombinant DNA technology …… 267
Ⅱ. The process of DNA recombination …… 270
Ⅲ. The application of recombinant DNA technology …… 272
Chapter 23 Oncogenes and tumor suppressor genes …… 274
Ⅰ. Oncogenes …… 274
Ⅱ. Tumor suppressor genes …… 276
Ⅲ. Growth factors and their receptors …… 278
Chapter 24 Gene Diagnosis and Gene therapy …… 281
Ⅰ. Gene diagnosis …… 281
Ⅱ. Gene therapy …… 284
Chapter 25 Genetically Engineered Drugs and Vaccines …… 293
Ⅰ. Protein and Peptide Drugs …… 294
Ⅱ. Genetically Engineered Vaccines …… 296

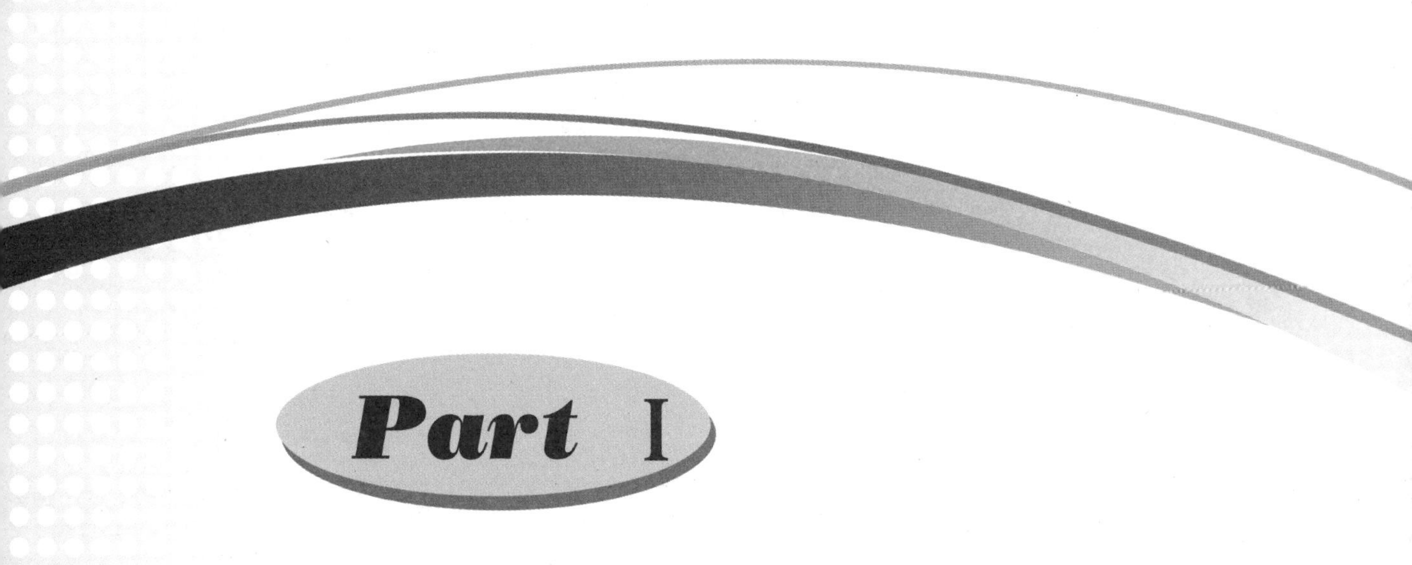

Part I

STRUCTURE AND FUNCTION OF BIOMOLECULE

Chapter 1

Structure and Function of Proteins

Examination Syllabus

1. **Amino acids and Peptide**: general structure, classification of amino acids; peptide bond, peptide chain.

2. **Protein structure**: four levels of protein structures, peptide unit, major types of secondary structures (especially α-helix and β-sheet), subunit, stabilizing forces.

3. **Structure-function relationship of proteins**: relationship between primary structure and protein function, higher order structure and protein function.

4. **Physico-chemical properties of proteins**: isoelectric point, denaturation, colloid property, precipitation, ultraviolet absorption.

Major points and practices

On average, the **nitrogen** composition in proteins is about 16% by weight. Thus 1 gram of nitrogen corresponds to 6.25 grams of proteins.

I. Amino acids and Peptides

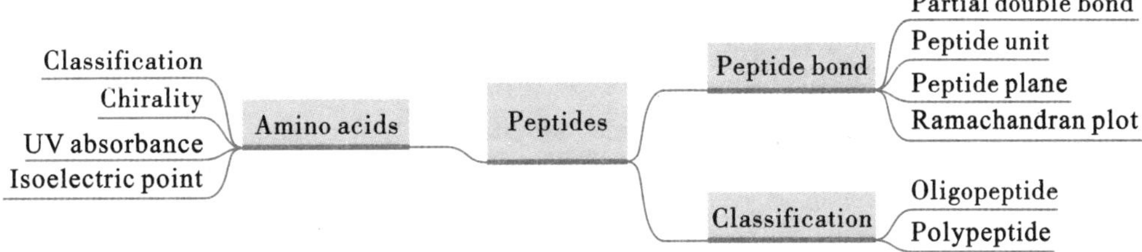

A. General Structure

1. Each amino acid has one amino group, one carboxyl group and an R group attached to the central alpha carbon atom.

2. Except glycine, all "standard" amino acids are *L*-α-amino acids.

B. Classification of amino acids

1. Aliphatic Amino Acids: Gly, Ala, Val, Leu, Ile, Met, Pro
2. Polar Uncharged Amino Acids: Ser, Thr, Cys, Met, Asn, Gln
3. AromaticAmino Acids: Phe, Tyr, Trp
4. Acidic Amino Acids: Asp, Glu
5. Basic Amino Acids: Lys, Arg, His

- *Notes*:

· Cysteine can be oxidized to form **disulfide bond**. Disulfide bonds can be broken (reduced) by reducing agents with free thiol groups (DTT, β-ME, GSH).

· Amino acids with hydroxyl groups (**Tyr**osine, **Ser**ine, and **Thr**eonine) can be **phosphorylated**. Phosphorylation often acts as a switch for protein function.

C. Physico-chemical Properties of Amino Acids

1. The **isoelectric point (pI) of an amino acid** is the pH where the net charge is zero on the amino acid.

2. Aromatic amino acids (mainly tryptophan and tyrosine) have **absorbance at ultraviolet wavelength** (around 280 nm). This can be used to quantify amino acids.

3. **Color Reaction**: When amino acids are heated together with a chemical called ninhydrin, a blue-purple color compound is produced. This color reaction can be used to detect amino acids qualitatively or quantitatively.

D. Peptides

1. **Peptide bond**: A substituted amide linkage between the α-amino group of one amino acid and the α-carboxyl group of another, with the elimination of a molecule of water.

2. Peptide bonds have partial double bond nature.

(1) The bond length is between a real single bond and a real double bond.

(2) This spatial arrangement of the peptide bond is trigonal (similar to double bonds), not tetrahedral.

3. **Peptide unit**: Six atoms of the peptide group ($C_{\alpha 1}$, $C_{\alpha 2}$, N, H, C, O) are on the same plane. This is called the **peptide plane or peptide unit**. The C-N bond cannot rotate due to its partial double bond nature.

4. The trans- configuration of the peptide bond is energetically favored, except when a proline residue is part of the peptide bond.

5. The N- C_α and C_α-C bond can freely rotate, described by two dihedral angles designated phi and psi. Not all phi and psi combinations are possible. Energetically allowed values are depicted in a plot with phi vs psi on the two axes, called **Ramachandran plot**.

6. **Peptide chain**: residues

(1) Backbone: N-C_α-C-N-C_α-C-N-C_α-C···

(2) Side chain: R group

(3) Terminus: N-terminus and a C-terminus.

(4) Direction: N→C

7. **Oligopeptides**: peptides formed by ≤10 amino acids.

Polypeptides: Peptides formed by >10 amino acids.

Proteins: polypeptides of > 50 amino acid residues.

8. Short peptides also have biological activities. One important example is **glutathione** (GSH), a natural reducing agent inside cells.

[Practices]

[A1 type]

[1] The ultraviolet absorbance of proteins is due to the presence of the following amino acids ________.

A. Alkyl amino acids

B. Aromatic amino acids

C. Positively charged amino acids

D. Negatively charged amino acids

E. Hydrophobic amino acids

[2] Which one of the following amino acids is not chiral? ________

A. Serine B. Tryptophan C. Glycine

D. Lysine E. Alanine

[3] Alanine has two dissociable groups, the pKa of which are 2.3 and 9.9. What's the pI of Alanine? ________

A. 6.1 B. 7.5 C. 2.3

D. 9.9 E. 4.5

[4] Which one ofthe following amino acids can be phosphorylated? ________

A. Tyrosine B. Alanine C. Aspartate

D. Methionine E. Glutamine

[5] Which one of the following amino acids is positively charged at physiological pH? ________

A. Leucine B. Lysine C. Aspartate

D. Serine E. Glutamate

[6] Which of the following statements is/are true concerning peptide ponds? ________

A. They are the only covalent bond formed between amino acids in polypeptide structures.

B. The angles between the participating C and N atoms are described by the values psi and phi.

C. They have partial double bond character.

D. A and C E. B and C

[7] The peptide bond in proteins is ________.

A. Planar, but rotates to three preferred dihedral angles

B. Non-planar, but rotates to three preferred dihedral angles

C. Non-planar, and fixed in a trans conformation

D. Planar, and usually found in a trans conformation

E. Planar, and a normal single bond

Ⅱ. Protein Structure

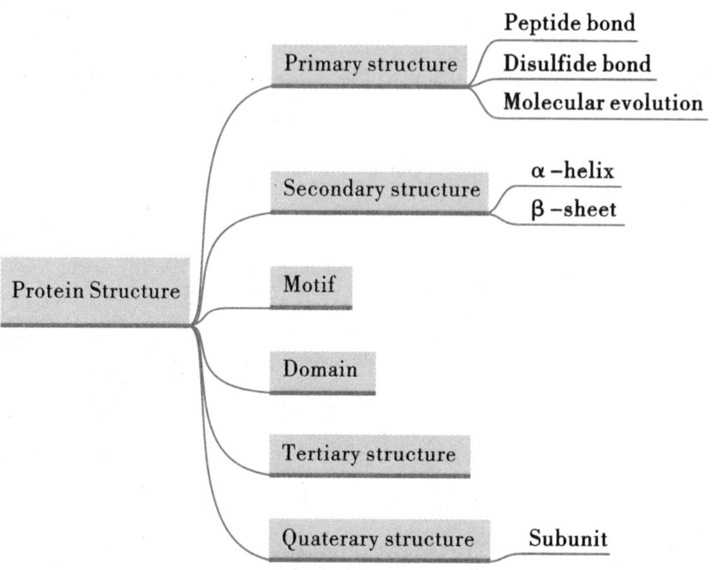

A. The Primary Structure

1. **The primary structure of a protein** is the covalent structure (**peptide bonds and disulfide bonds**) linking amino acid residues in proteins or the amino acid sequence in a protein.

2. The amino acid sequence of conserved proteins can be used to deduce the evolutional tree. This method is called **molecular evolution analysis**.

B. Higher Order Structures

1. **The secondary structure** is the local ordered arrangement of a protein's backbone atoms, without considering its side chain. Common secondary structures include **α-helix, β-sheet, β-turn, and Ω loop**. Secondary structures are maintained by **hydrogen bonds**.

2. **Motif** is a structural component of a protein molecule consisting of two or more secondary structural peptide fragment with specifie spatial conforwation and specifre function. Example: helix-loop-helix motif, zinc finger motif.

3. **Domains**: Some polypeptide chains fold into two or more compact regions that may be connected by a flexible segment of polypeptide chain. These compact globular units are called domains. Domains often perform discrete function and fold independently.

4. **The tertiary structure** is the three-dimensional structure of the polypeptide. Tertiary structure is maintained by **hydrophobic interactions (driving force), hydrogen bonds, electrostatic interactions (salt bridges), Van der Waals interactions.**

5. **Subunits**: Many proteins are composed of two or more polypeptide chains loosely, referred to as subunits.

6. **The quaternary structure** is defined by the interactions between different subunits in multi-subunit proteins. Quaternary structures are mainly maintained by hydrogen bonds, electrostatic interactions (salt bridges).

[*Practices*]

[A1 type]

[8] The primary structure of a protein is mostly maintained by____________.

A. Peptide bond B. Hydrogen bond C. Ionic bond

D. Hydrophobic bonds E. Van der Waals interactions

[9] Disulfide bonds are formed between which two amino acids ____________

A. Cysteine and cysteine B. Cysteine and cystine C. Cysteine and serine

D. Serine and threonine E. Serine and tyrosine

[10] Secondary structures in a protein refer to____________.

A. Linear sequence of amino acids joined together by peptide bond

B. Three-dimensional arrangement of all amino acids in polypeptide chain

C. Regular folding of local backbone of the polypeptide chain

D. Protein made up of more than one polypeptide chain

E. Independent and compact units inside a protein

[11] Which one of the following amino acid is an alpha helix terminator? ____________

A. Cysteine B. Alanine C. Proline

D. Isoleucine E. Histidine

[12] An oil drop with a polar coat is a metaphor referring to the three-dimensional structure of____________.

A. Fibrous proteins B. Collagen C. Globular proteins

D. Silk protein E. Ion channel

[13] What was the first protein whose complete tertiary structure was determined? ____________

A. Lysozyme B. Myoglobin C. Pancreatic ribonuclease

D. Pancreatic DNase E. Insulin

[14] The different orders of protein structure are determined by all the following bond types except ____________.

A. Peptide bonds B. Phospho-diester bonds C. Disulfide bonds

D. Hydrogen bonds E. Van der Waals interactions

[15] The secondary structure of a protein is primarily maintained by ____________.

A. Van der Waals force B. Hydrogen bond C. Ionic bond

D. Hydrophobic interactions E. Covalent bonds

[16] Which of the following best describes a protein domain? ____________

A. The α-helical portion of a protein

B. A discrete region of polypeptide chain that has folded into a self-contained three-dimensional structure

C. The β-pleated sheet portion of a protein

D. A feature that rarely occurs in globular proteins

E. A regular folding of local protein backbone

Ⅲ. Protein Function

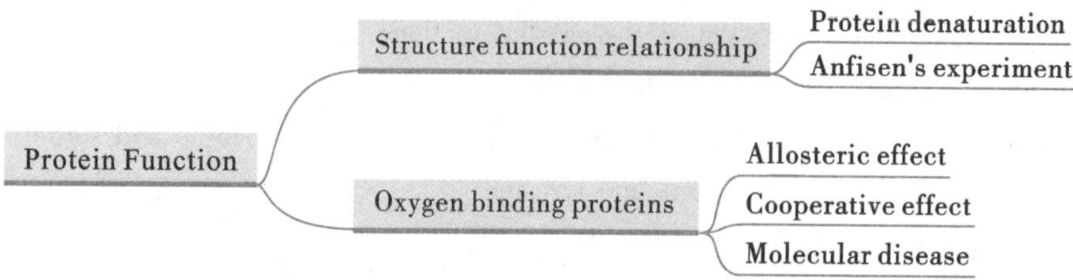

A. Structure-function relationship of proteins

1. **The primary structure of a protein determines its three-dimensional structure: Anfinsen's experiment.**

Sometimes, proteins need help from chaperones to fold into the correct conformation.

2. **Protein denaturation**: Partial or complete disruption of the natively spatial structure and loss of biological activity of a protein by physical or chemical factors is called protein denaturation. Protein denaturation disrupts the spatial conformation but not the primary structure of a protein.

Denaturants include strong acids or bases, organic solvents, heating, urea and reducing reagents such as DTT and β-ME.

3. **Change of essential amino acid residue in protein may lead to disease**: **Sickle cell anemia** is a type of **molecular disease**. People with sickle cell anemia have a glutamic acid to valine mutation in the 6th position of β subunit of adult hemoglobin resulting in a dysfunctional hemoglobin and sickle-shaped red blood cells.

4. **The higher order structure of a protein determines its function.**

B. Oxygen binding proteins

1. Myoglobin and hemoglobin are two oxygen binding proteins in the body. Myoglobin is the primary oxygen storage protein in muscle. Hemoglobin in red blood cells is responsible for oxygen transport from the lungs to the tissues for use in metabolism.

2. Heme group is the **prosthetic group** of myoglobin and hemoglobin that sequester iron at its center, which forms coordinate bond with an oxygen molecule.

3. Myoglobin is a monomer containing one heme group and one oxygen binding site. Hemoglobin is hetero-tetramer of 2 α subunits and 2 β subunits. Each subunit of hemoglobin is structurally similar to myoglobin. So hemoglobin has four heme groups and four oxygen binding sites.

4. **Cooperative effect**: Binding of a preceding ligand to one subunit of a protein affects the affinity of neighboring subunits for subsequent ligands. The oxygen saturation curve of hemoglobin is S-shaped, with increasing affinity of hemoglobin for oxygen as oxygen loading increases.

5. **Allosteric effect** is the regulation of a protein by binding a ligand at a site and results in conformational changes. The protein is allosteric protein and the ligand is allosteric effector.

[*Practices*]

[A1 type]

[17] Which of the following statements correctly describes the behavior of the hemoglobin protein in sickle

-cell disease? ____________.

A. The hemoglobin protein dissociates into four subunits

B. The hemoglobin protein lacks a heme group

C. Neighbouring hemoglobin proteins aggregate together

D. The hemoglobin protein possesses an iron atom in the Fe (Ⅲ) form rather than the normal Fe (Ⅱ) form

E. The hemoglobin adopts myosin like oxygen binding ability

[18] As hemoglobin binds oxygen molecules, its affinity for oxygen increases, driving the binding of further oxygen molecules. Which term best describes this phenomenon?____________.

A. Catalysis B. Saturation C. Cooperative effect

D. Isomerism E. Feedback

[19] In hemoglobin, allosteric effects occur____________.

A. only in humans

B. for maintaining Fe in the Fe^{2+} state

C. to minimize oxygen delivery to the tissues

D. to maximize oxygen delivery to the tissues

E. to regulate the stability of hemoglobin

[20] Which of the following statement is incorrect?____________.

A. Hemoglobin and myoglobin are the two oxygen binding proteins

B. Hemoglobin transports O_2 in the blood

C. Myoglobin stores O_2 in muscles

D. Myoglobin structure is similar to one subunit of hemoglobin

E. None of the above

[21] The oxygen in hemoglobin and myoglobin is bound to____________.

A. the iron atom in the heme group

B. the nitrogen atoms on the heme

C. histidine residues in the protein

D. lysine residues in the protein

E. the-OH group in the active center

[22] An allosteric activator of hemoglobin________.

A. increases the binding affinity of oxygen

B. decreases the binding affinity of oxygen

C. stabilizes the R state of the protein

D. stabilizes the T state of the protein

E. both (a) and (c)

Ⅳ. Working with proteins

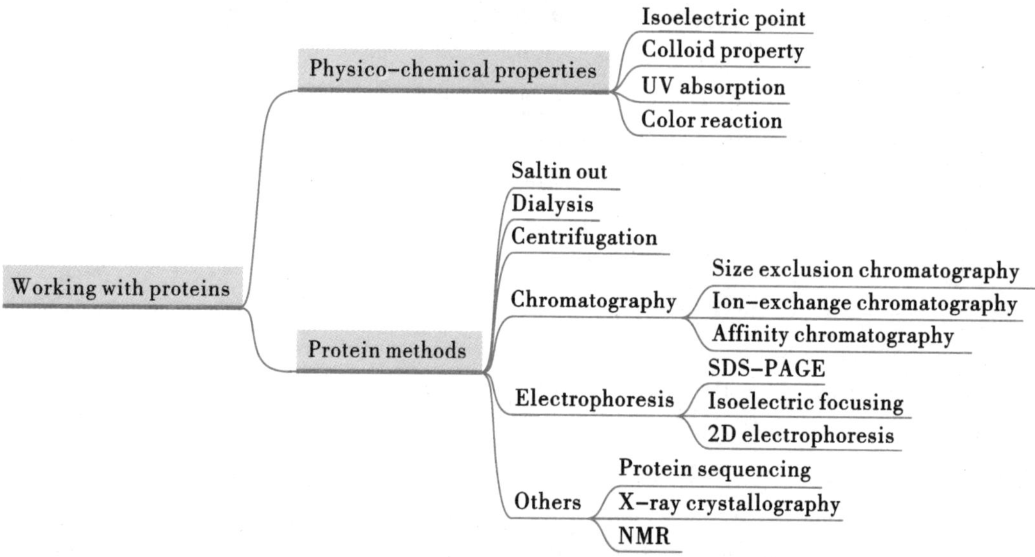

A. Physico-chemical properties of proteins

1. **Isoelectric point (pI)** of a protein: The pH at which the protein has zero net-charge is referred to as isoelectric point (pI) of the protein. Proteins are zwitterionic. The dissociable groups in a protein include its N terminal amino group, C terminal carboxyl group and some groups in the side chains of amino acid residues (if present).

(1) When **pH<pI**, the overall charge is **positive**.

(2) When **pH>pI**, the overall charge is **negative**.

2. **Colloid property**: The **hydration shell** and **electric repulsion** can stabilize proteins in solution.

3. **Ultraviolet absorption**: Proteins have maximal absorption at about **280nm** due to the presence of tryptophan and tyrosine, which can be used to determine protein concentration.

4. **Color reaction**: Proteins can react with Cu^{2+} to form pink color chelates at basic solution (Biuret reaction).

B. General methods of protein technology

1. Proteins can be separated and purified by **salting out**, **dialysis**, **centrifugation**, chromatography and some other methods.

2. **Chromatography**

(1) **Size exclusion chromatography** (also called gel filtration) separates proteins based on size differences. Larger proteins flow faster and are eluted earlier.

(2) **Ion-exchange chromatography** separates proteins according to net charge differences. There are cation exchange column and anion exchange column.

(3) **Affinity chromatography** separates proteins based on the high affinity of proteins to specific ligands. Affinity chromatography is the most specific one.

3. **Electrophoresis**

a. **SDS-PAGE** is a type of electrophoresis that separates proteins according to their differences on size. Binding of SDS renders proteins denatured and negatively charged, which makes the 3D structure and native charge of the proteins not important.

b. **Isoelectric focusing** separates proteins on the basis of isoelectric points. **Two-dimensional electrophoresis** combines SDS-PAGE and isoelectric focusing, which can analyze large amounts of proteins.

4. The amino acid sequence of a protein can be determined by **Edman Sequencing** or deducted from DNA sequence.

[*Practices*]

[A1 type]

[23] At the isoelectric pH of a tetrapeptide____________.

A. only the amino and carboxyl termini contribute charge

B. the amino and carboxyl termini are not charged

C. the total net charge is zero

D. two internal amino acids of the tetrapeptide cannot have ionizable R groups

E. the peptide does not have an ionized group

[24] In ion-exchange chromatography____________.

A. proteins are separated on the basis of their net charge

B. proteins are separated on the basis of their size

C. proteins are separated on the basis of their shape

D. proteins are separated on the basis of their affinity

E. either (b) or (c)

[25] Affinity chromatography uses the following facts for purification____________.

A. specific binding of a protein constituent for another molecule

B. protein - protein interaction

C. protein - carbohydrate interaction

D. protein - lipid interaction

E. none of the above

[26] During successful purification scheme, this may be expected that the____________.

A. specific activity increases

B. specific activity decreases

C. number of proteins in the sample decreases

D. number of proteins in the sample increases

E. both (a) and (c)

[27] Which of the following statements about SDS polyacrylamide gel electrophoresis is correct?

____________.

A. Wanted proteins can be tested for their biological activity after separation

B. Proteins are solubilized but not denatured when separated

C. SDS polyacrylamide gel electrophoresis separates proteins based on charge

D. SDS polyacrylamide gel electrophoresis separates proteins based on size

E. SDS polyacrylamide gel electrophoresis separates proteins based on pI

[B1 type]

No. 28 ~ 30 share the following suggested answers.

A. Salting out B. Size exclusion column. C. Ion Exchange column

D. Affinity column E. SDS-PAGE

A biochemist is attempting to separate a DNA-binding protein (protein X) from other proteins in a solution. Only three other proteins (A, B, and C) are present. The proteins have the following properties:

	PI (isoelectric point)	Size M_r	Bind to DNA?
protein A	7.4	82,000	yes
protein B	3.8	21,500	yes
protein C	7.9	23,000	no
protein D	7.8	22,000	yes

What type of protein separation techniques might she use to separate

[28] Protein X from protein A? ___________

[29] Protein X from protein B? ___________

[30] Protein X from protein C___________

Answers

1	2	3	4	5	6	7	8	9	10
B	C	A	A	B	C	D	A	A	C
11	12	13	14	15	16	17	18	19	20
C	C	B	B	B	B	C	C	D	D
21	22	23	24	25	26	27	28	29	30
A	E	C	A	A	E	D	B	C	D

Brief Explanation for Practices

1. The UV absorbance of amino acids is due to the aromatic ring.
2. Two substitution groups in glycine are exactly the same. So, it is not chiral.
3. Take the average of the two pKa.
4. Phosphorylation requires -OH group.
5. Lysine is positively charged at physiological pH.
6. Peptide bonds are not the only type of covalent bonds. There are also disulfide bonds. The participating C and N atoms in the peptide bond cannot be rotated.
7. The peptide bond is planar, and the trans- configuration is favored.
8. Primary structure is maintained by covalent bonds, mainly peptide bonds.
9. Cysteine has -SH group. Cystine does not.
10. Look for the key word "local".
11. The amino group in proline is a secondary amine.
12. In globular proteins, the hydrophobic residues are buried, like an oil droplet.
13. Myoglobin was the first protein whose structure was determined.
14. Phosphodiester bonds are found in nucleic acids, not proteins.
15. Secondary structures are maintained by H-bonds.

16. Look for key words that says "independent folding" etc.

17. In sickle cell disease, a glutamic acid to valine mutation in the β subunit of adult hemoglobin makes hemoglobin prone to aggregation.

18. Refer to the definition of coorperative effect.

19. Oxygen binding to hemoglobin change hemoglobin from T state to R state, which will have the maximal binding with oxygen.

20. Hemoglobin and myoglobin are two oxygen binding proteins. Hemoglobin transports oxygen and myoglobin stores oxygen.

21. Oxygen forms a coordinate bond with iron, which is sequestered in the center of the heme group.

22. Allosteric effect of hemoglobin will lead to increased affinity to oxygen and stabilize R state.

23. Refer to the definition of isoelectric point.

24. Ion-exchange chromatography separates proteins based on net charge differences.

25. Affinity chromatography separates proteins based on the high affinity of a protein to a specific molecule, which is called a ligand.

26. The purification process is expected to increase the activity of your interest protein and get rid of other unwanted proteins.

27. In SDS-PAGE, the proteins are denatured and lose activities. SDS-PAGE separates proteins based on size differences.

28. Size-exclusion (or gel filtration) chromatography (or column) to separate on the basis of size. Since both X and A can bind to DNA and pI of the X and A are close, we can't use affinity chromatography or ion-exchange chromatography.

29. Ion-exchange chromatography (or column) (or isoelectric focusing) to separate on the basis of charge. Because pI of B and X are 3.8 and 7.8 respectively, under certain pH condition (such as pH 7), B will be negatively charged and X will be positively charged.

30. Affinity chromatography (or column), using immobilized DNA since C can't bind to DNA.

[*Comprehensive Practices*]

[**A1 type**]

[1] Which pair of the following amino acids both contains -OH group? ____________

A. Glutamine and glutamate B. Serine and threonine C. Phenylalanine and tyrosine
D. Cysteine and cystine E. Leucine and isoleucine

[2] Which one of the following factors cannot denature proteins? ____________

A. Heating B. Strong acid and base C. Organic acid
D. Heavy metal E. Salting out

[3] Which one of the following statements is correct about pI value of a protein? ____________

A. The net charge of a protein is zero when pH equals to pI.
B. Proteins are denatured at pI.
C. When pH <pI, proteins are negatively charged.
D. Proteins are more stable at pI.
E. The value of pI has nothing to do with the number of acidic amino acid residues in a protein.

[4] The quaternary structure of a protein____________.

A. must have several identical subunits
B. must have several distinct subunits
C. must have equal number of distinct subunits

D. must have unequal number of distinct subunits

E. have unequal number and different type of subunits

[5] Which one of the followingstatements is true about protein subunits? ____________

A. One polypeptide chain folds into a helical structure.

B. Two or more polypeptide chain that folds into a beta sheet structure.

C. Two or more polypeptide chain form complex with a cofactor.

D. Every subunit has its own tertiary structure.

E. Every subunit can function on its own.

[6] The higher order structure of a protein is determined by____________.

A. hydrogen bonds in the polypeptide

B. peptide bonds in the polypeptide

C. the composition and order of amino acid residues in a polypeptide

D. the peptide planes in the polypeptide

E. the peptide units in the polypeptide

[7] Which one of the following is an acidic amino acid? ____________

A. Lysine　　B. Arginine　　C. Histidine

D. Aspartate　　E. Asparagine

[8] Which one of the following amino acid does not participate in protein? ____________

A. Homocysteine　　B. Cysteine　　C. Cystine

D. Methionine　　E. Histidine

[9] Which one of the following bonds is not broken during protein denaturation? ____________

A. Hydrogen bond　　B. Peptide bond　　C. Hydrophobic interaction

D. Ionic bond　　E. Disulfide bond

[10] Which one of the following can be used for protein precipitation during purification? ____________

A. Salting out　　B. Dialysis　　C. Gel filtration

D. SDS-PAGE　　E. None of the above

[11] All naturally occurring proteins must have____________.

A. α-helix　　B. β-sheet　　C. tertiary structure

D. quaternary structure　　E. cofactor

[12] Which one of the following is true about denatured proteins? ____________.

A. Hard to be digested by proteases

B. Molecular weight decreases

C. Solubility increases

D. Biological activity decreases

E. Peptide bonds are broken

[13] Which one of the following amino acid is made after protein synthesis? ____________

A. Cysteine　　B. Hydroxyproline　　C. Methionine

D. Serine　　E. Tyrosine

[14] Which happens when subunits dissociate in a protein? ____________

A. Disulfide bonds are broken.

B. The primary structure is disrupted.

C. The secondary structure is disrupted.

D. The tertiary structure is disrupted.

E. The quaternary structure is disrupted.

[15] Which happens when a protease digests a protein? ____________

A. Disulfide bonds are broken.

B. The primary structure is disrupted.

C. The secondary structure is disrupted.

D. The tertiary structure is disrupted.

E. The quaternary structure is disrupted.

[16] The primary structure of a protein is____________.

A. the arrangement of subunits B. α-helix C. β-sheet

D. amino acid sequence E. amino acid number

[17] Which one of the following carries net positive charge at pH = 7.5? ____________

A. Glu B. Lys C. Ser

D. Asn E. Tyr

[18] Hemoglobin has an S shaped oxygen binding curve because ____________.

A. Fe (II) in the heme group is oxidized by oxygen

B. The first subunit changes conformation after oxygen binding and leads to increase in oxygen binding of other subunits

C. Hemoglobin has stronger oxygen binding affinity than myoglobin

D. Hemoglobin has weaker oxygen binding affinity than myoglobin

E. None of the above

[19] Which one of the following proteins has quaternary structure? ____________

A. Insulin B. RNase C. Hemoglobin

D. Myoglobin E. Trypsin

[20] In a conjugated protein, a prosthetic group is____________.

A. a fibrous region of a globular protein

B. a nonidentical subunit of a protein with many identical subunits

C. a part of the protein that is not composed of amino acids

D. a subunit of an oligomeric protein

E. none of the above

Answers

1	2	3	4	5	6	7	8	9	10
B	E	A	E	D	C	D	A	B	A
11	12	13	14	15	16	17	18	19	20
C	D	B	E	B	D	B	B	C	C

[*Clinical Cases*]

[21] A child presents with severe vomiting, dehydration, and fever. Initial blood studies show acidosis with low bicarbonate. Preliminary results from the blood amino acid screen show two elevated amino acids, both with nonpolar side chains. A titration curve performed on one of the elevated species shows only two ionizable groups: one that is acidic and the other that is basic (i. e., no charged side chain). Which of the following pairs of elevated amino acids is most likely elevated?

A. Arginine and isoleucine B. Aspartic acid and glutamine C. Glutamic acid and threonine
D. Histidine and valine E. Leucine and isoleucine

[22] A 60-year-old man is brought to his physician from an institution for severe mental deficiency. The physician reviews his family history and finds he has an older sister in the same institution. Their parents are deceased but reportedly had normal intelligence and no chronic diseases. The man sits in an odd position as though he was sewing, prompting the physician to obtain a ferric chloride test on the man's urine. This test turns color with aromatic (ring) compounds, including certain amino acids, and a green color confirms the physician's diagnosis. Which of the following amino acids was most likely detected in the man's urine?

A. Glutamine B. Glycine C. Methionine
D. Phenylalanine E. Serine

[23] A 6-year-old black boy complains of acute abdominal pain that began after playing in a football game. He denies being tackled forcefully. He has a history of easy fatigue and several similar episodes of pain after exertion, with the pain usually restricted to his extremities. Microscopic evaluation of his blood would be expected to reveal which of the following cellular abnormalities?

A. Increased WBC count B. Deformed RBCs C. Decrease WBC count
D. Increase RBC count (erythrocytosis) E. Reduced platelet count

Answers

21	22	23
E	D	B

Brief Explanations for Clinical Cases

21. The two ionizable groups are amino group and carboxyl group, found in all amino acids. The two amino acids have nonpolar side chains. Only Leu and Ile fit the description.

22. The test shows aromatic ring amino acids. Only Phe fit the description.

23. The patient's symptoms are entirely consistent with an acute sickle cell crisis. These are brought on by exertion, which increases the levels of deoxyhemoglobin in RBCs and alter the shape of RBCs (sickle cells).

(*Zhang Yanling, Ren Fangfang, Su Xiong*)

Chapter 2

Structure and Function of Nucleic Acids

Examination Syllabus

1. **Nucleic acid**: chemical composition, types, molecular composition of nucleotides
2. **Structure and function of DNA**: rule of base composition of DNA (Chargaff's rule), primary structure, double helix, advanced structure, function
3. **Physical and chemical properties of DNA**: DNA denaturation and renaturation, nucleic acid hybridization, ultraviolet (UV) absorption
4. **Structure and function of RNA**: types (mRNA, tRNA, rRNA, other), structure, function

I. Nucleic acid

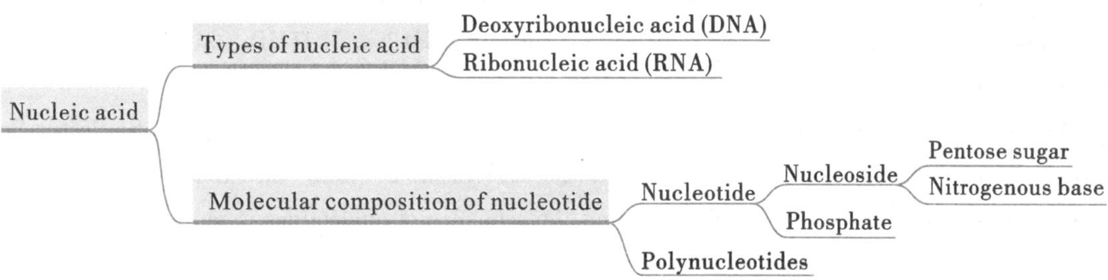

A. Types of nucleic acid

1. **Nucleic acid** is a biological macromolecule, composed of nucleotides, which are physical basis of heredity.

2. **Deoxyribonucleic acid (DNA)**, the carrier of the genetic information, is mainly located in the nucleus, whose small amount is in the mitochondria.

3. **Ribonucleic acid (RNA)** is mainly located in the cytoplasm, which can be involved in the expression of genetic information.

B. Molecular composition of nucleotide

1. **Nucleotide** is the basic unit of nucleic acid, which is composed of a **phosphate**, a **pentose sugar**

and a **nitrogenous base**. The basic unit of DNA is deoxyribonucleotide and that of RNA is ribonucleotide.

2. **Nitrogenous bases** are aromatic heterocyclic compounds found in nucleotides.

(1) The bases have two types, purines and pyrimidines.

(2) DNA and RNA contain the same purines adenine (A) and guanine (G) and the same pyrimidine cytosine (C).

(3) DNA contains thymine (T) whereas RNA contains uracil (U).

3. **Pentose sugars** are five carbon monosaccharides found in the nucleotides. RNA contains β-D-ribose while DNA contains β-D-2-**deoxy**ribose (Table 2.1).

Table 2.1 Comparison of molecular composition between DNA and RNA

Type	Purine	Pyrimidine	Pentose sugar	Phosphate
DNA	A,G	C,T	deoxyribose	phosphate
RNA	A,G	C,U	ribose	phosphate

4. **Nucleoside**is made of a pentose sugar and a nitrogenous base linked by **β-N-glycosidic bond**. That is, C-1′of the pentose sugar is linked to N-9 of a purine or N-1 of a pyrimidine.

Thus, nucleotide is an ester resulting from dehydration between hydroxyl group in pentose sugar of nucleoside and phosphate. Hydroxyl groups on the 2′, 3′ and 5′-carbon of the pentose sugar can be esterified with phosphates where 5′-hydroxyl is the most common.

5. **Nomenclature of nucleotides** indicates the molecular composition of the nucleotide.

(1) For example, adenosine monophosphate (AMP) contains adenine + ribose + phosphate.

(2) According to the number of phosphate groups attached, nucleotides are divided into three types, nucleoside 5′-monophosphate (NMP), nucleoside 5′- diphosphate (NDP) and nucleoside 5′-triphosphate (NTP).

(3) The prefix 'd' is used to indicated if the sugar is deoxyribose (e. g. dAMP).

(4) Four basic nucleotides in DNA are dAMP, dCMP, dGMP and dTMP, while those in RNA are AMP, CMP, GMP and UMP.

6. **Polynucleotides** refer to the polymer of nucleotides whose linkage is 3′,5′-**phosphodiester bond**. The polynucleotide chain has two ends, 5′-end and 3′-end, in the 5′-to-3′ direction.

As other components except the base are same, the sequence of nucleotides is represented with nitrogenous base.

[*Practices*]

[**A1 type**]

[1] The composition of backbone of polynucleotides includes__________.

A. base and pentose sugar B. base and phosphate C. base and base

D. pentose sugar and phosphate E. pentose sugar and pentose sugar

[2] The bases included in nucleic acid are__________.

A. 2 types B. 3 types C. 4 types

D. 5 types E. 6 types

[3] Which one is present neither in RNA nor in DNA? __________

A. Adenine B. Xanthine C. Guanine

D. Thymine E. Uracil

[4] For the products of DNA and RNA after complete hydrolysis, you may find that ________.

A. riboses are same but some bases are different.

B. bases are same but riboses are different.

C. both bases and riboses are different.

D. riboses are same but bases are different.

E. riboses and some bases are different.

[5] The maximal absorption peak of nucleic acid to ultra violet light is at ________.

A. 220 nm B. 230 nm C. 260 nm

D. 280 nm E. 300 nm

[6] The linkage between nucleotides in nucleic acid is ________.

A. 2′,3′- phosphodiester bond B. 3′,5′- phosphodiester bond C. 2′,5′- phosphodiester bond

D. 1′,5′- glycosidic bond E. hydrogen bond

Ⅱ. Structure and function of DNA

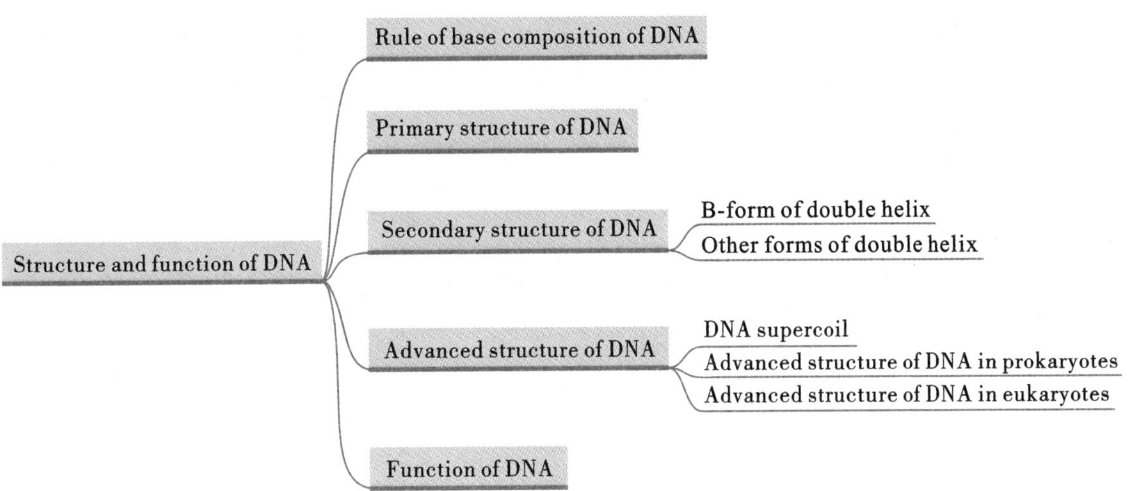

A. Rule of base composition of DNA refers to Chargaff's rule.

1. There is molar equivalence between the purines and pyrimidines in DNA.
2. The numbers of adenine residues are equal to those of thymine residues ([A]=[T]).
3. Those of guanine and cytosine residues are equal ([G]=[C]).

B. Primary structure of DNA

1. **Primary structure of DNA** refers to the sequence of the deoxynucleotides.
2. The linkage of the deoxynucleotides is 3′,5′-phosphodiester bond.
3. The direction is from 5′ end to 3′end. It is usually represented with either abbreviated type or sketched type.

C. Secondary structure of DNA

1. **Double helix** is secondary structure of DNA proposed by Watson and Crick, which is now known as B form DNA or B-DNA.

(1) The properties of double helix of B-DNA include: It consists of two polydeoxynucleotide chains, which are **antiparallel**, i.e., one strand runs in the 3′ to 5′ direction while the other in 5′ to 3′ direction. It is a **right-handed** helix and has major grooves and minor grooves. The deoxyribose-phosphate backbone is on the outside of the helix while the bases are stacked inside to form **complementary base pair** (bp), A═T, G≡C. Each turn of the helix is 3.54 nm with 10.5 bp, whose diameter is 2.37 nm.

(2) The factors stabilizing the helix include horizontal stabilizing factor, **hydrogen** bonds between complementary base pairs, and longitudinal stabilizing factor, **base-stacking interaction.**

(3) **Complementary base pairing rule** refers to adenine pairs with thymine by two hydrogen bonds (A═T), guanine pairs with cytosine by three hydrogen bonds (G≡C).

2. **Diversity of DNA double helix** is associated with conformational variants of DNA. Among these different forms of the double-helical structure of DNA, B, A, and Z forms are important (Table 2.2).

Table 2.2 Comparison of three conformations of DNA double helix

Features	A-DNA	B-DNA	Z-DNA
Helical sense	Right-handed	Right-handed	Left-handed
Base pairs per helical turn	11	10.5	12
Distance per helical turn (nm)	2.53	3.54	4.56
Diameter (nm)	2.55	2.37	1.84
Base tilt normal to the helix axis	19°	1°	9°

D. The higher order structure of DNA

1. **DNA supercoil** refers to the further structure based on the double helix. When the coiling direction is as same as that of the double helix, it can form **positively supercoiled** DNA. However, the opposite direction can form **negatively supercoiled** DNA. All natural DNA are negatively supercoiled.

2. **The higher order structure of DNA in prokaryotes**

The majority of DNA in prokaryotes is **circular covalently closed DNA** with double helix, which is further coiled in the cell to form a nucleoid structure.

3. **DNA organization in the nucleus in eukaryotes**

(1) **Nucleosome** is a basic unit of chromosome in eukaryotes, consisting of DNA and protein. The core includes DNA about 200 bp and two molecules of histones (H_{2A}, H_{2B}, H_3, and H_4). Nucleosomes are separated by spacer DNA where histone H_1 is attached. The continuous string of nucleosomes is termed as 10 nm fiber.

(2) This 10 nm fiber is further coiled to form 30 nm fiber, which has six nucleosomes in every turn, then further organized into loops. During the course of mitosis, the loops are further coiled; the chromosomes condense and become visible.

(3) Functional areas of eukaryotic chromosome contain telomeres and centromeres.

4. Mitochondrial DNA (mt DNA) is an exposed, circular and double-strandod DAN without any binding proteins, and very few non-coding regions.

E. Function of DNA

DNA can carry genetic information in the form of gene, which is the template of replication and transcription.

Gene is the sequence of the nucleic acids which is necessary for the synthesis of functional polypeptide chains or RNA. It's usually the DNA sequence (or RNA sequence in a RNA virus).

[*Practices*]

[A1 type]

[7] The rule of base composition of DNA refers to______.

A. [A]=[C]; [T]=[G] B. [A]+[T]=[C]+[G] C. [A]=[T]; [C]=[G]
D. ([A]+[T])/([C]+[G])=1 E. [A]=[G]; [T]=[C]

[8] Which is true about the base composition of DNA ?______

A. The number of adenines in DNA molecule is different from that of thymine.
B. The base composition in adulthood is different from that in adolescence in the same individual.
C. The base composition is different in the same individual at various nutritional status.
D. The base composition of various tissues in the same individual is different.
E. The base composition of DNA from various biological species is different.

[9] If the mole ratio of G in one DNA molecule is 32.8 %, that of A is ______.

A. 67.2 % B. 65.6 % C. 32.8 %
D. 17.2 % E. 34.4 %

[10] The primary structure of DNA is ______.

A. poly A structure B. nucleosome structure C. double helix structure
D. cloverleaf structure E. sequence of the deoxynucleotides

[11] Which statement is correct for the secondary structure of DNA?______

A. The sequence of the nucleotides in the nucleic acid chain
B. Cloverleaf structure of tRNA
C. DNA double helix
D. Supercoil structure of DNA
E. Nucleosome structure of DNA

[12] Which is **NOT** correct for B form of DNA double helix?______

A. Two deoxyribonucleotide chains are antiparallel.
B. The planes of the bases are perpendicular to the helix axis.
C. Each turn of the helix has 10.5 base pairs.
D. Each turn of the helix is 3.54 nm.
E. The deoxyribose and phosphate are stacked inside of the helix.

[13] Which is true for B form of DNA double helix? ______

A. One strand is left-handed helix while the other is right-handed.
B. Hydrogen bonds can longitudinally stabilize the structure of the double helix.
C. The ratio of [A+T]/ [G+C] is 1.
D. The bases between two strands are held together by covalent bonds.
E. The phosphate and deoxyribose form the backbone of the helix.

[X type]

[14] Which is true for B form of DNA secondary structure? ______

A. The double helix is right-handed.
B. Two deoxyribonucleotide chains are parallel.
C. Base pairing is found between two strands.

D. Each turn of the helix has 10.5 base pairs.

E. [A]=[G]; [T]=[C]

Ⅲ. Structure and function of RNA

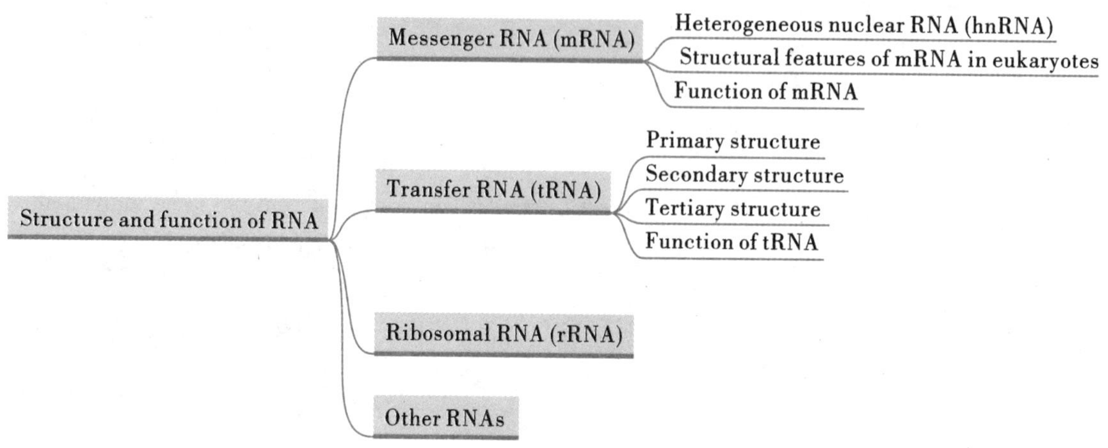

A. Types of RNA

1. RNAs contain two types: coding RNA and non-coding RNA.

2. **Coding RNA** refers to the RNA that can be transcribed from DNA and be translated into proteins, only including messenger RNA (**mRNA**).

3. **Non-coding RNA** cannot encode proteins. It can be divided into two classes.

a. **Constitutive non-coding RNA** is the RNA with basically constant abundance that can ensure the realization of basic biological functions, including transfer RNA (**tRNA**), ribosomal RNA (**rRNA**) and telomere RNA, small nuclear RNA (**snRNA**) and small nucleolar RNA (**snoRNA**), etc.

b. **Regulatory non-coding RNA** can play a vital role in the regulation of gene expression, can be divided into small non-coding RNA (**sncRNA**), long non-coding RNA (**lncRNA**) and circular RNA (**circRNA**). **sncRNA** contains microRNA (**miRNA**), small interfering RNA (**siRNA**), and **piRNA**.

B. Messenger RNA (mRNA)

1. **Heterogeneous nuclear RNA (hnRNA)** is the precursor of mRNA produced in the nucleus (in eukaryotes) with various size and very short half-life, whose molecular weight is larger than that of mature mRNA. hnRNA can be processed to mature mRNA.

2. **Structural features of mRNA in eukaryotes**: contain

(1) the cap sequence at 5′-terminal end (**5′-cap structure**), 7-methyl-guanosine triphosphate (m^7 Gppp),

(2) the polymer of adenylate residues at 3′-terminal end (**3′-poly (A) tail**).

(3) It is believed that both 3′- poly (A) tail and 5′- cap structure are responsible for mRNA transfer from nucleus to cytoplasm, stability of mRNA and regulation of the initiation of the translation.

3. **Function of mRNA** is the template of protein biosynthesis to transfer genetic information from DNA to protein. Three adjacent nucleotides in mRNA are considered as a group to specify a single amino acid, which is also called **codon**.

C. Transfer RNA (tRNA)

1. **Primary structure** of tRNA refers to the sequences about 70 ~ 90 nucleotides of tRNA. It has smallest molecular weight. tRNA contains 10% ~20% rare bases.

2. **Secondary structure** of tRNA refers to the cloverleaf structure. It contains mainly four arms, each arm with a base-paired stem.

(1) The **DHU arm** or D loop is so named due to the presence of dihydrouridine.

(2) The **anticodon arm** with the three adjacent nucleotide bases (anticodon) is responsible for the recognition of codon of mRNA.

(3) The **variable arm** is the most variable in tRNA.

(4) The **TψC arm** contains the sequence of thymine (T), pseudouridine (ψ) and cytosine (C).

(5) The **acceptor's arm** is capped with a sequence of CCA (5′ to 3′) at 3′-end to accept the activated amino acid.

3. **Tertiary structure** of tRNA is Inverted-L shape.

4. **Function of tRNA** is to transfer the activated amino acid during protein bio-synthesis.

D. Ribosomal RNA (rRNA)

1. rRNA has the maximal content, over 80 % of total RNA.

2. It can combine with proteins to form the ribosome, which is the factory of protein biosynthesis.

3. The ribosomes in prokaryotes are composed of two subunits, large (50S) and small (30S) subunits. The 50S subunit contains 5S rRNA and 23S rRNA while 30S subunit contains 16S rRNA.

4. The eukaryotic ribosomes are composed of large (60S) and small (40S) subunits. The 60S subunit contains 28S rRNA, 5S rRNA and 5.8S rRNA while 40S subunit contains 18S rRNA.

E. Other RNAs

Other RNAs refer to the RNAs present in the cell besides the above three RNAs. They play very important role in many aspects, including post-transcriptional modification of RNA and regulation of gene expression (Table 2.3).

Table 2.3 Cellular RNA and their functions

Type of RNA	Abbreviation	Function
Messenger RNA	mRNA	Template of protein synthesis
Transfer RNA	tRNA	Transfer amino acid
Ribosomal RNA	rRNA	Component of ribosome
Heterogeneous nuclear RNA	hnRNA	Precursor of mRNA
Small nuclear RNA	snRNA	Involved in mRNA processing

Continue to Table 2.3

Type of RNA	Abbreviation	Function
Small nucleolar RNA	snoRNA	Involved in rRNA processing
Small cytoplasmic RNA	scRNA	Involved in protein localization
Small interfering RNA	siRNA	Silence complementary target mRNA

[Practices]

[A1 type]

[15] tRNA contains ____________.

A. 3′-CCA-OH B. Cap sequence C. Codons

D. 3′- poly A tail E. Large subunit and small subunit

[16] Cap sequence refers to ____________.

A. m^3Gppp B. m^6Gppp C. m^5Gppp

D. m^4Gppp E. m^7Gppp

[17] Which is true for mRNA structure?________

A. 5′-end has poly (A) cap sequence.

B. 3′-end has mG tail.

C. The secondary structure is single-stranded coil and single-stranded helix.

D. Part of the chain can form double-stranded structure.

E. Three adjacent nucleotides form an anticodon.

[18] Which statement is wrong for RNA?________

A. It mainly has mRNA, tRNA and rRNA.

B. There are only mRNA and tRNA in the cytoplasm.

C. tRNA is the RNA with smallest molecular weight in the cell.

D. rRNA can combine with protein.

E. Not all of RNAs are single-stranded.

[19] Which one is involved in the formation of large subunit in prokaryotes?________

A. 28S rRNA B. 23S rRNA C. 16S rRNA

D. hnRNA E. 5.8S rRNA

[20] Which one can conform to the structural features of tRNA?________

A. 5′-end cap sequence B. 3′-end poly (A) tail C. Anticodon

D. Open reading frame E. 5′-CCA-OH

[21] If the anticodon of tRNA is GAU, the corresponding codon that can be recognized is ________.

A. AUC B. CUA C. CAU

D. AAG E. ACU

[22] Which statement is wrong for mRNA? ____________

A. It is spliced from hnRNA in the nucleus.

B. mRNA in eukaryotes has cap sequence and poly (A) tail.

C. The mRNAs are different with various sizes

D. It is the RNA with longest half-life.

E. It is the template of protein biosynthesis.

[B1 type]

No. 23 ~ 24 share the following suggested answers

A. rRNA B. mRNA C. tRNA

D. hnRNA E. snRNA

[23] RNA that contains the rarest bases is ____________.

[24] RNA that contains both introns and exons is ____________.

No. 25 ~ 27 share the following suggested answers

A. Sequence of nucleotides in the nucleic acid chain

B. Cloverleaf structure

C. DNA double helix

D. DNA supercoil

E. Nucleosome of DNA

[25] The primary structure of the nucleic acid is ____________.

[26] The secondary structure of tRNA is ____________.

[27] DNA in chromatin in eukaryotes is____________.

[X type]

[28] Which statement is true for clover leaf structure of tRNA? ____________

A. The first loop from 5′-end is DHU loop.

B. It has one codon loop.

C. It has one TψC loop.

D. 3′-end has identical sequence of CCA-OH.

E. It has one variable loop.

Ⅳ. Physical and chemical properties of nucleic acids and their applications

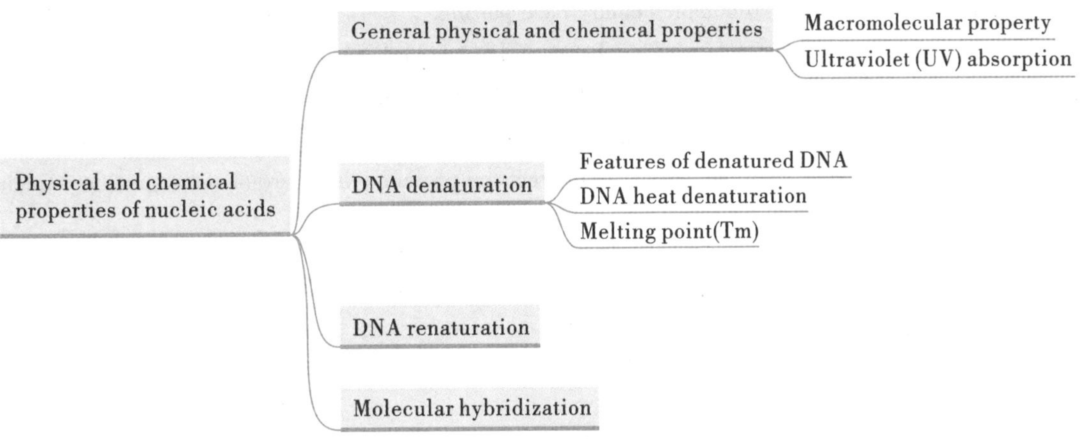

A. General physical and chemical properties of nucleic acid

1. **Macromolecular property** refers to the property of the molecules with large molecular weight. DNA is a linear polymer with very high viscosity while RNA is smaller with smaller viscosity.

2. **Ultraviolet (UV) absorption** refers to the nucleic acids have UV absorption at about 260 nm due

to the presence of the nitrogenous base.

B. DNA denaturation

1. **DNA denaturation** occurs when hydrogen bond between base pairs is disrupted and results in the separation of double helix and change of physicochemical properties due to many physical and chemical factors, such as acid, base, and high temperature. During the denaturation, the hydrogen bond is disrupted while 3′,5′-phosphodiester bond is not.

2. **Features of denatured DNA** include the reduction of viscosity and hyperchromic effect.

a. **Reduction of viscosity** is taken place due to the conversion from double helix to random coil.

b. **Hyperchromic effect** means the increased adsorption at 260 nm after DNA denaturation. It is resulting from the exposure of bases due to the disruption of hydrogen bond and then separation of double helix.

3. **DNA heat denaturation** refers to DNA denaturation is taken place when heated over 70-80℃.

4. **Melting point or melting temperature** (T_m) refers to the temperature when A_{260} value is increased to half of the maximum during DNA heat denaturation.

Tm is associated with high GC content in the molecule. Because cytosine pairs with guanine by three hydrogen bonds, higher GC content, higher interaction, higher temperature for disrupting hydrogen bonds.

C. DNA renaturation

1. **DNA renaturation** refers to the process that two single-stranded DNA separated previously can form a double helix again under proper conditions.

2. During renaturation, the **hypochromic effect** is occurred due to the decrease of A_{260} value.

3. **Annealing** refers to one type of renaturation of thermally denatured DNA. DNA can form double helix again if the temperature is slowly cooling down. It is generally believed that the temperature lower than Tm 5℃ is the optimal condition for DNA renaturation.

D. Hybridization

1. **Hybridization** refers to the process that two polynucleotide chains from different sources can anneal to form double chains according to complementary base pairing.

2. This phenomenon may be taken place between DNA and DNA from different sources, DNA and RNA, or RNA and RNA, which is called **nucleic acid hybridization**.

3. This principle can be used to study the location of a gene in DNA molecule, identify the sequence similarity between two nucleic acid molecules, and detect the presence or failure of some specific sequences in the sample.

[*Practices*]

[A1 type]

[29] The structural change during DNA denaturation is ____________.

A. disruption of phosphodiester bond

B. disruption of N-C glycosidic bond

C. disruption of C-C bond of pentose sugar

D. disruption of C-C bond of base

E. disruption of hydrogen bond between bases

[30] The phenomenon taken place during DNA heat denaturation includes that ____________.

A. polynucleotides are hydrolyzed into mononucleotide

B. the absorbance at 260 nm is increased

C. the base pair is linked by covalent bond

D. the viscosity of the solution is increased

E. the wavelength of the maximal absorption peak is changed

[31] T_m in the physicochemical properties of DNA means ____________.

A. the temperature during replication

B. the temperature during renaturation

C. the temperature when 50% of double strands are dissociated

D. the temperature when DNA changes from B form to A form

E. the temperature when DNA changes from A form to B form

[32] Whose T_m value is highest among the following DNA molecules? ____________

A. The content of [A+T] is 20%.

B. The content of [A+T] is 60%.

C. The content of [G+C] is 30%.

D. The content of [G+C] is 50%.

E. The content of [A+T] is 40%.

Answers

1	2	3	4	5	6	7	8	9	10
D	D	B	E	C	B	C	E	D	E
11	12	13	14	15	16	17	18	19	20
C	E	E	ACD	A	E	D	B	B	C
21	22	23	24	25	26	27	28	29	30
A	D	C	D	A	B	E	ACDE	E	B
31	32								
C	A								

Brief Explanations for Practices

1. In poly-nucleotide chain, both pentose sugar and phosphate form the backbone and bases are located on the side of the chain.

2. The bases mainly have five types, A, T, C, G and U.

3. The bases in DNA are A, G, C and T, while those in RNA are A, G, C and U. Xanthine is present in neither DNA nor RNA, which is the intermediate of nucleotide metabolism.

4. RNA and DNA contain same purines but various pyrimidines, where U is present in RNA while T is in DNA. RNA and DNA contain different pentose sugar, where ribose is in RNA while deoxyribose is in DNA.

5. The maximal absorption peak of nucleic acid is at 260 nm while that of protein is at 280 nm

6. The linkage between nucleotides in nucleic acid is 3′,5′- phosphodiester bond.

7. Chargaff's rule (Base pairing rule): [A] = [T], [G] = [C]

8. Chargaff's rule: The base compositions of DNA among various species are different, while those in different organs or tissues in the same individual are identical. Base pairing rule: [A] = [T], [G] =

[C]

9. The base composition in DNA is A (adenine), G (guanine), C (cytosine) and T (thymine), where A =T and G =C. It is known that G =32.8 %, so C =32.8 %, GC content is 65.6 %. AT content is the rest 34.4 %, thus A = (34.4%)/2=17.2%.

10. The primary structure of DNA is the sequence of polydeoxynucleotides or bases.

11. The primary structure is the sequence of nucleotides. Clover leaf structure is the secondary structure of tRNA. Both DNA supercoil and nucleosome are advanced structure.

12. DNA double helix is composed of two antiparallel deoxynucleotide chains, where hydrophilic deoxyribose-phosphate backbone is on the outside of the helix while bases are stacked inside.

13. DNA double helix is composed of two antiparallel deoxynucleotide chains, where two right-handed deoxyribose-phosphate chains form the backbone. The bases are located inside which are held together by hydrogen bonds (non-covalent bond). Hydrogen bonds between complementary bases can stabilize double helix horizontally and hydrophobic stacking interaction between base planes can stabilize it perpendicularly. In DNA molecule, A=T, G=C, A+G=T+C, therefore, (A+G)/(T+C)= 1.

14. For the secondary structure of DNA, two deoxynucleotide chains are antiparallel.

15. Transfer RNA contains 3′-CCA-OH which can attach activated amino acid. mRNA contains cap sequence, poly A tail and codon. tRNA can combine with protein to form large and small subunit of ribosome.

16. 5′-end of mRNA contains cap sequence, m^7Gppp (7-CH_3-guanosine triphosphate).

17. 5′-end of mRNA contains cap sequence, m^7Gppp, and that of 3′-end contains poly (A) tail. Three adjacent nucleotides following AUG can form one codon. Partial segment of the chain can form double-stranded structure not single-stranded helix.

18. Not only mRNA and rRNA, are present in the cytoplasm.

19. Large subunit in prokaryotes is composed of 23S rRNA, 5S rRNA and 31 proteins. Small subunit (30S) is composed of 16S rRNA and 21 proteins.

20. Answer C is the feature of tRNA while A, B and D are the features of mRNA. for tRNA, CCA-OH is at 3'-end not 5'-end.

21. Anticodon on tRNA can recognize the codon on mRNA due to base-pairing rule. For example, the anticodon on tRNA is GAU in the direction of 5′→3′, it can pair with the codon AUC on mRNA in the direction of 5′→3′. Note: Antiparallel complementation during pairing.

22. Half-life of mRNA is smallest among RNAs, some of which are just several minutes, even the longest of which is only a few hours, thus answer D is wrong.

23. Rare bases include DHU, methylated purine and pyrimidine, pseudouridine and ribothymidine. tRNA contains highest content of random bases in RNAs, about 10-20 %.

24. hnRNA is the primary product of transcription in the nucleus, which can completely pair with DNA template chain. Some nucleotide segments can conduct gene expression while others can not, so the former is called exon while the latter is intron. Thus hnRNA contains both introns and exons.

25. The primary structure of nucleic acid is the sequence of nucleotides.

26. DNA double helix is the secondary structure of deoxyribonucleotides. Clover leaf structure of tRNA belongs to the secondary structure of nucleic acid.

27. The chromatin in eukaryotes is composed of DNA and proteins, whose basic unit is nucleosome.

28. The 1st loop from 5′-end of clover leaf structure of tRNA is DHU loop. The 2nd loop is anticodon loop not codon loop because of three bases in the middle of it, that can complementary to the codon on mR-

NA. The 3rd loop is TψC loop with the feature of T and ψ. All of tRNA contains identical CCA-OH at 3′-end, which can attach the transferred amino acid.

29. During DNA denaturation, two chains are dissociated due to the disruption of hydrogen bond between base pairs.

30. A_{260} of DNA is increased due to more exposure of the conjugated double bonds during the dissociation and has certain proportional relationship with melting degree, which is called hyperchromic effect.

31. T_m is the melting temperature of DNA. It refers to the corresponding temperature when A_{260} is increased to half of the maximum during DNA heat denaturation. 50% of double-stranded DNA is dissociated at this temperature, thus the answer is C.

32. The melting temperature is associated with the base composition of DNA, i. e. , higher GC content, higher melting temperature. For answer A, AT content is smallest and GC content is highest, thus its T_m is highest.

Comprehensive Practices

[A1 type]

[1] Which one is present in RNA not in DNA?________

A. G B. C C. A D. U E. T

[2] If the content of Adenine in the DNA molecule is 15 % , the content of Cytosine should be ____________.

A. 30% B. 15% C. 40% D. 35% E. 25%

[3] Watson - Crick Model of DNA double helix refers to____________.

A. both the base plane and the ribose plane are parallel to the long axis of helix

B. both the base plane and the ribose plane are perpendicular to the long axis of helix

C. the base plane is perpendicular to the long axis of helix while the ribose plane is parallel to it

D. the base plane is parallel to the long axis of helix while the ribose plane is perpendicular to it

E. two chains are parallel in the same direction

[4] 5′-end of the majority of mRNA in eukaryotes has ____________.

A. poly (A) tail B. cap sequence C. stop codon

D. initiation codon E. CCA-OH

[5] Which statement is **NOT** true for tRNA?________

A. tRNA is usually composed of 70-90 nucleotides.

B. tRNA contains rare bases.

C. The secondary structure of tRNA is cloverleaf structure.

D. Anticodon loop contains the anticodon that is composed of three bases CCA.

E. Various tRNA are present in the cell.

[6] Which one is present on the anticodon loop of tRNA and can pair with various bases on mRNA? ________

A. I B. Ψ C. T D. X E. DHU

[7] Which statement is **NOT** true for RNA?________

A. tRNA contains more rare bases than other RNAs.

B. It is usually single-stranded molecule.

C. mRNA contains the genetic codon.

D. rRNA is the place for protein synthesis.

E. RNA contains regional double-stranded structure.

[8] Which statement is true for T_m? ________

A. Higher GC content, higher T_m

B. Higher AT content, higher T_m

C. Higher purity of nucleic acid, larger range of T_m

D. Smaller nucleic acid, larger range of T_m

E. The nucleic acid with higher T_m is usually RNA

[9] Which statement is **NOT** true for nucleic acid hybridization? ________.

A. Single-stranded DNA can also hybridize with RNA chain, which contains identical or nearly identical sequence of complementary bases, to form double helix

B. If two single-stranded DNA from different sources contain roughly identical sequence of complementary bases, they can form a newly hybridized DNA double helix

C. RNA can bind with its encoded polypeptide chain to form hybridized molecule

D. Hybridization can be applied in the research on the structure and function of nucleic acid

E. RNA can hybridize with RNA to form double-stranded structure

[B1 type]

No. 10 ~ 14 share the following suggested answers

A. double-stranded helix

B. regional double-stranded helix

C. supercoil

D. inverted L shape

E. poly (A) sequence

[10] The secondary structure of DNA is ____________.

[11] The tertiary structure of DNA is ____________.

[12] The secondary structure of RNA can form ____________.

[13] The tertiary structure of tRNA is ____________.

[14] 3′-end of mRNA has ____________.

No. 15 ~ 18 share the following suggested answers

A. tRNA B. mRNA C. rRNA

D. hnRNA E. siRNA

[15] The template of protein synthesis is ____________.

[16] The component of ribosome that can be involved in protein synthesis is ____________.

[17] The precursor of mature mRNA is ____________.

[18] The RNA that can carry amino acid and transfer it to ribosome is ____________.

No. 19 ~ 23 share the following suggested answers:

A. Hydrolysis of nucleic acid

B. Dissociation of nucleic acid

C. DNA denaturation

D. DNA renaturation

E. Hybridization

[19] Nucleic acid can form its components in the presence of relevant enzymes: ________

[20] DNA can be separated into two single chains during heating: ________

[21] Nucleic acid can have negative charge in the solution with certain pH value________

[22] Completely or partially complementary nucleic acid molecules can form double-stranded mole-

cule________

[23] Two complementary DNA chains can reform double helix structure by decreasing the temperature ________

[X type]

[24] The factors that can stabilize DNA double helix include: ________

A. Base stacking interaction

B. Phosphodiester bond

C. Ionic bond of phosphate residue

D. Hydrogen bond between bases

E. β-N-glycosidic bond between pentose sugar and base

[25] The molecular features of tRNA include: ________

A. It contains codon loop.

B. It contains amino acid arm.

C. 5′-end has the sequence of CCA.

D. The tertiary structure is Cloverleaf structure.

E. It contains DHU arm.

Answers

1	2	3	4	5	6	7	8	9	10
D	D	C	B	D	A	D	A	C	A
11	12	13	14	15	16	17	18	19	20
C	B	D	E	B	C	D	A	A	C
21	22	23	24	25					
B	E	D	AD	BE					

(***Peng Fan, Li Zhihong***)

Chapter 3

Structure and Function of Glycoconjugates

Examination Syllabus

1. **Glycans**: concept, function.
2. **Glycoprotein and proteoglycan**: concept.

[Major points]

Glycoconjugates (complex carbohydrates): glycoproteins, proteoglycans, and glycolipids.

Composition of glycoconjugates:

Glycoproteins and proteoglycans: covalently-linked proteins and glycans.

Glycolipids: glycans and lipids.

I. Glycans in glycoproteins

- Glycans in glycoprotein
  - Main monosaccharides
    - Glc, Gal, Man, Fuc
    - GlcNAc, GalNAC, NeuAc
  - General structures
    - N-linkedAsn-X-Ser/Thr
    - O-linkedSer/Thr
  - Functions
    - Stabilize the structure of the protein
    - Participate in the folding
    - Affect the targeting transport
    - Participate in intermolecular recognition

Glycans: the carbohydrate components of the complex carbohydrates, besides of single glycosyl group.

Glycoproteins are proteins that have oligosaccharide chains covalentlyattached to their polypeptide backbones.

A. The main monosaccharides types

glucose (Glc), galactose (Gal), mannose (Man), N-acetylgalactosamine (GalNAC), N-acetylglucosamine (GlcNAc), fucose (Fuc), and N-acetylneuraminic acid (NeuAc).

B. General structures of N-linked and O-linked glycoproteins

1. **Three main types of N-linked glycan**: hypermannose type, complex type, and hybrid oligosaccharides.

2. **Glycosylation site** of the N-linked glycoprotein is mainly **Asn-X-Ser/Thr**, and glycosylation site of the O-linked glycoprotein is directly on **Ser/Thr** residues.

3. The N-linked glycans are usually assembled with dolichol phosphate as the carrier, while no glycan carrier is needed for O-linked glycan synthesis.

4. Besides N-linked and O-linked glycan modification, β-N-acetylglucosamine glycosylation is a reversible monosaccharide modification (Figure 3.1).

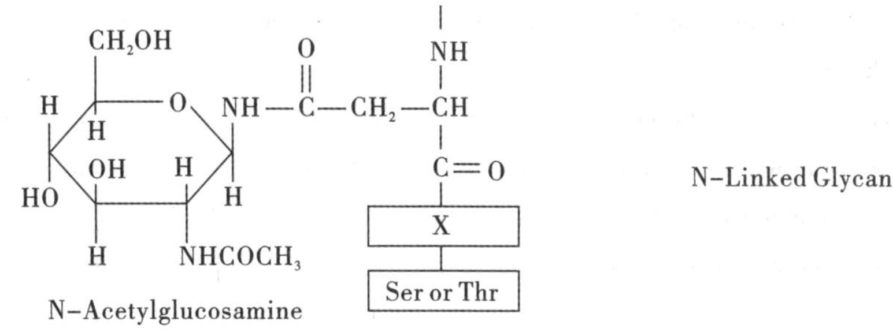

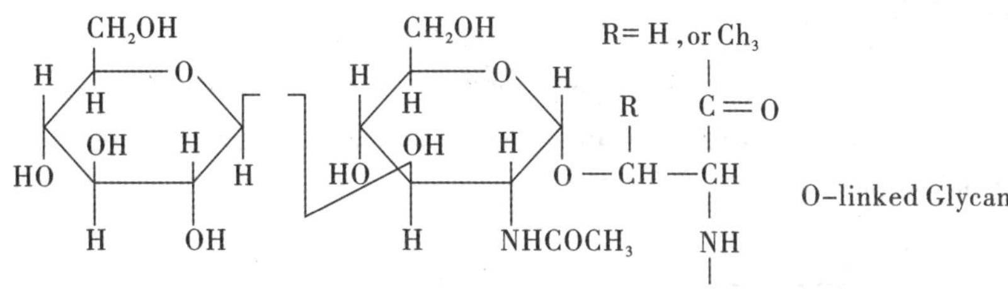

Figure 3.1 Structure of N-linked and O-linked glycoproteins.

C. Functionsof glycans in glycoprotein

1. Glycans can stabilize the structure and prolong the half-life of the protein;
2. Glycans participate in the folding or assembling of the newly generated peptide chains;
3. Glycans affect the targeting transport of glycoproteins in cells;
4. Glycans participate in intermolecular recognition.

Ⅱ. Glycosaminoglycans in proteoglycan molecules

- **Proteoglycan**
  - Glycosaminoglycans (GAGs)
    - hexuronic acid
    - hexosamine
  - Core proteinbinding domain
  - Functions
    - form the intercellular matrix
    - specific features

Proteoglycan is a very complex type of complex carbohydrates, mainly composed of glycosaminoglycans (**GAGs**) and their covalently linked to core proteins.

A. Glycosaminoglycans

1. Glycosaminoglycans are repeating **disaccharide units** composed of hexuronic acid and hexosamine.
2. **Six important glycosaminoglycans**: chondroitin 6-sulfate, keratan sulfate, dermatansulfate, hyaluronate, heparin, and heparan sulfate.

B. Core protein

The core protein contains a binding domain with spermine glycans.

1. All the core proteins contain the corresponding glycosaminoglycan substitution domains, and some of the proteoglycans are anchored on the cell surface or macromolecules of the extracellular matrix via core protein specific domains.
2. Glycosyl is attached to the core protein one by one during proteoglycan biosynthesis.
3. The monosaccharide UDP derivatives act as the monosaccharide donor, which are added one by one to the peptide chain but not as disaccharide unit.
4. Post modifications such as amination and sulfation can also occur after the completion of glycansynthesis.

C. FunctionsofProteoglycan

1. The **main function** of the proteoglycan is to form the intercellular matrix.
2. Diverse proteoglycans own specific features, such as anticoagulant, maintaining the mechanical properties of cartilage, and so on.

Ⅲ. Glycolipids

Glycolipids are compounds in which sugars are linked to lipids with glycosidic bonds through hemiacetal hydroxyl groups.

Glycolipids can be divided into **glycosphingolipids**, **glyceroglycolipids**, and steroid-derived glycolipids.

A. Glycosphingolipids

Glycosphingolipids are glycosylated ceramides.

1. Cerebroside is one kind of neutral glycosphingolipid containing no sialic acid.
2. Sulfatides are acidic glycosphingolipids with glycosyl moieties being sulfonated.
3. Gangliosides are acidic glycosphingolipids containing sialic acid.

B. Glyceroglycolipids

Myelin contains glyceroglycolipids. The most common glyceroglycolipids are monogalactosyl diacylglycerols and digalactosyl diacylglycerols, which are formed via the O-linked glycosidic bond between 3-OH of diacylglycerol and glycosyl residue.

Ⅳ. Functions of Glycan structure

Glycan structure indicates rich biological information.

1. Glycan isessential for glycoprotein acting diverse functions.

2. Glycan structural diversity contains abundant biological information.

(1) Space structure diversity of glycans is the fundamental basis for bio-information delivery.

(2) Space structure diversity of glycans is regulated by gene encoding of glycosyltransferases and glycosidases.

[*Practices*]

[**A1 type**]

[1] Which of the following monosaccharides are not involved in the sugar chain composition of glycoproteins?________

A. GalNAc B. Fuc C. NeuAc

D. Aluronic acid E. Gal

[2] Which of the following derivative is the activated glycosyl donor in the N-linked glycansynthesis?

A. UDP or GDP derivative B. UDP or CDP derivative C. ADP or GDP derivative

D. TDP or GDP derivative E. ADP or TDP derivative

[3] The glycosylation sites of N-linked oligosaccharide chains should be____________.

A. Asp-X-Ser/Thr B. Glu-X-Ser/Thr C. Asn-X-Ser/Thr

D. Gln-X-Ser/Thr E. None of the above

[4] Which of the following is NOT required for the synthesis and completion of N- linked glycans?

____

A. Glycosyltransferase B. Glucoside hydrolase C. Dolichol (dol) phosphate

D. Modification of sugar chains in nuclear

E. Being modifiedin the Golgi body

[5] Which of the following description about O-linked glycans and their synthesis is NOT true?

____________.

A. No glucose contained

B. The sugar chains are covalently bound with only Ser residues

C. Glycosyltransferase is required

D. No glycan carrier is required

E. Containing core disaccharide composed of N-acetylgalactosamine and galactose

[6] Which of the following description about proteoglycan is NOT true? ____________.

A. Proteoglycans are composed of glycosaminoglycans and core proteins via covalent linkage

B. The protein component is less than glycansin proteoglycan

C. The disaccharide units are firstly synthesized and repeated

D. Glycosaminoglycans consist of repeating disaccharide units

E. N-linked or O-linked glycans are contained in Proteoglycans

[7] N-linked glycoprotein contains ____________.

A. mannose B. glycosaminoglycan C. triple helix

D. two peptide chains E. three peptide chains

[X type]

[8] N-linked oligosaccharide chains ____________.

A. Containing core pentasaccharide.

B. Linked with the glutamine residue of the protein.

C. Divided into hypermannose type, complex type, and hybrid oligosaccharides.

D. N-Acetylneuraminic acid is contained.

E. The core pentasaccharides of the hybrid oligosaccharides contain 2-5 branched glycans being attached.

[9] Glycosaminoglycan____________.

A. Combining with the core protein with non-covalent binding to form proteoglycan.

B. Consisting of disaccharides units.

C. Consisting of hyaluronic acid, heparin, chondroitin sulfate, etc.

D. Consisting of glucuronic acid.

E. Consisting of serine.

Answers

1	2	3	4	5	6	7	8	9
D	A	C	D	B	C	A	ACD	BCD

(*Lin Li*)

Chapter 4

Enzyme

Examination Syllabus

1. **Enzyme catalysis**: molecular constituents and catalytic effects of enzymes, characteristics of enzymatic reaction, enzyme–substrate complex.

2. **Cofactors**: vitamins and coenzymes, roles of coenzymes, roles of metal ions.

3. **Enzyme kinetics**: the conception of K_m and V_{max}, optimum pH and optimum temperature.

4. **Inhibitors and activators**: irreversible inhibition, reversible inhibition, activators.

5. **Regulation of Enzyme Activity**: allosteric regulation, covalent modification, activation of enzyme precursors, isoenzymes.

6. **Ribozyme**: the definition of ribozyme.

Major points and Practices

Enzymes are biological catalysts that can accelerate the rate of a reaction. Enzymes can be made of proteins.

Ribozymes are RNA molecules that are capable of catalyzing specific biochemical reactions, usually the cleavage of other RNA molecules.

Ⅰ. General Properties of Enzymes

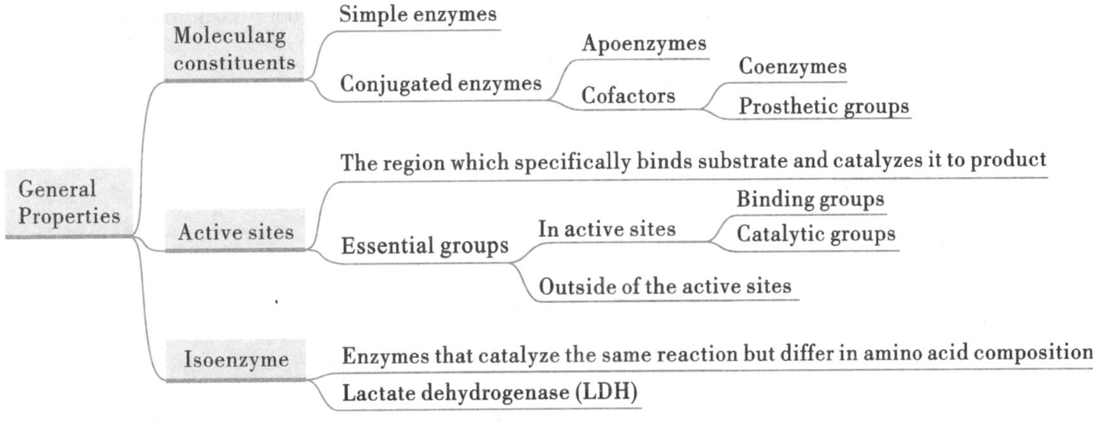

A. Molecular constituents of enzymes

1. Enzymes can be divided into **simple enzymes** and **conjugated enzymes** according to their composition. Simple enzymes are only composed of proteins. Conjugated enzymes contain **apoenzymes** (protein parts) and **cofactors** (nonprotein parts).

2. Neither apoenzyme nor cofactor alone has catalytic activity, only an enzyme together with its bound cofactor has its complete, catalytic activity and is called a **holoenzyme**.

3. **Cofactors** are nonprotein compounds that participate in the catalytic process. Commonly, most of the cofactors are **small organic molecules** or **metal ions**. Cofactors can be divided into two general classes: coenzymes and prosthetic groups.

(1) **Coenzymes** are small organic molecules that are required for the activity of certain enzymes. Coenzymes serve as recyclable shuttles or group transfer agents.

(2) **Prosthetic groups** are small inorganic ions that required for proper structure or to aid in catalysis for up to 70% of enzymes.

4. Many cofactors are derivatives of vitamins, especially the **water-soluble B vitamins** supply important components of numerous cofactors. Reactions that depend on such cofactors are blocked when the vitamin is deficient in the diet.

B. Active sites of enzymes

1. **Active site** is a three-dimensional, local region of the enzyme, the region is composed of several essential groups of amino acids, that has a special spatial structure which specifically binds substrate and catalyzes it to product.

2. **Essential groupsin active sites** include **binding groups** and **catalytic groups**, which are involved in binding substrates and transferring substrates to products. Essential groups **outside of the active** sites contribute to maintaining the spatial structure of active sites.

C. Isoenzyme

1. **Isozymes** are the multiple forms of enzymes that catalyze the same reaction at different location or times but differ in molecular structures, physical properties, and reaction kinetics.

2. **Lactate dehydrogenase (LDH)** is a tetramer containing two different subunits: H subunit and M subunit, which can form five different isozymes, LDH1, LDH2, LDH3, LDH4 and LDH5. Lactate dehydrogenase isoenzymes in different tissues have different expression profiles.

3. The characteristic tissue distribution of isoenzymes makes measurement of their concentrations in the plasma a useful diagnostic technique that can provide important information about tissue damage. For example, isozymes of LDH are used to detect myocardial infarctions (Figure 4.1).

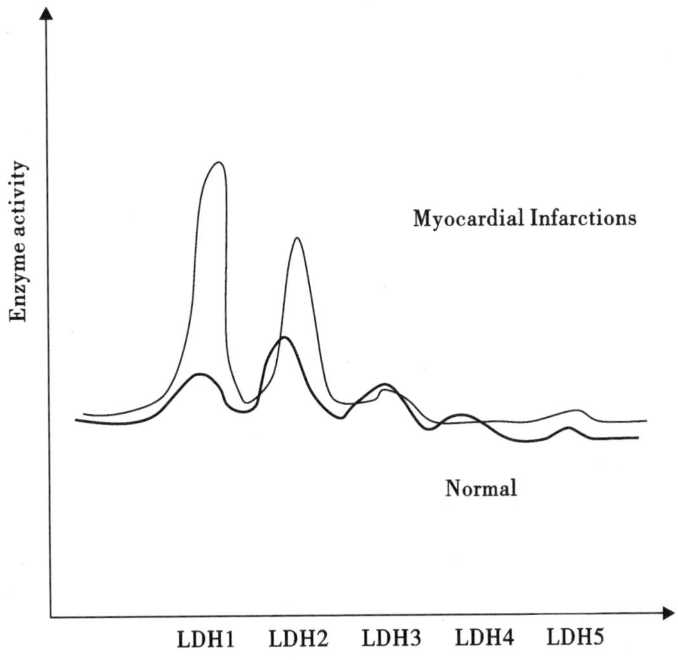

Figure 4.1 The changes of LDH is ozymes in myocardial infarctions

[*Practices*]

[A1 type]

[1]About the cofactors, which one is NOT true?____________.

A. Cofactors are nonprotein molecules required for an enzyme to be active

B. Cofactors loosely bind to the protein portion in enzyme molecules

C. Cofactors are usually (but not always) synthesized from vitamins

D. Cofactors are generally divided into coenzymes and prosthetic groups

E. Cofactors bind to the enzyme by covalent or noncovalent forces

[2]About the active site of enzymes, which one is NOT true?____________.

A. The active site cannot bind cofactors

B. The active site is formed by one or more regions of the polypeptide chain

C. The active site contains functional groups that participate in the reaction

D. Catalysis occurs at the active site

E. The active site generally takes the forms of a cleft or pocket

[3]Which of the following is correct about the isozymes? ____________.

A. They are same in molecule structure

B. They catalyze the same reaction but differ in molecular structure

C. They catalyze different reactions at different location or times

D. They catalyze the same reaction and have the same physical properties

E. They are encoded by the same genes

[4]About the holoenzyme, which one is NOT true?____________.

A. Holoenzyme is composed of apoenzyme and cofactor

B. Apoenzyme is responsible for the specificity of reactions

C. Coenzyme is responsible for the type and property of reactions

D. Only a holoenzyme has its completely catalytic activity

E. Apoenzyme alone has catalytic activity

[B1 type]

No. 5 ~ 7 share the following suggested answers

A. Vitamin B_1 B. Vitamin B_2 C. Vitamin B_{12}

D. Vitamin PP E. Pantothenic acid

[5] The vitamin contained in FAD is ___________.

[6] The vitamin contained in NAD^+ is ___________.

[7] The vitamin contained in CoA is ___________.

Ⅱ. Mechanism of Enzyme Action

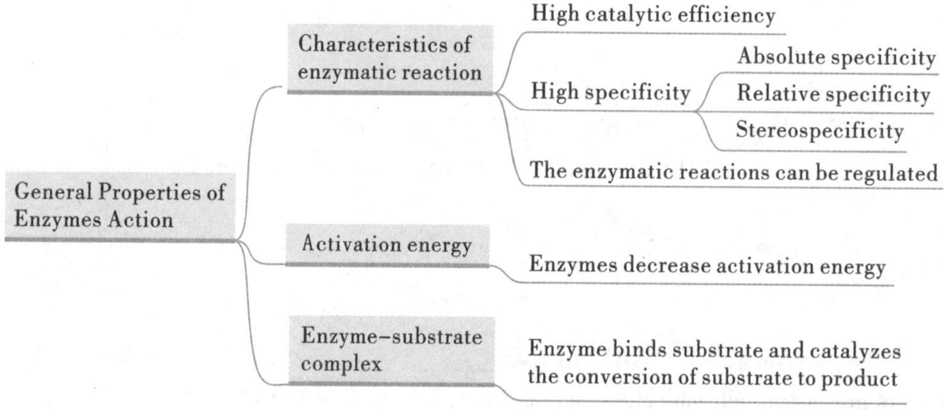

A. Characteristics of enzymatic reaction

1. Enzymes have **high catalytic efficiency**.

2. Enzymes have **high specificity** for their substrates.

(1) **Absolute specificity**

Some enzymes only react with a structurally specific substrate, catalyze a unique reaction to produce structurally specific products.

(2) **Relative specificity**

Some enzymes only catalyze one type of substrates or a specific chemical bond regardless of the substrates.

(3) **Stereospecificity**

Some enzymes are highly specific in stereoisomer which typically catalyze reactions only of one stereoisomer of a given compound.

3. The enzymatic reactions can **be regulated**.

The activities of enzymes can be regulated to maintain a stable intracellular environment.

B. Enzymes decrease the activation energy

1. **Activation energy** is the amount of energy required to convert 1 mol of substrate molecules from the ground state to the transition state.

2. The rate of reactions is determined by their activation energy. Enzymes can speed up the rates by **lowering the activation energy** without influencing the equilibrium of a reaction.

C. Enzyme-substrate complex

1. Enzyme binds substrate and catalyzes the conversion of substrate to product.
2. Enzyme catalysis has three basic steps.

(1) Binding of substrate: E + S↔ES

(2) Conversion of bound substrate to bound product: ES↔EP

(3) Release of product: EP↔E + P

Ⅲ. Enzyme kinetics

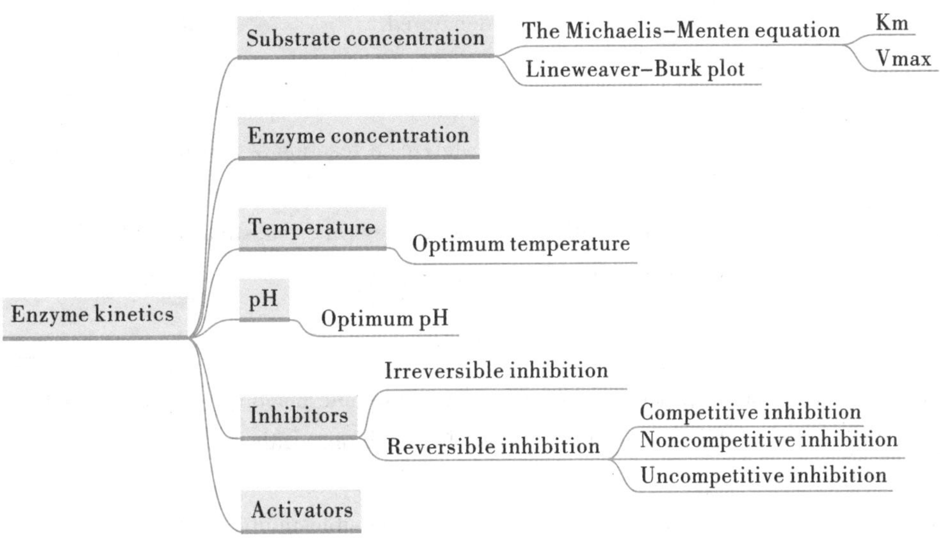

A. Substrate concentration

1. For a typical enzyme reaction (enzyme reaction is treated as if it had an only single substrate and a single product), as substrate concentration is increased, initial velocity (v_i) increases until it reaches a maximum value V_{max}.

2. **The Michaelis-Menten equation** illustrates in mathematical terms the relationship between initial velocity v_i and substrate concentration [S].

$$v_i = \frac{V_{max}[S]}{K_m + [S]}$$

(1) **The Michaelis constant K_m** is the substrate concentration at which v_i is half the maximal velocity ($V_{max}/2$) attainable at a particular concentration of enzyme. The K_m describes an enzyme's affinity for the substrate.

(2) V_{max} is the reaction velocity when enzyme is saturated by substrate.

3. A linear form of the Michaelis-Menten equation is used to determine K_m and V_{max}:

$$\frac{1}{v_i} = \frac{K_m}{V_{max}} \times \frac{1}{[S]} + \frac{1}{V_{max}}$$

In this equation, the relationship between $1/v_i$ and $1/[S]$ can be described as a straight line called a **double-reciprocal plot** or **Lineweaver-Burk plot**.

B. Enzyme concentration

When an enzymatic reaction reaches the maximal velocity, the maximal velocity can be proportionally accelerated by increasing enzyme concentration.

C. Temperature

1. **Optimum temperature** is the temperature at which an enzyme catalyzes a reaction at maximum efficiency.

The optimum temperature of most enzymes is around 25 ~ 40℃.

2. When the temperature is less than optimum temperature, raising the temperature increases the rate of enzymatic reactions. When the temperature is more than optimum temperature, the increasing temperature can decrease thevelocity, the enzyme is going to be denatured.

D. pH

1. **Optimum pH** is the pH at which an enzyme catalyzes a reaction at maximum efficiency. pH 5 ~ 8 is optimum for the most enzyme.

2. pH can affect the ionization of enzyme, substrate and enzyme-substrate complex. There is an optimum pH for each enzyme, at which enzyme activity is maximum.

E. Inhibitors

Inhibitors are compounds that decrease the rate of an enzymatic reaction. Many small molecules or ions known as inhibitors can alter the activity of an enzyme by combining with it.

1. **Irreversible inhibition**

Irreversible inhibitors bind covalently with enzymes. Irreversible inhibitors, form a covalent bond with or destroy a functional group on an enzyme that is essential for the enzyme's activity. Ordinarily, irreversible inhibition can be overcome mainly by the synthesis of new enzyme.

2. **Reversible inhibition**

Reversible Inhibitors do not react covalently with enzymes but undergo rapid, equilibrium binding. The inhibition is therefore instantaneous (**Figure** 4.2, **Table** 4.1).

(1) **Competitive inhibition**

Competitive inhibitor competes with the substrate by binding to the active site of the enzyme. Competitive inhibitors typically resemble substrates. Competitive inhibitors increaseK_m but do not affect V_{max}. Many drugs are enzyme competitive inhibitors.

(2) **Noncompetitive inhibition**

In noncompetitive inhibition, binding of the inhibitor does not affect the binding of substrate. Noncompetitive inhibitors bind enzymes at sites distant from the substrate-binding site. Noncompetitive inhibitors lowerV_{max} but do not affect K_m.

(3) **Uncompetitive inhibition**

Uncompetitive inhibitors bind only to the enzyme-substrate complex but not the free enzyme. Uncompetitive inhibitors reduce both K_m and V_{max}.

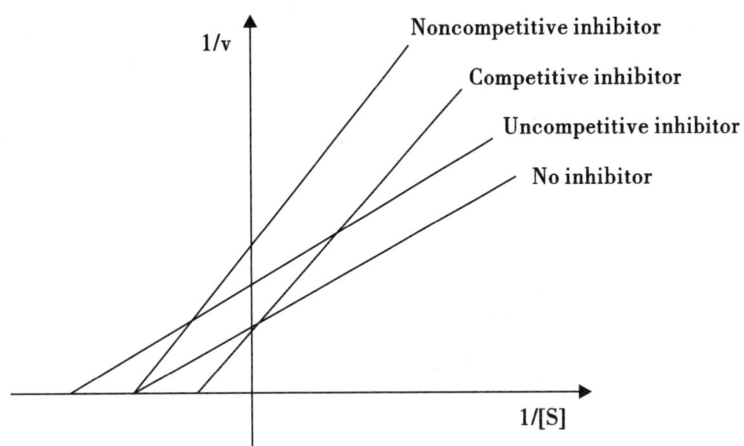

Figure 4.2 Lineweaver-Burk plot of reversible inhibition

Table 4.1 Comparison of the three reversible inhibitions

Characteristics	Competitive inhibition	Noncompetitive inhibition	Uncompetitive inhibition
Inhibitor binds with	Enzyme	Enzyme, Enzyme-substrate	Enzyme-substrate
Kinetic parameters			
K_m	↑	–	↓
V_{max}	–	↓	↓

↑, increased; ↓, decreased; –, no change

F. Activators

Activators are compounds that can alter the activity of an enzyme or increase the rate of an enzymatic reaction. Many of the activators are metal cations.

[Practices]

[A1 type]

[8] Enzymes accelerate reaction rates due to which one of the following?________.

A. Increase the effective concentration of the substrate

B. Reduce the frequency of collisions between the substrate molecules

C. Lower the energy of activation required to reach a transition state

D. Increase the free energy level of final state of the reaction

E. Decrease the free energy level of the initial state of the reaction

[9] About the K_m, which one is NOT true?________.

A. The K_m is the dissociation constant of ES complex

B. The K_m is the substrate concentration at which v_i is the half of V_{max}

C. The K_m describes an enzyme's affinity for the substrate

D. The K_m is related to the structure of enzyme

E. The K_m is related to the properties of substrate

[10] In physiological conditions, when the substrate is sufficient, the rate of enzymatic reaction is determined by____________.

A. Enzyme concentration

B. Sodium concentration

C. Temperature

D. pH value

E. Coenzyme

[11] About the inhibitors, which one is true?________.

A. All the inhibitors bind to enzymes by covalent bound

B. Many drugs and toxins act by inhibiting enzymes

C. The effect of noncompetitive inhibitors are overcome by raising the [S]

D. Competitive inhibitors bind to the ES complex

E. Uncompetitive inhibitors bind to the free enzyme

[B1 type]

No. 12 ~ 13 share the following suggested answers

A. is a competitive inhibitor

B. is a noncompetitive inhibitor

C. increases the V_{max}

D. decreases the K_m

E. none of the above choices are correct.

[12] Curve B depicts the effect of an inhibitor on the system described by curve A. This inhibitor ____________.

[13] Curve C depicts the effect of a different inhibitor of the system described by curve A. This second inhibitor ____________.

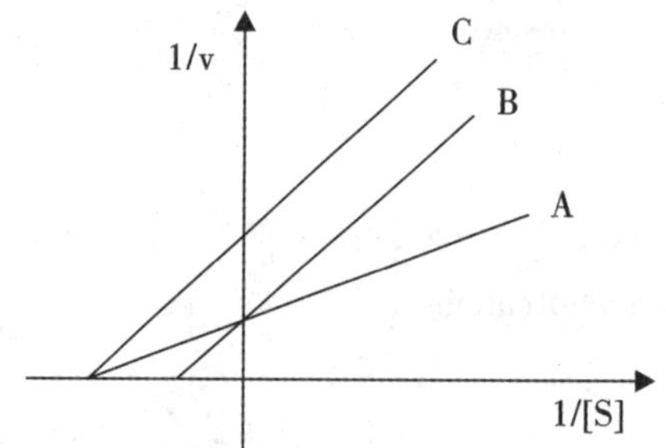

[X type]

[14] About the effect of temperature on enzyme activity, which ones are correct?________.

A. Temperature influences the rate of enzymatic reactions

B. The optimum temperature of most enzymes is around 25 ~ 40℃

C. Raising the temperature always increases the rate of enzymatic reactions

D. At excessively high temperature, most enzymes become denatured

E. The optimum temperature is the constant of enzyme

Ⅳ. Regulation of enzyme

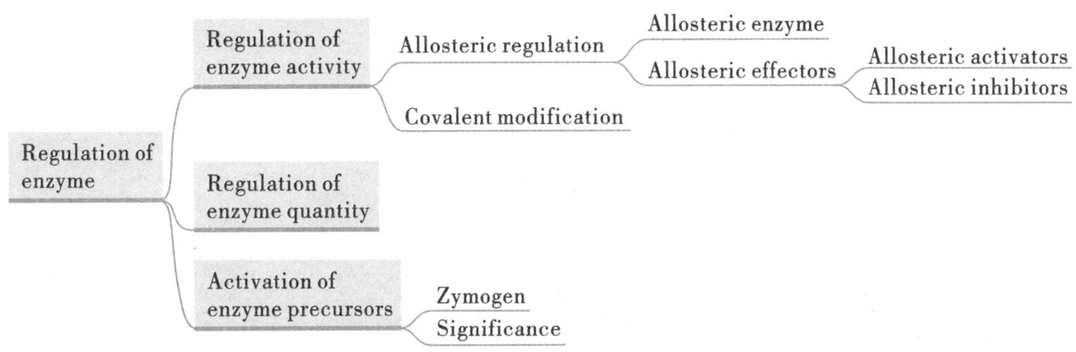

A. Regulation of enzyme activity

1. Allosteric regulation

(1) **Allosteric enzyme** is enzyme whose activity is regulated by effectors through altering its conformations. Allosteric enzymes do not obey Michaelis–Menten kinetics.

(2) Allosteric enzymes are regulated by molecules called **allosteric effectors** that bind to regulatory site, distant from the active site.

(3) Allosteric effectors which can enhance the affinity of the enzyme for substrate are **allosteric activators**. On the contrary, allosteric effectors which can decrease the enzyme's affinity for substrate are **allosteric inhibitors**.

2. Covalent modification

(1) Many enzymes may be regulated by covalent modification by adding (or removing) a functional group to (or from) the enzymes, such as phosphoryl, adenylyl, uridylyl, adenosine diphosphate ribosyl, and methyl group.

(2) Enzyme **phosphorylation** and **dephosphorylation** are the common way found in metabolic regulation, gene regulation, and signaling pathways.

B. Regulation of enzyme quantity

The quantity of an enzyme depends on the enzyme synthesis and degradation. In human, changes in protein level serve long-term adaptive requirement, whereas changes in catalytic efficiency are best suited for rapid and transient alterations in metabolite flux.

C. Activation of enzyme precursors

1. Some of enzymes are synthesized in vivo as an inactive precursor, **zymogen**, which can be converted into active enzyme by cleavage of some covalent bonds in their peptide chains.

2. This type of regulation or control of the activity of enzyme is very important for the protection of the cells that produce the digestive enzymes. The synthesis and secretion of proteases as catalytically inactive proenzymes protect the tissues of origin from autodigestion.

[Practices]

[A1 type]

[15] An allosteric inhibitor of an enzyme usually ____________.

A. binds to the active site

B. participates in feedback regulation

C. denatures the enzyme

D. causes the enzyme to work faster

E. is a hydrophobic compound

[16] About the allosteric enzyme, which one is NOT true?________.

A. Most of them are key enzymes in metabolic pathways

B. Their binding to allosteric effects is reversible

C. Containing catalytic subunit and binding subunit

D. The activity is regulated by effectors through altering its conformations

E. They also obey Michaelis–Menten kinetics

[17] Which one does not belong to covalent modification?________.

A. Phosphorylation or dephosphorylation

B. Acetylation or deacetylation

C. Methylation or demethylation

D. Adenylation or deadenylation

E. Hydrogenation and dehydrogenation

[18] About the zymogen, which one is NOT true?________.

A. Zymogen do not have catalytic activity

B. In human, many enzymes are released as inactive precursors

C. Zymogen can protect the tissues of origin from autodigestion

D. The activation of zymogen is achieved through covalent modification

E. Pepsinogen can be activated by gastric acid

[X type]

[19] About the regulation of enzyme, which ones are correct?________.

A. All the enzymes can be regulated

B. Enzyme activity is mainly regulated through allosteric regulation and covalent modification

C. The covalent modification involves only the addition function groups to the enzymes

D. Changes in enzyme level requirea comparatively long time

E. Allosteric regulation is a rapid way to regulate the enzyme activity

Answers

1	2	3	4	5	6	7	8	9	10
B	A	B	E	B	D	E	C	A	A
11	12	13	14	15	16	17	18	19	
B	A	B	ABD	B	E	E	D	BDE	

Brief Explanations for Practices

1. Cofactors can be divided into two general classes: coenzymes and prosthetic groups. Coenzymes bind

in a transient, dissociable manner either to the enzyme or to a substrate. Prosthetic groups are distinguished by their tight, stable incorporation into a protein's structurc by covalent or noncovalent forces.

2. Usually, the cofactors participate in the composition of the enzyme activity center.

3. Isoenzymes are the multiple forms of enzymes that catalyze the same reaction at different location or times but differ in molecular structure, physical properties, and reaction kinetics. They are encoded by different genes.

4. Only a holoenzyme has its completely catalytic activity. The apoenzyme is required for catalytic activity, determining the substrate specificity.

8. Enzymes increase the rate of the reaction by decreasing the activation energy (the amount of energy required to convert 1mol of substrate molecules from the ground state to the transition state). The enzyme does not change the initial energy level of the substrate or the final energy level of the products.

9. The K_m value is not always constant, which is related to enzyme structure, substrate structure, pH, temperature and so on, but not the enzyme concentration.

10. In physiological conditions, both temperature and pH value are stable. When the substrate is sufficient, the rate of enzymatic reaction is influenced by the enzyme concentration.

11. Irreversible inhibitors bind covalently with enzymes, but reversible inhibitors do not react covalently with enzymes. The effect of competitive inhibitors can overcome by raising the [S]. Competitive inhibitors bind to the free enzyme. Uncompetitive inhibitors bind to ES complex.

12. A competitive inhibitor of an enzyme increases K_m but does not affect V_{max}. So the intercept on y ($1/V_{max}$) doesn't change, the intercept on x ($-1/K_m$) increases in a double-reciprocal plot.

13. A noncompetitive inhibitor of an enzyme decreases V_{max} but does not affect K_m. So the intercept on y ($1/V_{max}$) increased, the intercept on x ($-1/K_m$) doesn't change in a double-reciprocal plot.

14. When the temperature is less than optimum temperature, raising the temperature increases the rate of enzymatic reactions. When the temperature is more than optimum temperature, the increasing temperature can decrease the velocity, the enzyme is going to be denatured.

15. Allosteric effectors bind to regulatory site, distant from the active site of enzyme. Allosteric effectors do not denature enzymes they bind. Allostericinhibitors can decrease the enzyme's affinity for substrate.

16. Allosteric effectors can change the enzyme's affinity for substrate (K_m). Allosteric enzymes do not obey Michaelis-Menten kinetics.

17. The covalent modification involves the addition to and removal from the enzyme, by separate enzymes, of the groups such as phosphoryl, adenylyl, uridylyl, adenosine diphosphateribosyl, and methyl group.

18. Zymogen can be converted into active enzyme by cleavage of some covalent bonds in their peptide chains.

19. Usually, the key enzymes in metabolic pathways can be regulated. The covalent modification involves the addition to and removal from the enzyme of the groups.

[Comprehensive Practices]

[A1 type]

[1] The relationship between an enzyme and a reactant molecule can best be described as ________.

A. atemporary association

B. an association stabilized by a covalent bond

C. one in which the enzyme is changed permanently

D. apermanent mutual alteration of structure

E. completely complementary binding

[2]The pancreatic enzymes, trypsin, chymotrypsin and elastase all have ____________.

A. the same catalytic triad at their active sites

B. similar sequences and tertiary structures

C. the same catalytic mechanism

D. similar processing pathways from inactive zymogens

E. all of the above choices are correct

[3]The active site of an enzyme ____________.

A. is formed only after addition of a specific substrate

B. is directly involved in binding of allosteric inhibitors

C. is found at the center of globular enzymes

D. binds competitive inhibitors

E. remains rigid and does not change shape

[4]The factors related to the kinetics of enzyme reaction are included, EXCEPT____________.

A. pH and temperature B. [Enzyme] C. [Substrate]

D. inhibitor E. time

[5]The effect of noncompetitive inhibitor on enzyme-catalyzed velocity is ____________.

A. K_m increased, V_{max} unchanged

B. K_m decreased, V_{max} decreased

C. K_m unchanged, V_{max} decreased

D. K_m decreased, V_{max} increased

E. K_m decreased, V_{max} unchanged

[6]Which of the following is NOT the mechanism for enzyme regulation in cells?

A. Binding of regulatory peptides via disulfide bonds

B. Proteolysis

C. Covalent modification

D. Induced changed in conformation

E. Protein synthesis

[7]The physiological significance of zymogen activation is________.

A. to accelerate metabolism

B. to resume enzyme activity

C. to facilitate growth

D. to avoid self damage

E. to protect enzyme activity

[8]The difference between coenzymes and prosthetic groups is ____________.

A. coenzymes are organic molecules, and prosthetic groups are metal ions

B. coenzymes incorporate into enzyme by covalent forces

C. coenzymes loosely bind to the enzyme and easily to be removed

D. coenzymes but not prosthetic groups participate in enzymatic reactions

E. prosthetic groups do not contain vitamins

[9]The meaning of K_m is ____________.

A. the affinity of enzyme binding to substrate

B. the optimum enzyme concentration

C. the rate of enzymatic reaction

D. the maximal velocity of enzymatic reaction

E. the substrate concentration at the maximal velocity

[10]The common feature of allosteric regulation and covalent modification is ____________.

A. inducing the conformation changes of enzyme

B. a rapid and transient alteration

C. as low mechanism for regulating enzyme

D. the enzyme is chemically modified

E. the effect can be amplified

[11]Which of the following is correct about the ribozymes?________.

A. The ribozymes do not have catalytic activity

B. The essence of ribozymes is protein

C. The essence of ribozymes is ribonucleic acid

D. The ribozymes are nuclease

E. The coenzyme of the ribozymes is CoA

[12]In a one-substrate enzymatic reaction, when $[S] \ll K_m$, ________.

A. the velocity of reaction is maximum

B. the velocity of reaction is too slow to measure

C. the velocity of reaction is proportionate to initial [S]

D. the velocity of reaction doesn't change when increasing the [S]

E. the velocity of reaction decreases when increasing the [S]

[B1 type]

No. 13 ~ 14 share the following suggested answers

A. monomeric enzyme

B. oligomeric enzyme

C. multifunctional enzyme

D. simple enzyme

E. conjugated enzyme

[13]The enzymes which are located in the same peptide chain but have different catalytic functions are ____________.

[14]The enzymes consisting of apoenzyme and cofactor are____________.

No. 15 ~ 16 share the following suggested answers

A. irreversible inhibition

B. feedback inhibition

C. competitive inhibition

D. noncompetitive inhibition

E. uncompetitive inhibition

[15]Inhibition of phosphate pesticide on choline esterase is ____________.

[16]Inhibition of sulfanilamide on dihydrofolate synthetase is____________.

[X type]

[17]About the enzyme, which ones are correct?____________

A. Enzymes do not invent new reaction.

B. Enzymes just make reactions occur faster.

C. Each enzyme usually catalyzes a specific biochemical reaction.

D. The enzyme forms an enzyme-substrate complex in its active site.

E. The enzymes can affect equilibrium of their catalyzed reactions.

[18] Which of the following cofactors involve in transferring of hydrogen ions?____________.

A. NAD^+ B. FAD C. FMN

D. Tetrahydrofolic acid E. Coenzyme A

[19] About the covalent modification, which ones are correct?____________.

A. The enzymes that can be regulated by covalent modification have active and inactive forms

B. The covalent modification is a process catalyzed by other enzyme molecules

C. The effect of covalent modification can be amplified

D. This is a slow mechanism for regulating enzyme activity

E. Enzyme phosphorylation and dephosphorylation are the common way found in covalent modification

Answers

1	2	3	4	5	6	7	8	9	10
A	D	D	E	C	A	D	C	A	B
11	12	13	14	15	16	17	18	19	
C	C	C	E	A	C	ABCD	ABC	ABCE	

(*Lai Mingming*)

Chapter 5

Vitamins

[*Examination Syllabus*]

1. **Lipid-soluble vitamins**: function, deficiency
2. **Water-soluble vitamins**: function, deficiency

Ⅰ. Lipid-soluble vitamins

A. Characteristic of Lipid-soluble vitamins

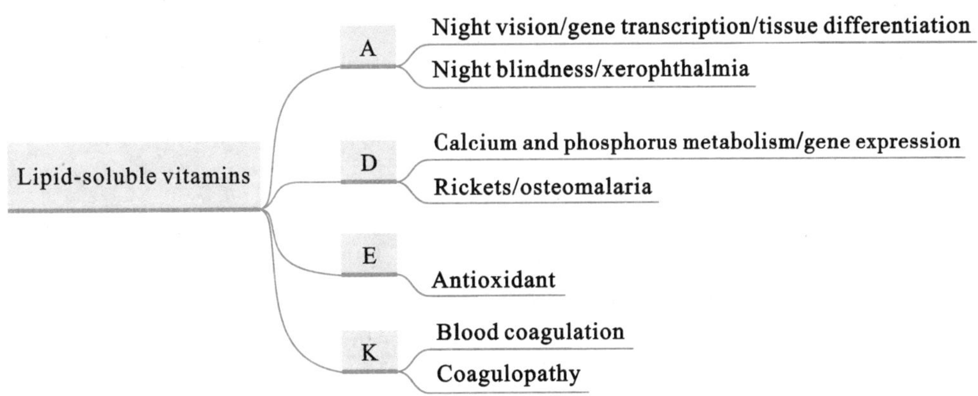

(1) Lipid-soluble vitamins are absorbed through the intestinal tractwith the help of lipids.

(2) They are transported in the blood in lipoproteins or attached to specific binding proteins.

(3) They are not readily excreted, and significant quantities are stored in the liver and adipose tissue.

(4) Excessive intake can result intoxic effect.

B. Functions of Lipid-soluble vitamins

Lipid-soluble vitamins have diverse functions:

(1) directly taking part in metabolism.

(2) regulating related metabolism.

(3) modulating gene expression (Table 5.1).

Table 5.1 Lipid-soluble vitamins

Vitamin	Active form	Function	Deficiency	Signs and symptoms
A	Retinol Retinal Retinoic acid	Night vision Promotion of growth Gene expression Differentiation and maintenance of epithelial tissues	Night blindness Xerophthalmia	Increased visual threshold Dryness of cornea
D	1,25-$(OH)_2D_3$	Calcium and phosphorus metabolism Gene expression	Rickets Osteomalacia	Soft, pliable bones
E	–	Antioxidant	Rare	–
K	Menadione Menaquinone Phylloquinone	γ-Carboxylation of glutamate residues in clotting and other proteins	coagulopathy	Bleeding

Ⅱ. Water-soluble vitamins

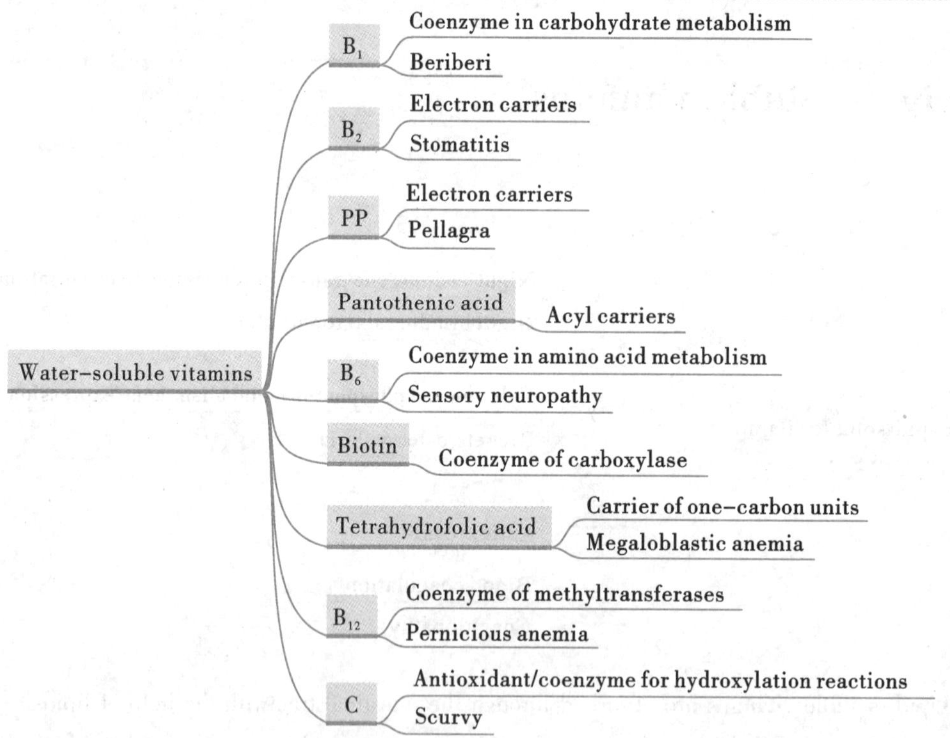

1. **Water-soluble vitamins** could dissolve easily in water and travel freely in the blood. They are not stored in the tissue and need intake from food frequently. They Disorder of absorption or long-term insufficient in food can result in related deficiency disease.

2. **Water-soluble vitamins** function mainly as enzyme cofactors (Table 5.2).

Table 5.2 Water-soluble vitamins

Vitamin	Active form	Function	Deficiency	Signs and symptoms
B_1	TPP	Coenzyme in carbohydra temetabolism: Pyruvate→acetyl CoA α-ketoglutarate→Succinyl CoA Succinyl CoA	Beriberi	Tachycardia, vomiting, convulsions
B_2	FMN, FAD	Electron transfer	R are	Dermatitis Angular stomatitis
PP	NAD^+, $NADP^+$	Electron transfer	Pellagra	Dermatitis Diarrhea Dementia
Pantothenic acid	CoA ACP	Acyl carrier	R are	-
B_6	Pyridoxal phosphate	Coenzyme for enzymes, particularly in amino acid metabolism	R are	Glossitis Neuropathy
Biotin	-	Carboxylation reactions	R are	-
Folic acid	FH_4	Transfer one-carbon units; synthesis of methionine, purines, and thymidine monophosphate	Megaloblastic anemia	Anemia
B_{12}	Methylcobalamin Deoxyadenosylcobalamin	Coenzyme for reactions: Homocysteine→methionine Methylmalonyl CoA→succinyl CoA	Pernicious anemia	Megaloblastic anemia
C	Ascorbic acid	Antioxidant Coenzyme for hydroxylation reactions	Scurvy	Sore, spongy gums Loose teeth Poor wound healing

[*Practices*]

[**A1 type**]

[1] The active form of vitamin D is________.

A. 25-(OH)D_3 B. 1,25-$(OH)_2D_3$ C. 1,24,25-$(OH)_2D_3$

D. 24,25-$(OH)_2D_3$ E. Not above all

[2] The active form of folic acid is________.

A. F B. FH_2 C. FH_4 D. NAD^+ E. FAD

[3] In vivo, ____________ can be converted into NAD^+ and $NADP^+$.

A. Vitamin B_1 B. Vitamin B_{12} C. Vitamin B_6

D. Vitamin B_2 E. Vitamin PP

[4] ____________ is the composition of CoA.

A. Riboflavin B. Folic acid C. Pantothenic acid

D. Biotin E. Niacin

[5] ________ can act as a prosthetic group directly without conversion.

A. Vitamin B_1 B. Vitamin B_2 C. Folic acid

D. Biotin E. Pantothenic acid

[6]The deficiency of____________ will cause anemia.

A. Vitamin B_6 B. Vitamin B_{12} C. FH_4

D. Vitamin B_1 E. Not above all

[A2 type]

[7]A 55-year-old woman presents with fatigue of several months' duration. Blood studies reveal a macrocytic anemia, reduced levels of hemoglobin, elevated levels of homocysteine, and normal levels of methylmalonic acid. Which of the following is the most likely deficient in this woman?____________

A. Folic acid B. Folic acid andvitamin B_{12} C. Vitamin C

D. Vitamin B_6 E. Biotin

[B1 type]

No. 8 ~ 11 share the following suggested answers.

A. Riboflavin B. Vitamin B_6 C. Pantothenic acid

D. Nicotinic acid E. Vitamin D

[8]NAD^+ is conver fecl from ____________.

[9]FMN is the actlue form of ____________.

[10]Coenzyme A is syutles ized from ____________.

[11]Pyridoxal phosphate is the acti oe form of ____________.

No. 12 ~ 15 share the following suggested answers.

A. Vitamin A B. Vitamin C C. Vitamin D

D. Vitamin K E. Vitamin B_1

[12]Which is involved in blood clotting____________.

[13]Which is involved in calcium metabolism____________.

[14]Which is involved in collagen synthesis____________.

[15]Which is involved in vision____________.

[X type]

[16]Enzymes that have TPP as a coenzyme are____________.

A. Pyruvate dehydrogenase complex

B. Lactate dehydrogenase

C. Isocitrate dehydrogenase

D. α-ketoglutarate dehydrogenase complex

E. Transketolase

[17]Which of the following vitamins belong to lipid-soluble vitamins?____________

A. Vitamin A B. Vitamin C C. Vitamin D

D. Vitamin K E. Vitamin E

[18]Which of the following vitamins belong to water-soluble vitamins?____________

A. Vitamin A B. Vitamin C C. Vitamin PP

D. Vitamin K E. Folic acid

[19]Which vitamins of the following are the composition of pyruvate dehydrogenase's cofactors?____________

A. Thiamine B. Nicotinic acid C. Riboflavin

D. Folic acid E. Pantothenic acid

Answers

1	2	3	4	5	6	7	8	9	10
B	C	E	C	D	B	A	D	A	C
11	12	13	14	15	16	17	18	19	
B	D	C	B	A	ADE	ACDE	BCE	ABCE	

[*Clinical Case*]

[20] An 11-month-old girl is being evaluated for the bowed appearance of her legs. The parents report that the baby is still being breastfed and takes no supplements. Radiologic studies confirm the suspicion of vitamin D-deficient rickets. Which one of the following statements concerning vitamin D is correct?

A. A deficiency results in an increased secretion of calbindin.

B. Chronic kidney disease results in over production of 1,25-dihydr ixycholecalciferol.

C. 25-hydroxycholecalciferol is the active form of the vitamin.

D. It is required in the diet of individuals with limited exposure to sunlight.

E. Its opposes the effect of parathyroid hormone.

Answers: D

Brief Explanations

Deficiency of vitamin D results in a decreased secretion of calbindin. Chronic kidney disease results in down production of 1,25-dihydrixycholecalciferol. The active form of Vitamin D is 1,25-$(OH)_2D_3$. It has the same effect of parathyroid hormone.

(***Li Meining, Guo Rui***)

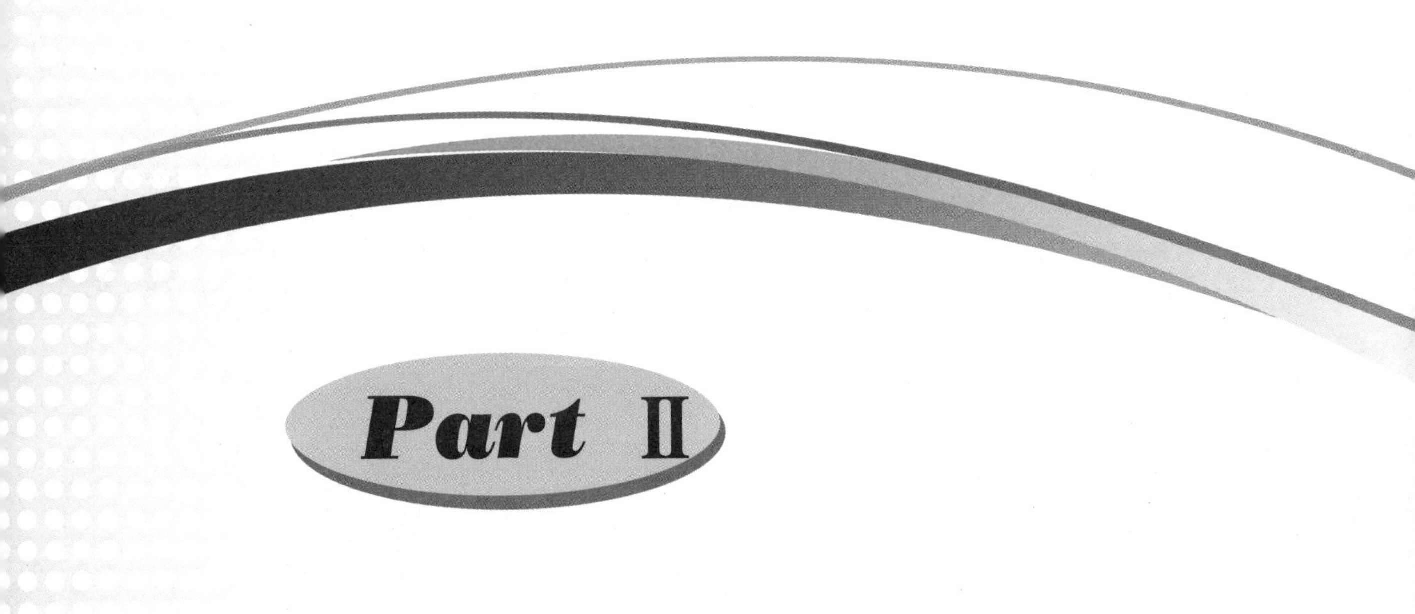

Part Ⅱ METABOLISM AND REGULATION

Chapter 6

Carbohydrate metabolism

[Examination Syllabus]

1. **Catabolism of carbohydrate**: main pathways, key enzymes and physiological significance of glycolysis; main pathways and energy supply of aerobic oxidation of glucose; physiological significance of the tricarboxylic acid cycle

2. **Glycogenesis and glycogenolysis**: glycogenesis in the liver; glycogenolysis in the liver.

3. **Gluconeogenesis**: main pathways, key enzymes of gluconeogenesis; physiological significance of gluconeogenesis; lactate Cycle

4. **Pentose phosphate pathway**: key enzymes and important productions; physiological significance of pentose phosphate pathway

5. **Blood glucose and regulation**: the concentration of blood glucose; regulation by insulin, glucagon, and glucocorticoid

I. The anaerobic oxidation of glucose

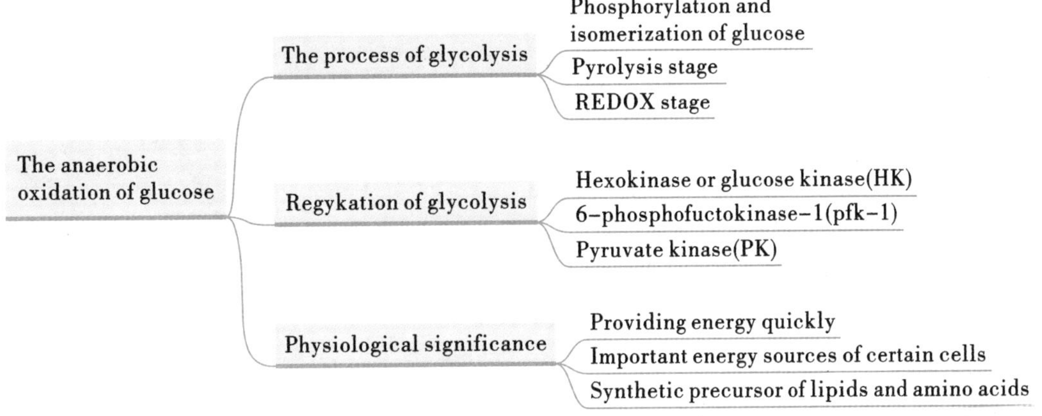

A. Definition

Anaerobic oxidation of glucose (latic acid fermentation): In the case of hypoxia, glucose is catalyzed by a series of enzymes to produce the lactic acid.

The **location** of anaerobic oxidation is in the **cytosol**.

B. The process of anaerobic oxidation

1. **The first stage**: The glucose is broken down into two wcole cales of pyruvate(glycolysis).

(1) **Phosphorylation and isomerization of glucose**

①Glucose (Glc) → glucose-6-phosphate (G-6-P):

Enzyme: hexokinase (HK).

Thefirst key steps, irreversible, consume 1 molecule ATP.

②G-6-P → Fructose- 6-phosphate (F-6-P): phosphohexose isomerase.

③F-6-P → Fructose 1,6 -biphosphate (F-1,6-BP)

Enzyme: phosphofructose kinase-1(PFK-1)

Thesecond key step, irreversible, consume 1 molecule ATP.

The mostimportant regulatory point in the process of glucose oxidation.

(2) **Pyrolysis stage**

④ F-1,6-BP → 2 molecular phosphotriose (dihydroxyacetone phosphate and glyceraldehyde 3-phosphate), which can be converted into each other.

(3) **REDOX stage**

⑤ Glyceraldehyde 3- phosphate →1,3-biphosphoglycerate

Oxidation: NAD^+→ NADH + H^+.

Enzyme: Glyceraldehyde 3- phosphate dehydrogenase (G-3PD)

Forming a **high-energy phosphate bond.**

⑥ 1,3-biphosphoglycerate → 3-phosphoglycerate

Phosphorylation: ADP → ATP.

Enzyme: Phosphoglycerate kinase.

The first substrate-level phosphorylation: the energy of a metabolite molecule is transferred directly to ADP (or GDP) to generate ATP (or GTP).

⑦3-phosphoglycerate → 2-phosphoglycerate

⑧2-phosphoglycerate → phosphoenolpyruvate

Enzyme: enolase

Forming a **high-energy phosphate bond.**

⑨Phosphoenolpyruvate → pyruvate

Enzyme: pyruvate kinase (PK)

Phosphorylation: ADP → ATP.

The second substrate-level phosphorylation

Thethird key step, irreversible.

2. **The second stage**: anaerobic conditions

⑩ Pyruvate → lactate

Enzyme: lactate dehydrogenase (LDH)

NADH + H^+→ NAD^+

Reduction reaction, NADH + H^+ comes from the fifth (⑤) step of the dehydrogenation reaction.

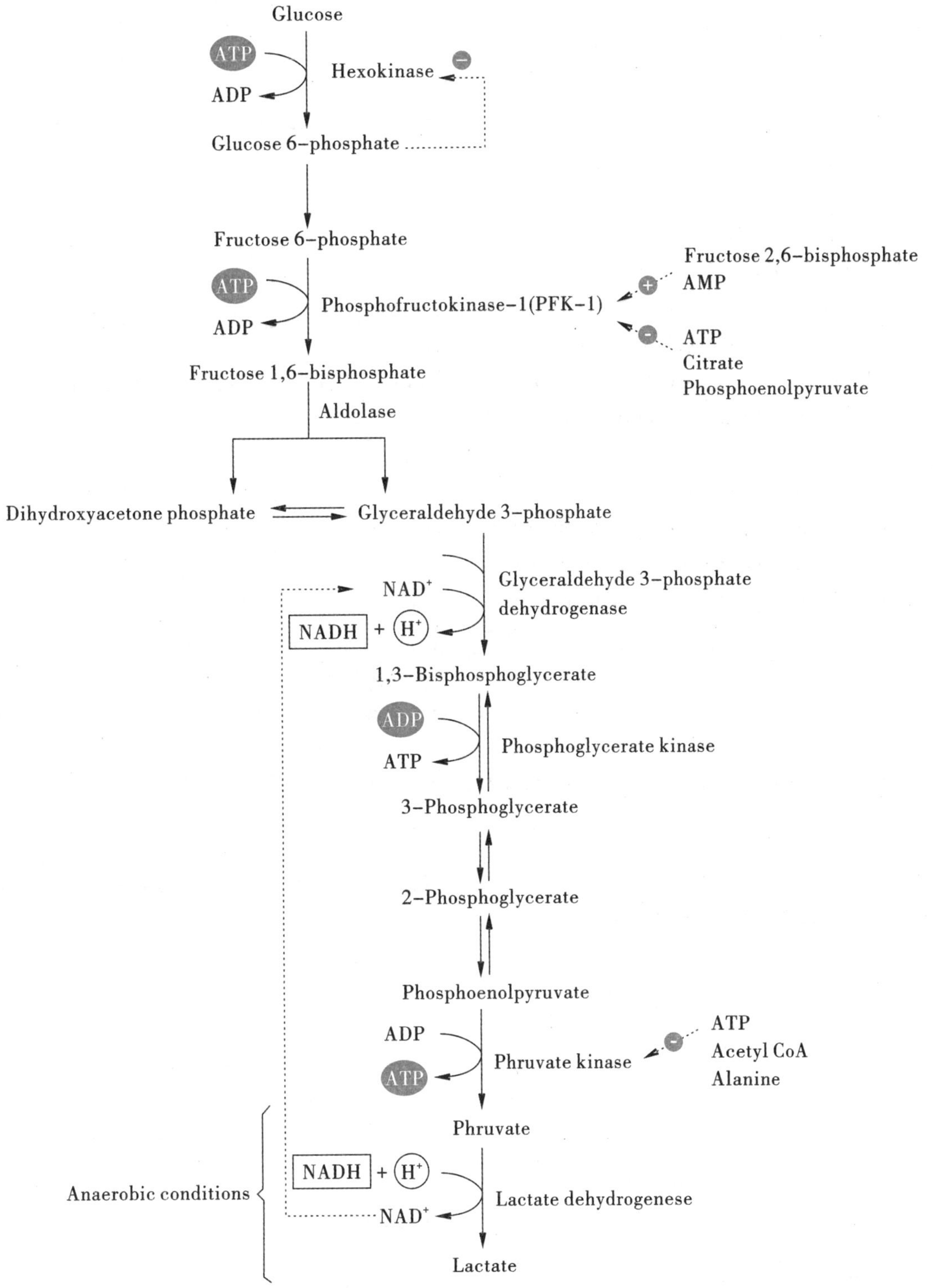

Figure 6.1 The steps of glycolysis

"1,2,3" of glycolysis

"1"——One dehydrogenation reaction.

"2"——Two substrate-level phosphorylation reactions.

"3"——Three irreversible reactions and three key enzymes.

A molecule of glucose can produce2 molecules of ATP by anaerobic glycolysis.

C. Regulation of glycolysis

Three key enzymes:

Hexokinase (HK) or glucose kinase, phosphofructokinase-1 (PFK-1), Pyruvate kinase (PK).

Regulation mode: allosteric regulation, covalent modification.

1. HK

(1) The activator: Mg^{2+}.

(2) The feedback inhibitor: glucose-6-phosphate.

2. PFK-1

(1) Allosteric inhibitors: citrate, ATP, phosphoenolpyruvate.

(2) Allosteric activators: ADP, AMP, F-1,6-BP (positive feedback), F-2,6-BP (the strongest).

(3) Insulin can induce its formation.

3. PK

(1) Allosteric inhibitors: ATP

(2) Allosteric activators: ADP andfructose 1,6-biphosphate.

(3) Phosphorylation modification decreased the activity.

(4) Insulin can induce PK generation.

D. Physiological significance

1. Provide energy quickly: an effective way of the body to gain energy in anaerobic or anaerobic conditions.

2. Important energy sources of certain cells (e. g., red blood cells, white blood cells, etc.) under normal oxygen supply.

Ⅱ. Aerobic oxidation

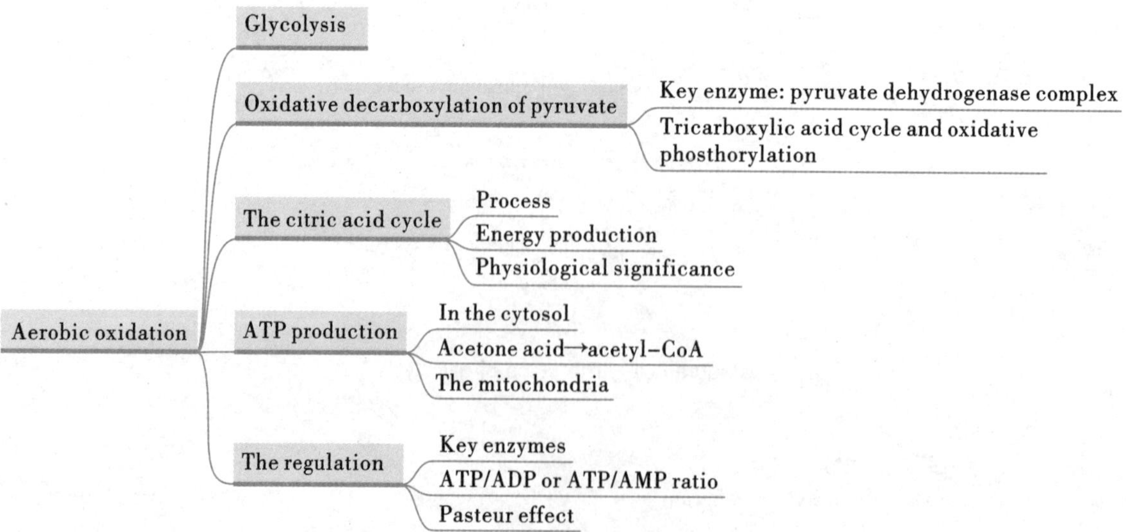

The process of aerobic oxidation include: glycolysis, oxidative decarboxylation of pyruvate and the citrici acid cycle.

A. Glycolysis

Glucoseis decomposed into pyruvate (cytoplasm).

B. Oxidative decarboxylation of pyruvate to produce acetyl-CoA (mitochondria)

1. **Key enzyme**: pyruvate dehydrogenase complex (table 6.1), the reaction is irreversible.

Table 6.1 The composition of pyruvate dehydrogenase complex

Enzyme	Coenzyme
E_1: pyruvate dehydrogenase	TPP
E_2: dihydrolipol transacetylase	lipoate, HS CoA
E_3: dihydrolipoyl dehydrogenase	FAD、NAD^+

2. The reaction formula

CH_3–CO–COOH + CoA ~ SH + NAD^+ → CH_3–CO ~ CoA + CO_2 + NADH + H^+

C. Tricarboxylic acid cycle and oxidative phosphorylation (mitochondria)

The citric acid (Krebs) cycle

1. Overall reaction equations:

Acetyl-CoA + 3NAD^+ + FAD + GDP + Pi + 2H_2O → 2CO_2 + 3NADH + 3H^+ + $FADH_2$ + GTP + HSCoA

2. Process (Figure 6.2)

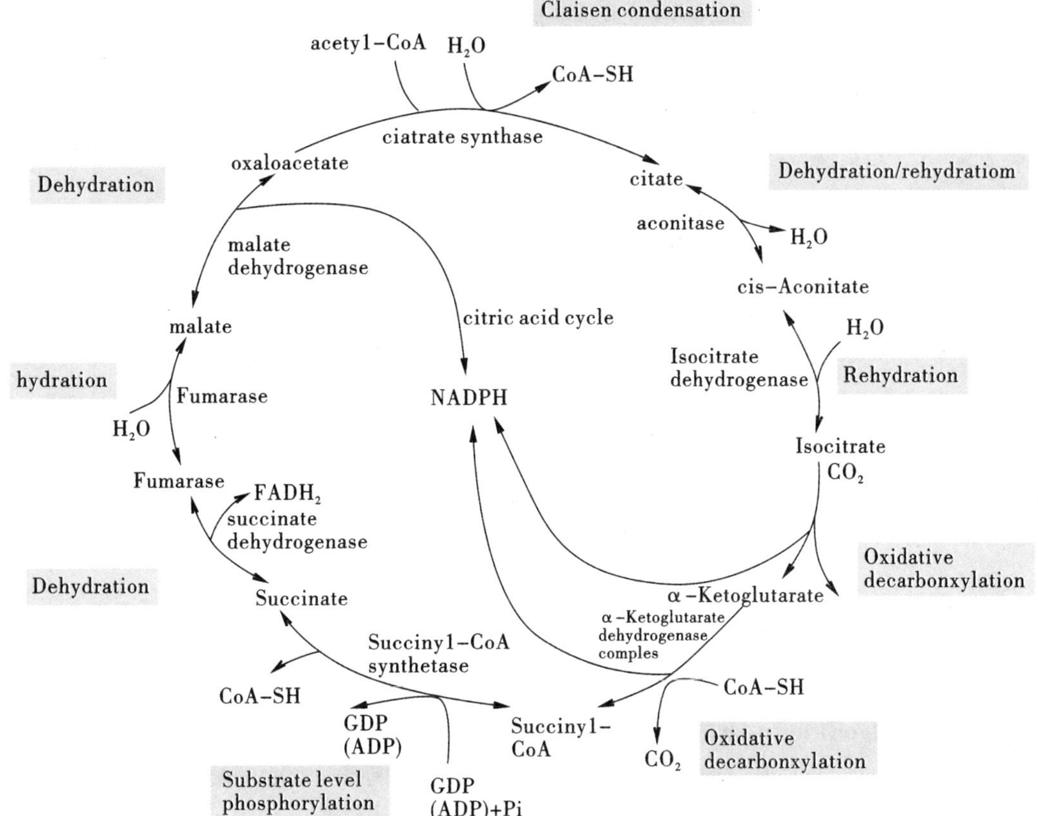

Figure 6.2 The steps of the citric acid

"1,2,3,4"of aerobic oxidation:

"1": one substrate-level phosphorylation reaction produces 1 molecule GTP.

"2": twice oxidation decarboxylation to produce 2 molecules CO_2

β-oxidation decarboxylation: catalyzed by isocitrate dehydrogenase;

α-oxidative decarboxylation: catalyzed byα-ketoglutarate dehydrogenase complex.

"3": three irreversible reactions and three key enzymes (citrate synthase, isocitrate dehydrogenase, α-ketoglutarate dehydrogenase complex)

"4": four dehydrogenations (3 molecules $NADH+H^+$, 1 molecule $FADH_2$)

3. Energy Production:

1 molecules ATP from the substrate-level phosphorylation

9 molecules ATPs through 4 dehydrogenation and oxidative phosphorylation

The total is 10 molecules ATPs.

4. The physiological significance of the tricarboxylic acid cycle.

a. TCA cycle is the common catabolic pathway of three major nutrients (carbohydrate, fat and protein).

b. TCA cycle provides reducing equivalents (H^++ e) for the respiratory chain.

c. Components of TCA cycle are important biosynthetic intermediates.

D. ATP production (Table 6.2)

Table 6.2 ATP produced by aerobic oxidation

Phase	Location	Reaction	coenzyme	ATP
Phase 1	Cytosol	Glucose→ glucose 6-phosphate		-1
		Fructose 6-phosphate → Fructose 1, 6- biphosphate		-1
		Glyceraldehyde 3-phosphate→1, 3-biphosphoglycerate	NAD^+	2.5(or1.5)×2
		1, 3-biphosphoglycerate→3-phosphoglycerate		1 × 2
		Phosphoenolpyruvate → pyruvate		1 × 2
Phase 2	Mitochondria	Pyruvate → acetyl-CoA	NAD^+	2.5 × 2
Phase 3	Mitochondria	Isocitrate → α-ketoglutarate	NAD^+	2.5 × 2
		α- ketoglutarate → succinyl-CoA	NAD^+	2.5 × 2
		Succinyl-CoA →Succinate		1 × 2
		Succinate → fumarate	FAD	1.5 × 2
		Malate→ oxaloacetate	NAD^+	2.5 × 2
	Total			30 or 32

Physiological significance: it is the most important way of the body's energy production. Not only does it have high productivity, but also energy utilization is also high because of the gradual release of the energy produced.

E. Regulation of aerobic oxidation

1. **Key enzymes and regulation** (Table 6.3)

(1) Glycolytic pathway: hexokinase (HK); Pyruvate kinase (PK); phosphofructo kinase-1 (PFK-1).

(2) Pyruvate oxidation decarboxylation: pyruvate dehydrogenase complex.

(3) Tricarboxylic acid cycle: citrate synthase; Isocitrate dehydrogenase; α-ketoglutarate dehydrogenase complex.

Table 6.3 Regulation of aerobic oxidation

Enzyme	Activator	Inhibitor	Regulation
Pyruvate dehydrogenase complex	AMP, CoA, NAD^+, Ca^{2+}	ATP, acetyl-CoAand NADH	Allosteric regulation, covalent modification.
Citrate synthase	ADP	NADH, citrate, succinyl CoA, ATP	High activity, maybe not accelerate TCA.
Isocitrate dehydrogenase	Ca^{2+}, ADP	ATP	Mainregulation point, feedback inhibition.
α - ketoglutarate dehydrogenase complex	Ca^{2+}	ATP, NADH, succinyl CoA, NADH	Feedback inhibition

Note: The regulation of key enzymes in the glycolytic pathway is shown in the last section.

2. **Regulation characteristics**

(1) Regulation of aerobic oxidation is to meet the needs of the body or organ for energy.

(2) The regulation of aerobic oxidation is achieved by regulating the key enzyme.

(3) ATP/ADP or ATP/AMP ratio all affects the rate of aerobic oxidation.

F. Pasteur effect

aerobic oxidation inhibits glycolysis in aerobic conditions.

Ⅲ. Pentose phosphate pathway

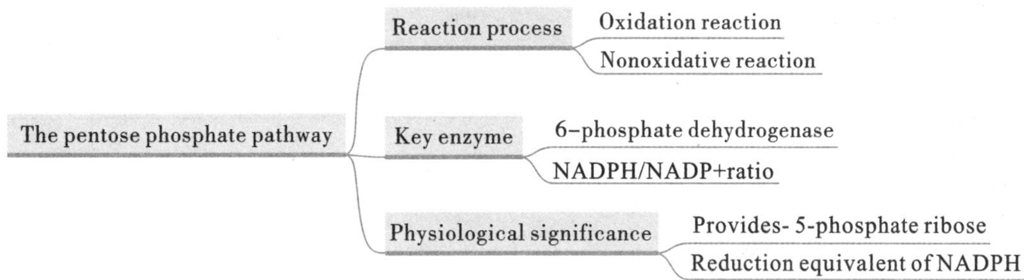

The pentose phosphate pathway (PPP) is another important way of glucose metabolism, its main purpose is not to produce ATP, but to generate **NADPH** required by many biosynthetic and **ribose-5-phosphate** for nucleotide. It occurs in the cytosol, in the liver, breast, red blood cells and other tissues.

A. Reaction process

Phase 1: oxidative phase produces pentose phosphate, NADPH and CO_2.

Key enzyme: glucose 6-phosphate dehydrogenase (G6PD), the enzyme activity is mainly affected by theNADPH/$NADP^+$ ratio.

Phase 2: non-oxidative phase recycles pentose phosphates to glucose 6-phophate and generates **ribose 5-phosphate** (the interconversion of sugars).

B. Physiological significance

1. It provides ribose 5-phosphate for the biosynthesis of nucleotides and nucleic acids.

2. It provides NADPH as the hydrogen donor involved in various metabolic reactions, and maintains the reduction state of glutathione (GSH).

[*Summary of glucose catabolism pathway*]

Table 6.4 Comparison of metabolic pathways of glucose

Reaction	Location	Conditions	Key enzyme	Production	Significance
Glycolysis	Cytosol	Hypoxia, NAD^+	HK, PFK-1PK	lactate, 2ATP	Quick energy supply; energy sources of RBC and other tissue
Aerobic Oxidation	Cytosol mitochondria	O_2, NAD^+, FAD	HK, PFK-1PK; Pyruvate dehydrogenase complex; Citrate synthase; Isocitrate dehydrogenase; α-ketoglutarate dehydrogenase complex	30 or 32 ATP, CO_2, H_2O	Main energy productionpathway; the linkage of three major biomolecule metabolism.
Pentose phosphate way	Cytosol	$NADP^+$	G6PD	NADPH; ribose 5-phosphate	NADPHis provided for anabolic metabolism. The raw material of nucleotide synthesis;
Glucuronate pathway	Cytosol	NAD^+ $NADP^+$	UDPG Pyrophosphorylase; UDPG dehydrogenase	UDPGA; NADPH; Vit C	Participate in biotransformation; saccharide source of proteoglycans; Vit C synthesis in Plants

IV. Glycogen metabolism

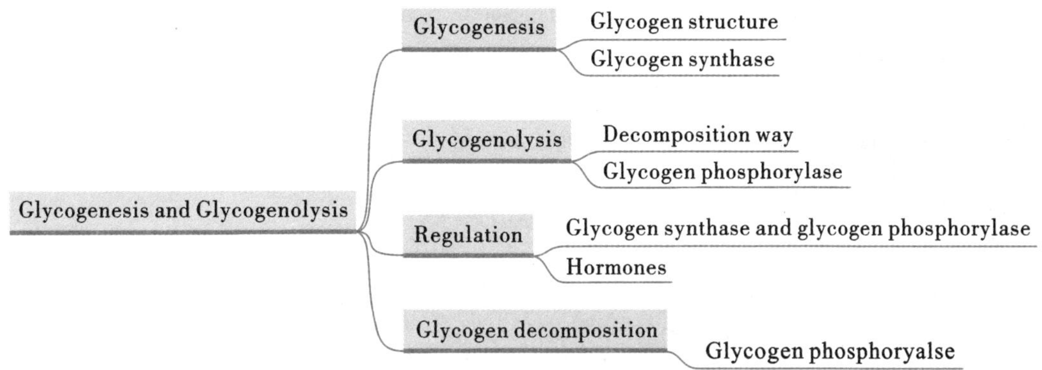

Glycogen is the storage form of glucose main found in liver and muscle.

A. Glycogenesis

Glycogensis is the synthesis of glycogen from glucose (Figure 6.3).

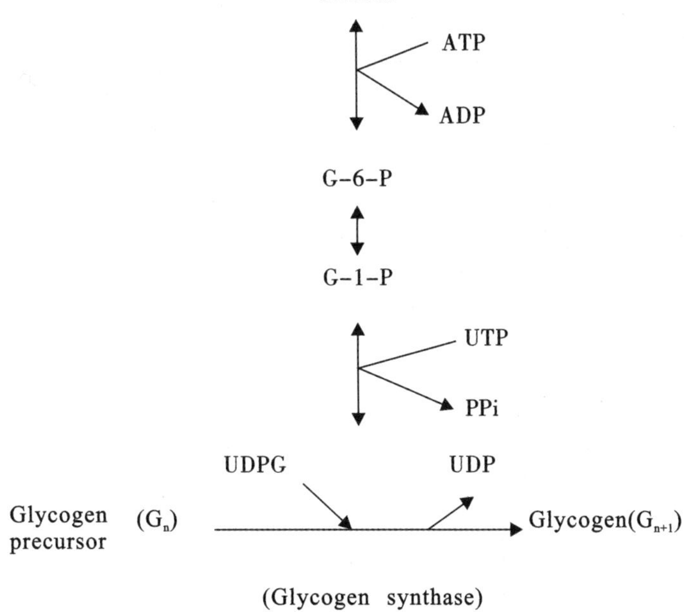

Figure 6.3 The steps of glycogenesis

Glycogen synthase catalyzes the addition of glucose from UDPG to the end of a glycogen molecule, forming an α-1,4 linkage main chain. **Branching enzyme** removes at least 6 units and add them to another chain to form α-1,6 branch.

Reaction features

1. It mainly occurs in the cytoplasm of liver and muscle cells.
2. The UDPG (UDP-glucose) is the donor of the active glucosyl residues.
3. **Glycogen synthase** is the key enzyme in glycogen synthesis and is regulated by various factors such as G-6-P.

4. The synthesis of glycogen is a process ofenergy consumption, and the synthesis of 1 molecule glycogen requires 2 ATPs.

B. Glycogenolysis

Glycogenolysis is the process of glycogen decomposition into free glucose (Figure 6.4). Key enzyme: **glycogen phosphorylase.**

Glycogen phosphorylase only catalyzes the α-1,4-glycosidic bonds, thus decomposing the linear glycogen chain till four glucose residues remain on either side of α-1,6 branch. The **debranching enzyme** catalyzes the α-1,6-glycosidic bonds and produce free glucose.

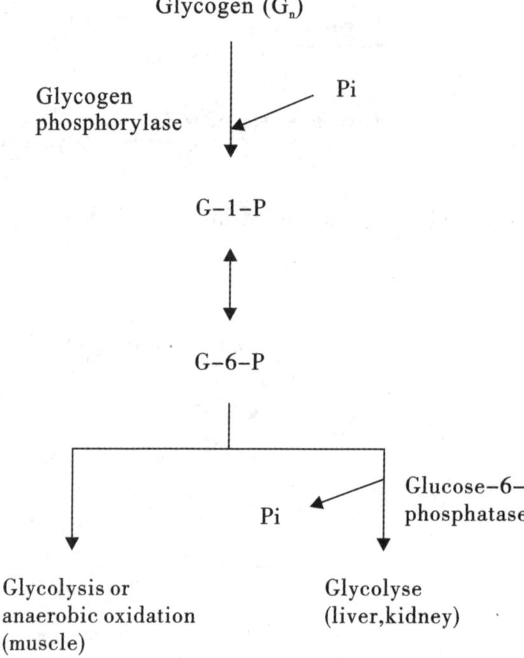

Figure 6.4 The steps ofglycogenolysis

C. Regulation of glycogensis and glycogenolysis

The synthesis and decomposition of glycogen are affected by hormones such as epinephrine, glucagon, and insulin.

Glycogen synthase and glycogen phosphorylase are the key enzymes of glycogensis and glycogenolysis, and the activity is regulated by allosteric regulation and covalent modification.

V. Gluconeogenesis

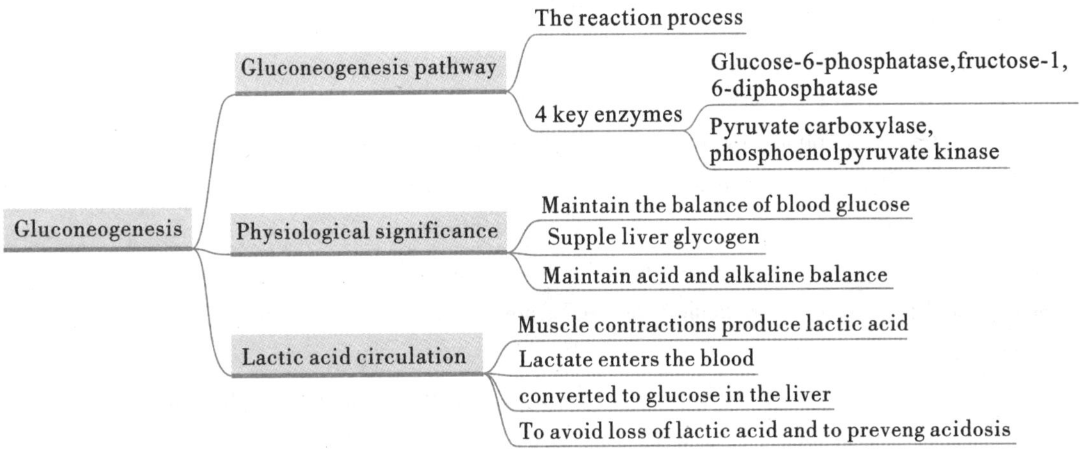

Gluconeogenesis refers to the conversion of noncarbohydrate (such as lactate, glycerol, and certain amino acids) to glucose or glycogen. The main organ of gluconeogenesis is the liver.

A. Gluconeogenesis pathway (Figure 6.5)

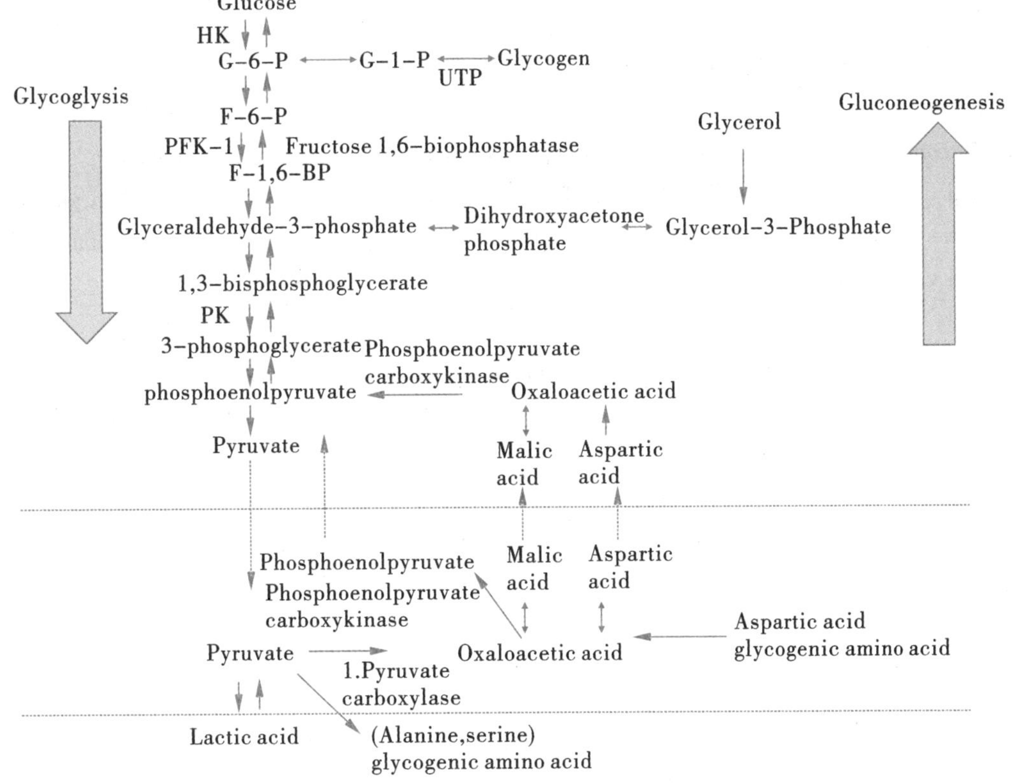

Figure 6.5 The major pathway of glycolysis and gluconeogenesis

1. **Reaction features**

(1) **The reaction process**: the process of glucose from pyruvate, which is the reverse process of gly-

colysis, but not exactly reversible.

(2) 4 **key enzymes**: glucose-6-phosphatase, fructose-1,6-diphosphatase, pyruvate carboxylase, phosphoenolpyruvate carboxykinase.

(3) **Regulation**: through two substrate cycles:

substrate cyclebetween F-6-P and F-1,6-BP.

substrate cyclebetween phosphoenolpyruvate and pyruvate.

2. "1,2,3" of gluconeogenesis

"1": one ATP consumption (pyruvic acid +ATP→oxaloacetate);

one GTP consumption (oxaloacetate +GTP→ phosphoenolpyruvate)

"2": **two mechanisms of oxaloacetic acid transportation**:

Malic acid shuttle mechanism (with alanine as a raw material for gluconeogenesis);

Aspartic acid generation by glutamic oxalacetic transaminase mechanism (lactic acid as a raw material for gluconeogenesis)

"3": **three bypass reaction** (throughout to the bypass of irreversible glycolytic reactions):

pyruvate→ oxaloacetic acid→ phosphoenolpyruvic acid;

F-1,6-BP→ F-6-P;

G-6- P→ glucose.

B. Physiological significance

1. The **main** physiological significance of gluconeogenesis is to maintain the balance of blood glucose.
2. Supply liver glycogen.
3. Maintain acid-base balance.

C. Lactate cycle

Muscle contraction produces lactic acid through glycolysis. Gluconeogenesis activity in muscle is low, so lactic acid can diffuse into the liver by way of blood, and it is converted to glucose. When glucose is released into the blood, it can be taken up by the muscle, thus forming a cycle called the lactate cycle, also known as the Cori cycle (Figure 6.6).

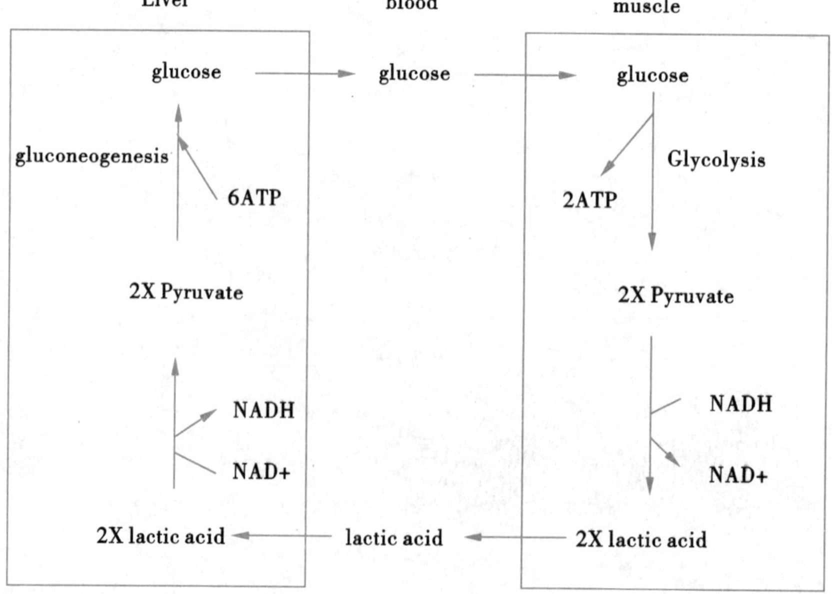

Figure 6.6 Lactate Cycle

1. Physiological significance: To recover the lactic acid and avoid lactic acidosis caused by acid accumulation.

2. Lactic acid cycle is the process of energy consumption, converting 2 molecules of lactate to yield glucose need to consume 6 molecules of ATP.

Ⅵ. Blood glucose and the regulation

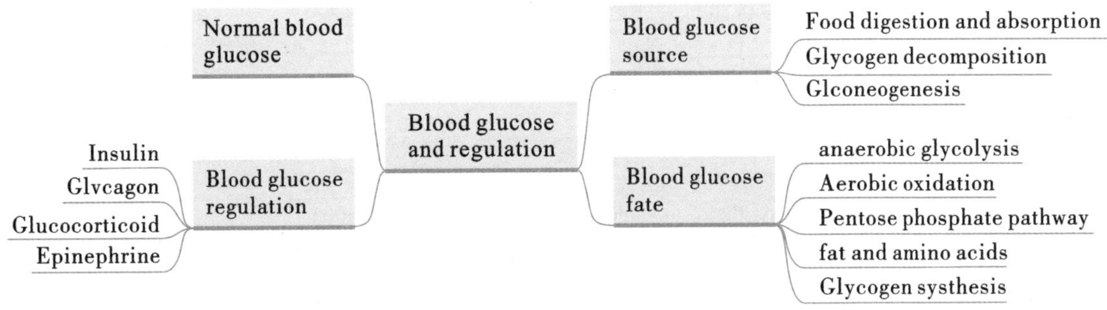

A. Normal blood glucose: 3.89 ~ 6.11 mmol/L

1. **Blood glucose source**: food digestion and absorption; Glycogen decomposition; gluconeogenesis

2. **Blood glucosefate**: oxidation; glycogen systhesis; pentose phosphate pathway; convert into fats and amino acids (Figure 6.7).

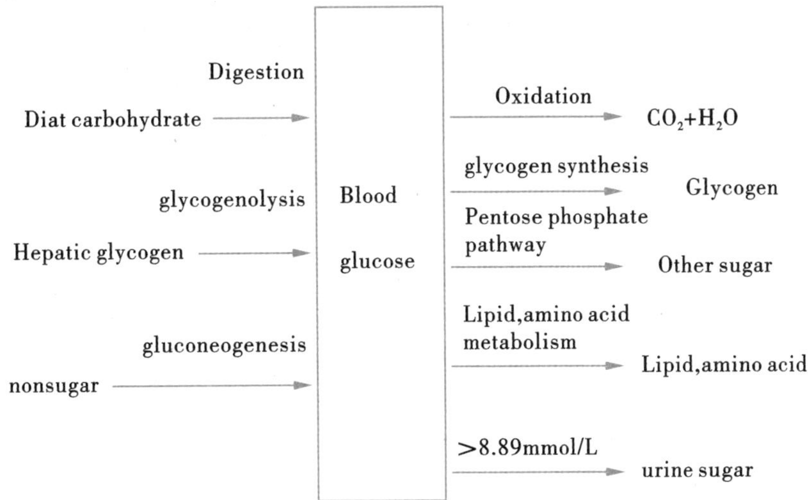

Figure 6.7 Source and fate of blood glucose

B. Blood glucose regulation

1. Enzymatic regulation is the most basic pathway.

2. Main hormones: insulin (lower blood glucose).

Glucagon, glucocorticoid, epinephrine (increase blood glucose).

The regulation mechanism of these hormonesis shown in table 6.5.

Table 6.5 Blood glucose regulation

Hormone	Regulation mechanism	Effect
Insulin	(1) Promote muscle and fat cells to uptake glucose by glucose transporters (2) Accelerate glycogen synthesis and inhibit glycogenolysis (3) Accelerate the aerobic oxidation of glucose (4) Inhibit hepatic gluconeogenesis (5) Inhibit fat mobilization	lower blood glucose
Glucagon	(1) Promote glycogenolysis and inhibit glycogen synthesis (2) Inhibit glycolysis and accelerate gluconeogenesis (3) Promote fat mobilization	increase blood glucose
Glucocorticoid	(1) Promote the degradation of muscle protein, accelerate gluconeogenesis (2) Inhibit the uptake and utilization of glucose by extrahepatic tissues and inhibits the oxidative decarboxylation of pyruvate (3) Enhance the effect of other hormones that promote fat mobilization	increase blood glucose
Epinephrine	Trigger cAMP-dependent phosphorylation cascades in liver and muscle cells and accelerate glycogenolysis.	increase blood glucose

[*Practices*]

[A1 type]

[1] The final product of glycolysis in vivo is ____________.

A. propionic acid B. lactic acid C. pyruvate

D. ethanol E. acetone

[2] The key enzyme of glycolysis pathway is ____________.

A. phosphoglycerate kinase B. enolase enzyme C. pyruvate kinase

D. pyruvate dehydrogenase complex E. pyiuvate carboxylase

[3] Which substance is involved in the substrate-level phosphorylation during glycolysis?

A. Phosphoenolpyruvate B. Fructose-1,6- biphosphate C. 2,3-diphosphoglycerate

D. Glyceraldehyde-3-phosphate E. 3-phosphoglycerate

[4] The production of lactic acid is conducive to the continuous process of glycolysis ____________.

A. the catalytic reaction is reversible

B. lactic acid is acidic

C. lactate dehydrogenase has 5 isozymes

D. reoxidation of $NADH+H^+$ to NAD^+

E. provide ATP

[5] How manynumbers of substrate-level phosphorylation are in the glycolysis process of 1 molecule of glucose ?____________

A. 1 B. 2 C. 3

D. 4 E. 5

[6] When 1 mol of glucose residues in glycogen molecules are lysed into lactic acid, How manymol number

of ATP are produced? ____________

A. 1 mol B. 2 mol C. 3 mol D. 4 mol E. 5 mol

[7] What is the strongest allosteric activator of phosphofructokinase-1? ____________

A. Fructose 2,6-biphosphate B. ATP C. ADP
D. AMP E. Fructose 1,6-biphosphate

[8] Activators of 6-phosphofructokinase-1 does not include ____________.

A. Citric acid B. Fructose 2,6-biphosphate C. AMP
D. Fructose 1,6-biphosphate E. ADP

[9] The number of ATP is produced after 1 mol acetyl CoA through tricarboxylic acid cycle oxidation is ____________.

A. 8 mol B. 9 mol C. 10 mol D. 11 mol E. 12 mol

[10] Which one is the product of 1 molecule acetyl CoA through tricarboxylic acid cycle oxidation? ____________

A. CO_2 and H_2O B. Oxaloacetate and CO_2 C. Citric acid
D. Oxaloacetate E. CO_2 and 4 molecules restore the equivalent.

[11] Where in the suborganelle does the tricarboxylic acid cycle occur? ____________.

A. Nucleus B. Cytosol C. Microsomal
D. Golgi apparatus E. Mitochondria

[12] When 1 mol pyruvate is completely oxidized to generate carbon dioxide and water, the mol number of ATP product is ____________.

A. 12 B. 12.5 C. 13 D. 13.5 E. 15

[13] What is the high-energy compound generated through substrates level phosphorylation in the tricarboxylic acid cycle? ____________

A. ATP B. GTP C. UTP D. CTP E. TYP

[14] In the tricarboxylic acid cycle and associated respiration, the most abundant phase of ATP production is ____________.

A. Citric acid → isocitrate acid
B. Isocitrate acid → α-ketoglutaric acid
C. α-ketoglutaric acid → succinic acid
D. Succinic acid → malic acid
E. Malic acid → oxaloacetate

[15] When one molecule of succinate is dehydrogenated to fumarate, How many molecules of ATP are generated ? ____________

A. 1 B. 1.5 C. 2 D. 2.5 E. 3

[16] Which kind of enzyme is allostericly activated by acetyl-CoA? ____________

A. Glycogen phosphorylase B. Pyruvate carboxylase C. Phosphofructokinase-1
D. Citrate synthase E. Isocitrate dehydrogenase

[17] The process of substrate level phosphorylation in the tricarboxylic acid cycle is? ____________

A. Isocitrate acid→α-ketoglutaric acid
B. Citric acid→isocitrate acid
C. α-ketoglutaric acid→succinic acid
D. Malic acid→oxaloacetate
E. Succinate→fumarate

[18]The physiological significance of the glucose pentose phosphate pathway is not included____________.

A. Product large amounts of ATP for energy

B. Participate in the body hydroxylation

C. Maintain the reduction state of glutathione

D. Provide ribose for nucleic acid synthesis

E. As hydrogen donor of anabolic

[19]The decomposition products of muscle glycogen cannotbe converted to ____________.

A. UDPG　B. 1-phosphate dextran　C. Glucose

D. Lactic acid　E. Glucose-6-phosphate

[20]How many ATP mols are generated when 1 mol glucose first synthesizes glycogen, and then decomposes into lactic acid? ____________

A. 0　B. 1　C. 2　D. 3　E. 4

[21]How many numbers of high-energy phosphate bondare consumed when adding 1 molecule glucose residue to glycogen synthesis? ____________

A. 1　B. 2　C. 3　D. 4　E. 5

[22]The direct product of hepatic glycogen is ____________.

A. glucose-6-phosphate and glucose

B. glucose-1-phosphate

C. glucose

D. glucose-1-phosphate and glucose

E. 6-phosphate fructose

[23]The key enzyme of glycogen synthesis is ____________.

A. hexokinase　B. UDPG-pyrophosphorylase　C. glucokinase

D. glycogen synthase　E. phosphoglucomutase

[24]The key enzyme of glycogen decomposition is ____________.

A. phosphorylase　B. branching enzymes　C. glucose-6-phosphatase

D. debranching enzyme　E. glucose phosphate mutase

[25]Hepatic glycogen can directly supply blood sugar because of ____________.

A. glucose oxidase　B. glucose-6-phosphate dehydrogenase　C. glucokinase

D. glucose-6-phosphatase　E. phosphohexose isomerase

[26]The key enzyme of gluconeogenesis is ____________.

A. phosphoglycerate kinase　B. enolase　C. pyruvate kinase

D. pyruvate dehydrogenase complex　E. pyruvate carboxylase

[27]Among the enzyme involved in glucose metabolism, which kind of enzyme-catalyzed reaction is reversible?____________

A. Glycogen phosphorylase　B. Hexokinase　C. Fructose diphosphatase

D. Pyruvate kinase　E. Phosphoglycerate kinase

[28]Which of the following descriptions is incorrect when enough ATP is available in the liver cells?____________.

A. Pyruvate kinase is inhibited

B. Phosphofructokinase-1 is inhibited

C. Isocitrate dehydrogenaseis inhibited

D. Fructose bisphosphatase is inhibited

E. The Acetyl-CoA is decreased after entering the tricarboxylic acid cycle

[29] Which descriptionabout oxaloacetate is wrong in the gluconeogenesis from lactic acid about oxaloacetate?______.

A. Generated in mitochondria

B. Phosphoenolpyruvate cannotbe directly produced

C. Does not go through the mitochondrial membrane

D. The production of oxaloacetate from pyruvate consumes one ATP

E. Generating phosphoenolpyruvate from oxaloacetate depletes one GTP

[30] Which of the following hormones cannot make blood sugar level to rise ?______

A. Growth hormone B. Adrenaline C. Insulin

D. Glucagon E. Glucocorticoids

[31] Which of the following is not a diabetic disorder of glucose metabolism?______.

A. Glycogen synthesis reduced and decomposition accelerated

B. Blood sugar increased

C. The conversion of glucose-6-phosphate to glucose weakened

D. Glycogen glycolysis and aerobic oxidation weakened

E. The speed of glucose through muscle and fat cells slowed down

[32] In long-term hunger, blood sugar mainly come from ______.

A. the amino acids of muscle protein degradation

B. glycogenolysis

C. glycerin gluconeogenesis

D. the amino acids of hepatic protein degradation

E. muscle glycogen degradation

[B1 type]

No. 33 ~ 34 shared the following suggested answers

A. Glucose phosphate dehydrogenase B. Pyruvate carboxylase

C. Pyruvate kinase D. Isocitrate dehydrogenase E. Glycogen synthase

[33] What is the key enzyme of glycolysis?______

[34] What is the rate-limiting enzyme of tricarboxylic acid cycle?______

No. 35 ~ 37 shared the following suggested answers

A. 10 mol ATP B. 12.5 mol ATP C. 15 mol ATP

D. 32 mol ATP E. 33 mol ATP

[35] When 1 mol glucose is thoroughly oxidized, Howmany mol ATP are produced ?______

[36] When 1 mol acetyl CoA is thoroughly oxidized, Howmany mol ATP are produced______.

[37] When 1 mol pyruvate is thoroughly oxidized? Howmany mol ATP are produced ______

[X type]

[38] Which is the correct description of the mechanism during which the small intestinal transit absorbs glucose ?______

A. Need to use ATP to provide energy

B. The absorbing speed of glucose is higher than pentose sugar

C. Dependent on Na^+ concentration gradient of intracellular and extracellular

D. The absorption process is not a simple spread

E. Dependent on K^+ concentration gradient of intracellular and extracellular

[39] Which is the key enzyme of glycolysis?____________

A. 6-Phosphofructokinase-1

B. Pyruvate dehydrogenase complex

C. Pyruvate kinase

D. Hexokinase

E. UDPG-pyrophosphatase

[40] What is the important function of pentose phosphate pathway ?____________.

A. It is the metabolic center of glucose lipid and amino acids

B. Provides NADPH for fatty acid synthesis

C. Provide raw materials for the synthesis of nucleic acids

D. Provide raw materials for the synthesis of cholesterol

E. Provides FADH for fatty acid synthesis

[41] Which reaction is capable of substrate level phosphorylation? ____________

A. Phosphoenolpyruvate → Pyruvate

B. Fructose-6-phosphate → Fructose-1,6-biphosphate

C. 1,3-biphosphoglycerate → 3-phosphoglycerate

D. Glyceraldehyde-3-phosphate→3-phosphoglycerate

E. Fructose-1,6-biphosphate→ Fructose-6-phosphate

[42] Which description is correct about glycogen synthase regulation? ____________.

A. The enzyme is inactivated by the action of phosphoprotein phosphatase-1

B. Glycogen synthase has activity and inactivity two forms

C. PKA can activate phosphoprotein phosphatase inhibitors

D. cAMP can inactivate glycogen synthase through dependent on protein kinase

E. It is the rate-limiting enzyme of the glycogen synthesis pathway

[43] Which description is correct about glycogen structure?____________.

A. There is only one non-reductive terminal

B. Glycogen exists in the cytosol

C. Glycogen contains α -1,4-glucosidic bond andα- 1,6 -glucosidic bond

D. Synthesis and decomposition occur at multiple reductiveterminals

E. The major structure unit of glycogen is α-D-glucose

[44] Which isirreversible during the gluconeogenesis reaction?____________

A. Lactic acid→Pyruvate

B. Fructose-1,6-biphosphate→Fructose-6-phosphate

C. Glucose-6-phosphate→Glucose

D. Pyruvate→Phosphoenolpyruvate

E. Pyruvate→Oxaloacetate

[45] The physiological significance of lactate cycle is including____________.

A. avoid fuel waste

B. avoid lactic acidosis

C. replenish blood sugar

D. prevent lactic acid accumulation

E. lactic acid discharged from urine in the end

[46] After ATP/AMP ratio is increased in intracellularly, what can be inhibited?____________

A. Hexokinase

B. 6-Phosphofructokinase-1

C. Glyceraldehyde-3-phosphate dehydrogenase

D. Pyruvate dehydrogenase complex

E. Pyruvate decarboxylase

[47] Which description is correct about glucagon to regulate glucose metabolism________.

A. Activate glycogen synthase

B. Promote PEP carboxylase synthesis

C. Activate hormone-sensitive lipase

D. Inhibit fructose-2,6-bisphosphate synthesis

E. The mechanismis increased glucose source and reduces glycogen consumption

Answers

1	2	3	4	5	6	7	8	9	10
B	C	A	D	D	B	A	A	C	E
11	12	13	14	15	16	17	18	19	20
E	B	B	C	B	B	C	A	C	B
21	22	23	24	25	26	27	28	29	30
B	D	D	A	D	E	E	D	B	C
31	32	33	34	35	36	37	38	39	40
C	A	C	D	D	A	B	ABCD	ACD	BCD
41	42	43	44	45	46	47			
AC	BCD	BCE	BCD	ABCD	BD	BCD			

(*Zhang Yuzhe*, *Zhou Ti*)

Chapter 7

Biological Oxidation

Examination Syllabus

1. **Oxidative phosphorylation**: definition of oxidative phosphorylation, the composition of two respiratory chains, ATP synthase, regulation of oxidative phosphorylation and its influencing factors.

2. **ATP and other high energy compounds**: ATP cycle and high energy phosphate bond, utilization of ATP, other high-energy phosphate compounds.

[*Major points*]

The **biological oxidation** is the oxidation of substances presenting in living organisms.

Ⅰ. Electron transport chain

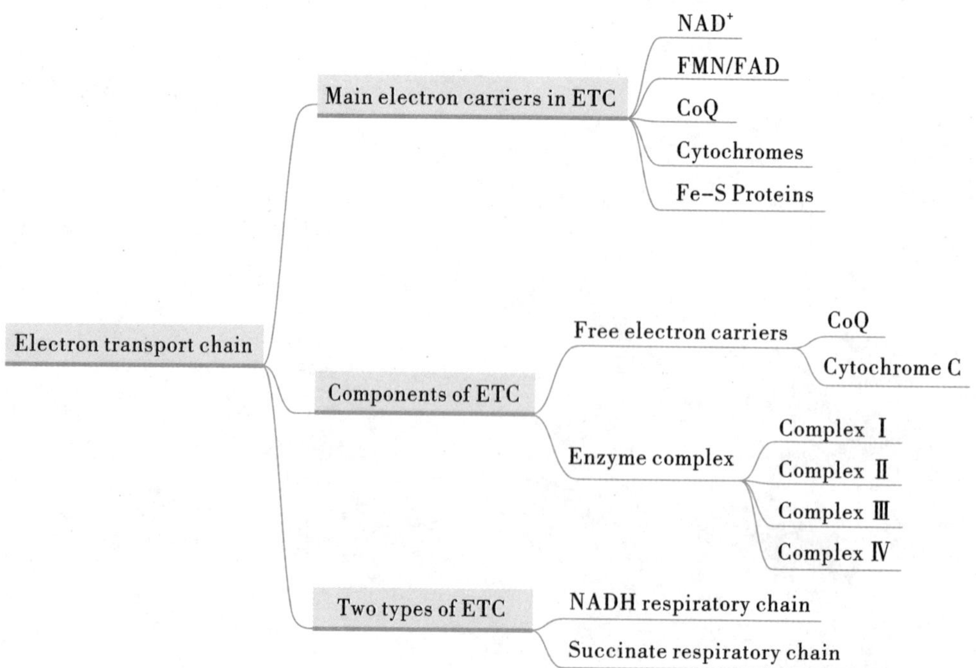

Electron transportsystem, electron transport chain (ETC), or the respiratory chain: a series of mitochondrial enzymes and redox carrier molecules which ferry reducing equivalents from substrates to oxygen.

This system captures the free energy available from substrate oxidation so that it may later be applied to the synthesis of ATP.

A. Components ofthe electron transport chain

1. Multiprotein complexes

The mitochondrial respiratory chain consists of a series of electron carriers, most of which are membrane proteins with prosthetic groups capable of transferring electrons. The electron carriers in the inner mitochondrial membrane are assembled in the form of four multiprotein complexes, named complex Ⅰ, Ⅱ, Ⅲ, and Ⅳ.

2. Free electron carriers

(1) **Ubiquinone** is small, lipid-soluble component of the electron transport chain, and freely diffusible in the lipid bilayer of the inner mitochondrial membrane. It moves between complexes I or II and III as a freely fat-soluble compound.

(2) **Cytochrome C**, as a freely water-soluble protein, is mobile on the outer surface of the inner membrane. It moves between complexes III and IV.

B. Electron carriers of the respiratory chain (Table 7.1)

Table 7.1 Electron carriers of the respiratory chain

Two-Electron Carriers			One-Electron Carriers		
Carriers	Important component	Function	Carriers	Important component	Function
NAD^+	VitPP	transfer 1e or 2e	Cyt	Heme	transfer only 1e
FMN/FAD	$VitB_2$	transfer 1e or 2e	Fe-S proteins	Fe-S centers	transfer only 1e
CoQ	—	transfer 1e or 2e			

C. Two types of the respiratory chain

There are two types of the respiratory chain in mitochondria. The CoQ is an intersection of the two types of the respiratory chain.

1. NADH respiratory chain

NADH →Complex Ⅰ →CoQ → Complex Ⅲ→Cyt c → Complex Ⅳ→O_2

2. $FADH_2$ respiratory chain (Succinate respiratory chain)

Succinate → Complex Ⅱ →CoQ →Complex Ⅲ→Cyt c → Complex Ⅳ→O_2

Ⅱ. Oxidative phosphorylation and ATP synthesis

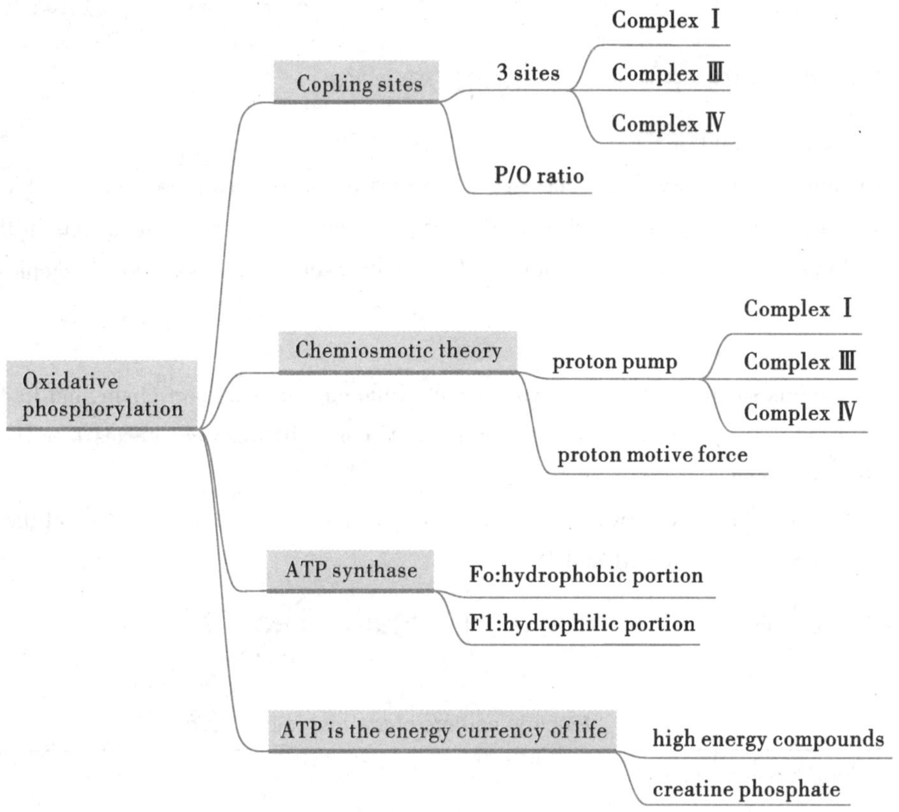

A. Two types of phosphorylation

1. **Oxidative phosphorylation** is the term used to describe the synthesis of ATP, which is energetically coupled to the oxidation of reducing equivalents to water, in the mitochondria. The oxidative phosphorylation is the main source of energy in the aerobic cell.

2. **Substrate-level phosphorylation** is referred to the phosphorylation of ADP or GDP coupled to a dehydrogenation (or dehydration) of an organic substrate, independent of the respiratory chain.

B. The sites coupling proton flux with phosphorylation

1. There are 3 sites in the respiratory chain where ATP is formed by oxidative phosphorylation (Figure 7.1).

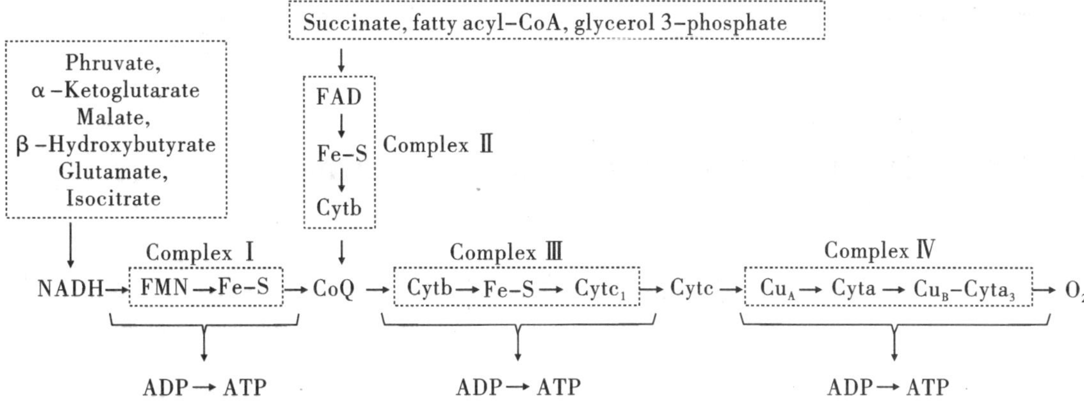

Figure 7.1 Sites that couple the electron transport chain with phosphorylation of ADP to form ATP

2. **P/O ratio**, or the **phosphate/oxygen ratio**, refers to the amount of ATP produced from the transfer of two electrons through an electron transport chain, terminated by reduction of an oxygen atom.

(1) NADH can be oxidized through the entire electron transfer chain to produce a ratio of 2.5 (the production of 2.5 mole ATP).

(2) Succinate ($FADH_2$) bypasses the first phosphorylation site, so the P/O ratio would be 1.5 (the production of 1.5 mole ATP).

C. Chemiosmotic theory

The chemiosmotic theory which is proposed by Peter Mitchell describes the coupling of electron transport and oxidative phosphorylation. The flow of electrons along the electron transport chain allows protons to be pumped from the matrix of the mitochondria to the intermembrane space. The protons are pumped across the inner mitochondrial membrane at three sites in the respiratory chain to produce a proton gradient. When protons move back through proton channels of the ATP synthase in the inner mitochondrial membrane, ATP is synthesized.

D. ATP synthase

ATP synthase, also called complex Ⅴ, is composed of two parts (F_1 and F_o).

1. Hydrophilic portion: F_1($\alpha_3\beta_3\gamma\delta\varepsilon$ subunits).
2. Hydrophobic portion: F_o($a_1b_2c_{9-12}$ subunits).

A model of the F_1 and F_o components of the ATP synthase can be thought of as a rotating molecular motor. The a, b, α, β, and δ subunits form the stator of the motor, and the c, γ, and ε subunits compose the rotor. The flow of protons through the proton channel in F_o turns the rotor and drives the conformational changes of α and β subunits that synthesize ATP.

E. ATP: the energy currency of life

1. **High energy compound** is those that contain one or more high-energy bonds, which release more than 25 kJ/mol of free energy when hydrolyzed. The two phosphoanhydride bonds in ATP, which link the three phosphate groups, each has a larger negative ΔG *for hydrolysis. Hydrolysis of these bonds, especially the terminal one, drives many energy-requiring reactions in biological systems.*

2. **Creatine Phosphate**

Besides ATP, there is still another high-energy phosphate compound called creatine phosphate (CP), also known asphosphocreatine. It acts as a rapidly mobilizables torage form of energy in skeletal muscle and brain.

Ⅲ. Regulation of oxidative phosphorylation

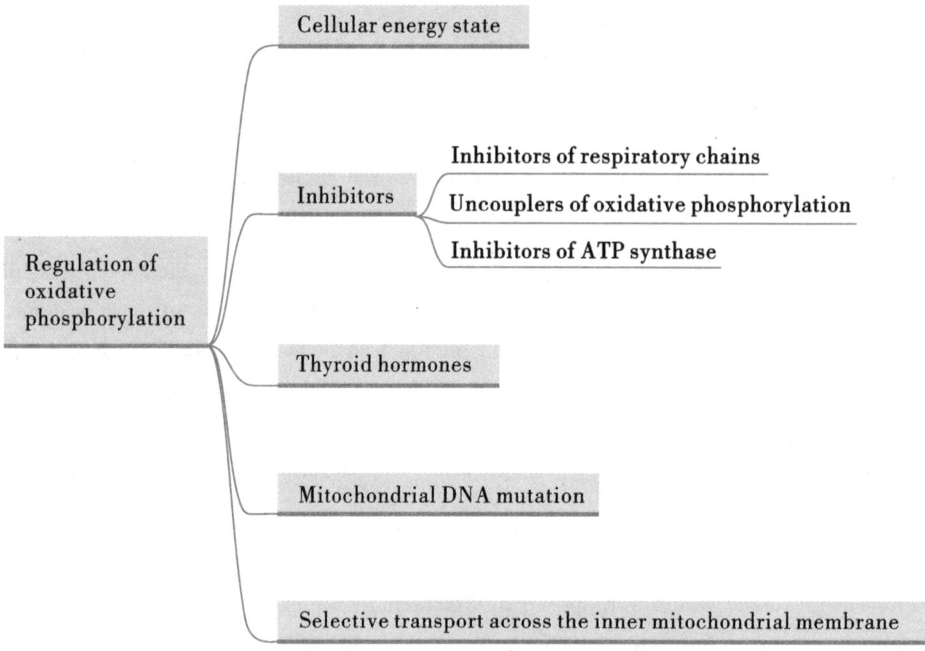

A. Effect of ADP

The concentration of intracellular ADP is one measure of the energy status of cells. The ATP/ADP ratio can rapidly indicate the change of cellular energy demand and is an important regulatory factor for oxidative phosphorylation.

B. The factors affecting ETC

Much information about the respiratory chain has been obtained by the use of inhibitors. For the descriptive purpose, they may be divided into three types.

1. **Inhibitors of the electron transport chain** (**Table** 7.2)

Table 7.2 Inhibitors of the electron transport chain

Sites	Inhibitors	Functions
Complex Ⅰ	Rotenone: A fish poison and also insecticide. Amobarbital Piericidin A, an antibiotic that resembles coenzyme Q.	Inhibits transfer of electrons through complex Ⅰ Inhibits electrons transfer through complex Ⅰ Blocks electron transfer by competing with CoQ
Complex Ⅱ	Carboxin Malonate	Specifically, inhibit the transfer of reducing equivalent from complex Ⅱ Acompetitive inhibitor of complex Ⅱ

Continue to Table 7.2

Sites	Inhibitors	Functions
Complex Ⅲ	Antimycin A Phenformin Dimercaprol	Blocks electron transfer from cyt b to cyt c_1
Complex Ⅳ	Cyanide (CN^-) Azide (N_3^-)	Inhibits terminal transfer of electrons to molecular O_2
	Hydrogen sulfide (H_2S) Carbon monoxide (CO)	Inhibits cytochrome oxidase by combining with O_2 binding site

2. **Uncouplers of oxidative phosphorylation**

They are compounds that allow oxidation of substrates to continue without corresponding synthesis of ATP. 2,4-Dinitrophenol and uncoupling proteins in brown adipose tissue of newborns are classic examples of uncouplers.

3. **Inhibitors of ATP synthase**

Oligomycin prevents both the stimulation of oxygen uptake by ADP and the phosphorylation of ADP to ATP.

C. Thyroid hormones

Thyroid hormones regulate Na^+,K^+-ATPase activity in the cell membrane and accelerate the hydrolysis of ATP to ADP and Pi, which enhances the oxidative phosphorylation. T3 induces uncoupling protein gene expression which causes the increase of the metabolite oxidation release energy and the rate of heat production. The basal metabolic rate is higher in patients with hyperthyroidism.

D. mtDNA mutation

Mutations in mitochondrial genes have a profound impact on oxidative phosphorylation and cause a decrease in the generation of ATP.

Ⅳ. Selective transport across the inner mitochondrial membrane

A. Oxidation of cytosol NADH

The NADH dehydrogenase of the inner mitochondrial membrane of animal cells can accept electrons only from NADH in the matrix. Given that the inner membrane is not permeable to NADH, how can the NADH in the cytosol be reoxidized to NAD^+ by O_2 via the respiratory chain? Two shuttle systems have been considered to permit this process.

1. **Glycerolphosphate shuttle** mainly exists in the brain and skeletal muscle.
2. **Malate aspartate shuttle** is mainly found in liver, kidney and heart.

B. Adenine nucleotide transporter

The inner mitochondrial membrane is generally impermeable to charged species, but two specific systems which are:

1. **Adenine nucleotide translocase** transport ADP and Pi into the matrix and ATP out to the cytosol.

2. **Phosphate translocase**, A second membrane transport system essential to oxidative phosphorylation, promotes symport of one $H_2PO_4^-$ and one H^+ into the matrix.

V. Other biological oxidations without ATP producing

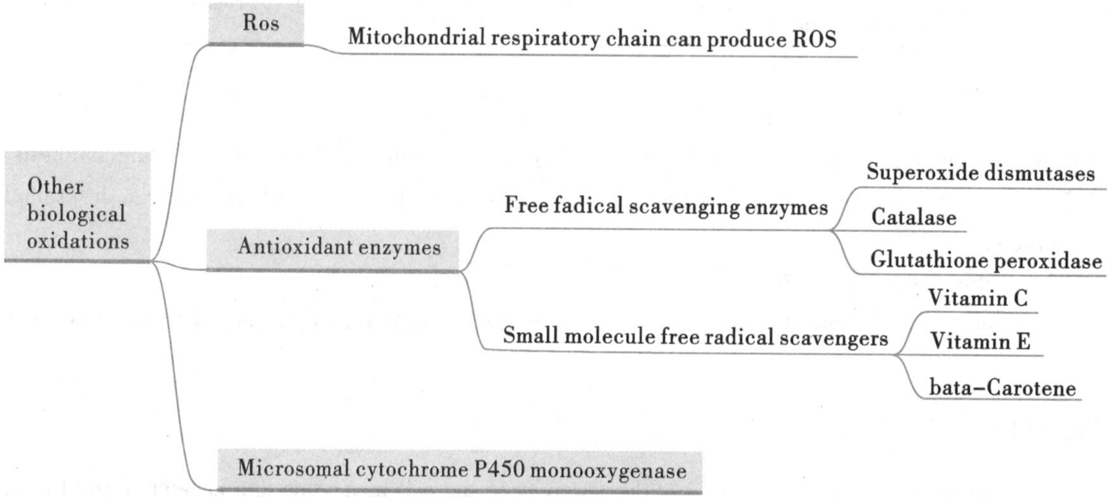

Besides the mitochondrial oxidation system, there are other oxidative systems in microsomes, peroxisomes and other parts of cells, which are involved in the oxidation process outside the respiratory chain. It does not produce ATP and is mainly involved in the biotransformation of metabolites, drugs and poisons. Peroxisomes contain catalase and peroxidase, and microsome contains monooxygenase.

A. ROS

The mitochondrial respiratory chain can also produce reactive oxygen species (ROS)

B. Antioxidant enzymes

scavenging ROS, peroxisomes

Catalase

Peroxidase: Glutathione peroxidase

Superoxide dismutases (SOD)

C. Microsomal cytochrome P450 (Cyt P450)

monooxygenase, mixed-function oxidase or hydroxylase

catalyzes hydroxylation of substrate molecules

$$RH + NADPH + H^+ + O_2 \rightarrow ROH + NADP^+ + H_2O$$

[Practices]

[A1 type]

[1] The major molecule that the human body generates and uses for energy is ________.

A. glucose B. fatty acid C. creatine phosphate

D. ATP E. GTP

[2] Eukaryotic respiratory chain exists in the ____________.

A. cytoplasm B. nucleus C. mitochondria

D. peroxisome E. microsome

[3] Which of the following is the main form of energy storage in muscle tissue? ____________

A. ATP B. GTP C. UTP

D. CTP E. Creatine phosphate

[4] Which of the following statements about cytochrome is true? ____________.

A. They are all electron carriers

B. The coenzymes of various cytochromes are identical

C. All cytochromes bind tightly to the mitochondrial inner membrane

D. They are all present in mitochondria

E. All cytochromes are inhibited by CN^- and CO

[5] Which one of the following carriers from the electron transfer chain transfer electrons directly to oxygen? ____________

A. Cytochrome aa_3 B. Cytochrome b C. Cytochrome c

D. Cytochrome c_1 E. Iron-sulfur protein

[6] The most important factor in the regulation of oxidative phosphorylation is____________.

A. O_2 B. $FADH_2$ C. NADH

D. cytochrome aa_3 E. ADP/ATP

[7] The hormone that regulates oxidative phosphorylation is ____________.

A. thyroxine B. adrenaline C. glucocorticoid

D. insulin E. glucagon

[8] Which of the following statements about the respiratory chain is not true?____________

A. They couple with ADP phosphorylation in the transfer of hydrogen and electrons.

B. CO inhibits electron transports of the respiratory chain.

C. Hydrogen carriers are also electron carriers.

D. Electron carriers are all hydrogen carriers.

E. Electrons transfer from carriers of lower E'° to higher E'°.

[9] Which of the following statements about uncoupling oxidation and phosphorylation in mitochondria is true? ____________

A. ADP phosphorylation accelerates the utilization of oxygen.

B. Phosphorylation of ADP continues, but the utilization of oxygen is stopped.

C. ADP phosphorylation stops, but the utilization of oxygen continues.

D. Phosphorylation of ADP has no change, but the utilization of oxygen is stopped.

E. Both ADP phosphorylation and utilization of oxygen cease.

[10] Which one of the following statements about mammalian energy metabolism is true? ____________.

A. ATP is only formed in the absence of O_2

B. ATP hydrolysis is an exergonic reaction

C. ATP is only formed in the presence of O_2

D. The heat produced by ATP hydrolysis specifically drives other reactions

E. $FADH_2$ cannot be utilized to form ATP

[11] Inhibition of the synthesis of ATP by oligomycin during oxidative phosphorylation is thought to be due

to____________.

A. blocking of the proton gradient between NADH dehydrogenase and reductive CoQ

B. blocking of the proton gradient between cytochrome c and cytochrome oxidase

C. uncoupling of electron transfer between NADH and flavoprotein

D. dissociating of CoQ from mitochondrial membranes

E. inhibiting of mitochondrial ATP synthase

[12] The connection between oxidation phosphorylation and electron transport is best described by____________.

A. existence of a higher pH in the cisternae of the endoplasmic reticulum than in the cytosol

B. synthesis of ATP as protons flow into the mitochondrial matrix along a proton gradient that exists across the inner mitochondrial membrane

C. symmetric distribution of the ATPase of the inner mitochondrial membrane

D. dissociation of electron transport and oxidative phosphorylation

E. absence of ATPase in the inner mitochondrial membrane

[13] Which one of the following statements about cytochrome c is true? ____________

A. A two-electron carrier

B. Covalently bound to complex II

C. An iron-sulfur protein

D. A single electron carrier

E. None of the above

[14] Which of the following statements about CoQ (Ubiquinone) is true? ____________

A. Transfers only one electron

B. Is lipid-soluble compound

C. Reduces oxygen directly

D. Requires ATP

E. Is a small protein

[15] The sequence of electron transport in the electron transfer chain is____________.

A. NADH → CoQ → Complex II → Complex III → Complex IV

B. $FADH_2$ → Complex II → CoQ → Complex III → Complex IV

C. $FADH_2$→Complex II→CoQ→Complex III→Cytochrome c→Complex IV

D. NADH→CoQ→Complex II→Complex III→Cytochrome c→ Complex IV

E. NADH→CoQ→ComplexI→ComplexIII→Cytochrome c→ Complex IV

[16] Oxidative phosphorylation occurs in the ____________.

A. cytosol B. mitochondrion C. endoplasmic reticulum

D. microsome E. nucleus

[17] The respiratory chain transfer a pair of reducing equivalents generated by which of the following substances, its P/O ratio is about 2.5? ____________

A. α-Glycerophosphate B. Succinate C. β-Hydroxybutyrate

D. Ascorbic acid E. Fatty acyl CoA

[18] Which of the following enzymes catalyzes substrate-level phosphorylation during glycolysis? ____________

A. Pyruvate kinase

B. Phosphate triose isomerase

C. Phosphoglycerate mutase

D. *L*–Lactate dehydrogenase

E. Glyceraldehyde–3–phosphate dehydrogenase

[19] Which of the following has been shown to be an uncoupling agent for oxidative phosphorylation? ____________

A. 2–Deoxyglucose B. Uncoupling prctein C. Arsenate

D. Thyroxine E. 2,3–Bisphosphoglycerate

[20] The increase of basal metabolic rate in patients with hyperthyroidism is due to____________.

A. uncoupling protein gene expression enhancement

B. ATP synthesis increase

C. decrease of adenine nucleotide translocase activity

D. decrease of Na^+, K^+–ATPase activity in the cell membrane

E. increase of phosphate translocase activity

[21] Which of the following compounds is the best for the rapid ATP replenishment? ____________

A. Glycogen B. Creatine C. Myoglobin

D. Fatty acids E. Phosphocreatine

[22] Which of the following statements is about the physiological significance of the malate–aspartate shuttle? ____________.

A. Transports oxaloacetate into the mitochondrial matrix for oxidation

B. Transports FAD to the mitochondrial matrix

C. Transamination of glutamate and oxaloacetate

D. Providing adequate oxaloacetate for the tricarboxylic acid cycle

E. Transports reducing equivalents from cytosolic NADH into the mitochondrion

[23] Two distinct shuttle mechanisms have been defined which are capable of moving electrons from NADH in the cytosol into the mitochondrion. These two shuttles produce different amounts of ATP per pair of electrons from oxidative phosphorylation because____________.

A. they operate at different efficiencies depending on the substrates available

B. some tissues have only one of the shuttles, while other tissues may have both

C. only some sources of NADH can enter the malate–aspartate shuttle

D. the glycerol–3–phosphate shuttle mechanism uses exclusively lactate

E. the two shuttles transfer electrons to different points in the respiratory chain

[**A2 type**]

[24] A comatose laboratory technician is rushed into the emergency room. She dies while you are examining her. Her most dramatic symptom is that her body is literally hot to your touch, consistent with an extremely high fever. You find out that her lab has been working on metabolic inhibitors and that there is a high likelihood that she accidentall yingested one. Which one of the following is the most likely culprit?

A. Barbiturates B. Piericidin A C. Rotenone

D. 2,4–Dinitrophenol E. Cyanide

[25] In 1959, a 30–year–old Swedish woman arrived at Luft's clinic and had the highest metabolic rate ever recorded. From muscle biopsies, Luft came to the conclusion that her condition was caused by abnormal mitochondria, abnormal production of ATP, and uncontrolled muscle metabolism. She is the first patient that was diagnosed with Luft disease. Which of the following descriptions of the patient is true?

__________.

A. Abnormal increase of ATP level in mitochondria

B. Electron transfer rate in mitochondria is very low

C. Increased proton gradient across the mitochondrial inner membrane

D. Hydride does not inhibit electron transfer

E. Patients showed high basal metabolic rate and elevated body temperature

[**B1 type**]

No. 26 ~ 27 shared the following suggested answers

A. $NADH^+/NADH+H^+$

B. $FAD/FADH_2$

C. Cytb Fe^{3+}/Fe^{2+}

D. Cyta Fe^{3+}/Fe^{2+}

E. Cytc Fe^{3+}/Fe^{2+}

[26] Which one of the above-mentioned redox couples have the highest oxidation-reduction potential in the respiratory chain? __________

[27] Complex II from the respiratory chain is composed of which of the above-mentioned redox couples? __________

No. 28 ~ 29 shared the following suggested answers

A. Succinyl CoA synthase B. Citrate synthase C. Isocitrate dehydrogenase

D. Succinate dehydrogenase E. Fumarase

[28] Which of the enzymes above catalyzes substrate-level phosphorylation during the TCA cycle? __________

[29] Which of the above is the NAD^+-linked enzyme during the TCA cycle? __________

No. 30 ~ 31 shared the following suggested answers

A. Complex I B. Complex II C. Complex III

D. Complex IV E. Complex V (ATP synthase)

[30] Succinate dehydrogenase is located in which complex of the mitochondrial respiratory chain? __________

[31] The only complex which actually uses oxygen in the mitochondrial respiratory chain is__________.

No. 32 ~ 33 shared the following suggested answers

A. Cytochrome b B. Coenzyme Q C. Cytochrome c

D. Cytochrome a_3 E. Cytochrome c_1

[32] If the mitochondrial electron carriers are artificially oxidized and NADH is then added to the system, the last carrier to become reduced is__________.

[33] Which one of the above electron carriers in the electron transport chain is not membrane-bound? __________

No. 34 ~ 35 shared the following suggested answers

A. Coupling of oxidation and phosphorylation

B. The effect of CO on electron transfer

C. Storage and utilization of energy

D. The combination of $2H^+$ and 1/2 O_2

E. The reaction catalyzed by lactate dehydrogenase

[34] Which of the above processes is related to the interconvert of ADP and ATP with each other? __________

______

[35] Which of the above processes is associated with ATP generation? ____________

[X type]

[36] Choose all that applies: which of the following reactions occur in mitochondria? ____________

A. Glycolysis
B. Citric acid cycle
C. Fatty acid β-oxidation
D. Synthesis of palmitic acid
E. Oxidative phosphorylation

[37] Which of the following compounds contain vitamin B, and are also components of the respiratory chain? ____________

A. Flavin adenine dinucleotide
B. Nicotinamide adenine dinucleotide
C. Pyridoxal phosphate
D. Coenzyme Q
E. Tetrahydrofolic acid

[38] Which of the following coenzymes transfer both hydrogen and electrons? ____________

A. NAD^+
B. FMN
C. FAD
D. Cytochrome
E. CoQ

[39] Which of the following enzymes have prosthetic groups containing iron-sulfur centers? ____________

A. NADH-ubiquinone oxidoreductase
B. Succinic-ubiquinone reductase
C. Ubiquinone-cytochrome c reductase
D. Cytochrome c oxidase
E. ATP synthase

[40] Which of the following are proteins with heme prosthetic groups? ____________

A. Hemoglobin
B. Myoglobin
C. Catalase
D. Cytochrome
E. Iron-sulfur protein

[41] Which of the following reactions produce high-energy phosphoric compounds by substrate-level phosphorylation? ____________

A. 1,3-Bisphosphoglycerate→3-phosphoglycerate
B. Oxaloacetate→phosphoenolpyruvate
C. Isocitrate→α-ketoglutarate
D. Phosphoenolpyruvate→pyruvate
E. SuccinylCoA→succinate

[42] Which of the following compounds contain high energy bonds? ____________

A. UTP
B. Creatine phosphate
C. Glucose-6-phosphate
D. Acetyl CoA
E. 1,3-Bisphosphoglycerate

[43] Which of the following processes can release sufficient energy for ATP formation during transferring electrons by the electron transport chain? ____________

A. NADH→CoQ
B. CoQ→Cytc
C. FAD→CoQ
D. Cytc→$Cytaa_3$
E. $Cytaa_3$→O_2

[44] Which of the following compounds inhibit electron transport in the respiratory chain? ____________

A. CO
B. CN^-
C. N_3^-
D. Antimycin A
E. 2,4-Dinitrophenol

[45] The factors contributing to the acceleration of oxidative phosphorylation are ____________.

A. ADP/ATP ratio increased
B. increased ATP/ADP ratio
C. thyroxine
D. insulin
E. O_2

[46] NADH that is produced in the cytoplasm is transported into the mitochondrial matrix by which system

for oxidation? ____________

A. Glycerol-3-phosphate shuttle B. Alanine-glucose cycle C. Malate aspartate shuttle

D. Lactate cycle (Cori cycle) E. Citrate pyruvate cycle

[47] Which of the following substances can enter and exit the mitochondria through the translocase in the mitochondrial inner membrane? ____________

A. NADH B. Phosphate C. ATP

D. Acetyl CoA E. Oxaloacetate

[48] Which of the following statements about monooxygenase are true? ____________

A. It catalyzes hydroxylation of substrates

B. The reaction catalyzed by monooxygenase requires cytochrome P450

C. NADPH is required to provide reduction equivalent

D. The reaction product is hydrogen peroxide.

E. The reaction catalyzed by monooxygenase relates to the biotransformation of drugs or poisons

Answers

1	2	3	4	5	6	7	8	9	10
D	C	E	A	A	E	A	D	C	B
11	12	13	14	15	16	17	18	19	20
E	B	D	B	C	B	C	A	B	A
21	22	23	24	25	26	27	28	29	30
E	E	E	D	E	D	B	A	C	B
31	32	33	34	35	36	37	38	39	40
D	D	C	C	A	BCE	AB	ABCE	ABC	ABCD
41	42	43	44	45	46	47	48		
ADE	ABDE	ABE	ABCD	AC	AC	BC	ABCE		

Brief Explanations for Practices

1. ATP, a high energy compound, is the currency of life. It can directly provide energy for various physiological activities of cells. A: Glucose is a type of carbohydrate. B: Fatty acids are components of fat. C: Creatine phosphate is the form in which energy is storedin muscle tissue. E: GTP is a high-energy phosphate compound.

2. Electron transport chain, also be called as the respiratory chain, consists of a series of mitochondrial enzymes and redox carrier molecules that transfer reducing equivalents from substrates to oxygen.

3. Creatine phosphate acts as a rapidly mobilizable storage form of energy in skeletal muscle and brain. Creatine kinase transfers the terminal phosphate group of ATP to creatine, generating creatine phosphate.

4. Cytochromes are a group of electron transfer proteins with heme prosthetic groups in which iron interconverts between Fe^{3+} and Fe^{2+} during oxidation and reduction.

5. B: Cytochrome b transfers electrons to cytochrome c_1. C: Cytochrome c transfers electrons to cytochrome aa_3. D: Cytochrome c_1 transfers electrons to cytochrome c. E: Components of complexes I, II and III participate in the electron transfer in the respiratory chain.

6. Oxidative phosphorylation is the most important way to synthesize ATP. The concentration of intra-

cellular ADP is a measurement of the energy status incells. The ADP/ATP ratio can indicate the change of cellular energy demand and is an important regulatory factor for oxidative phosphorylation.

7. Thyroxine regulates Na^+, K^+–ATPase activity in the cell membrane and accelerates the hydrolysis of ATP, which enhances oxidative phosphorylation. It also induces uncoupling protein gene expression, which causes metabolite oxidation to increase energy release and the rate of heat production.

8. Electron carriers in the respiratory chain are not all hydrogen carriers. For example, cytochromes are proteins with heme prosthetic groups in which the iron atom oscillates between Fe^{3+} and Fe^{2+} during oxidation and reduction. They are single electron carriers and don't transfer hydrogen.

9. Uncoupling agents specifically increase the proton permeability in the inner mitochondrial membrane and abolish the linkage between the respiratory chain and the phosphorylation of ADP to ATP.

10. In mammalian systems, ATP may be synthesized in the presence or the absence of O_2. ATP is formed either of two ways of phosphorylation, by substrate–level phosphorylation in glycolysis or by the oxidation phosphorylation in mitochondria. ATP hydrolysis is an exergonic reaction and drives many energy–requiring reactions in biological systems.

11. Oligomycin binds to a 23kd polypeptide (OSCP) in the F_0 baseplate of mitochondrial ATP synthase and thus blocks utilization of energy derived from electron transport for ATP synthesis. Oligomycin has no effect on coupling but inhibits mitochondrial phosphorylation so that both oxidation and the phosphorylation cease in its presence.

12. The chemiosmotic theory which is proposed by Peter Mitchell describes the coupling of electron transport and oxidative phosphorylation. The flow of electrons along the electron transport chain allows protons to be pumped from the matrix of the mitochondria to the intermembrane space. The protons are pumped across the inner mitochondrial membrane at three sites in the respiratory chain to produce a proton gradient. When protons move back through proton channels of the ATP synthase in the inner mitochondrial membrane, ATP is synthesized.

13. Cytochromes are proteins with heme prosthetic groups in which the iron atom oscillates between Fe^{3+} and Fe^{2+} during oxidation and reduction. Cytochromes are one–electron carriers.

14. CoQ is ubiquitous, a lipid–soluble component of the electron transport chain found in the inner mitochondrial membrane. It is one of a group of benzoquinone derivatives that have an isoprenoid side chain of varying length.

15. See NADH respiratory chain and $FADH_2$ respiratory chain.

16. Oxidative phosphorylationrefers to the synthes is of ATP, which is energetically coupled to the oxidation of reducing equivalents to water, in the mitochondria.

17. β–Hydroxybutyrate is catalyzed by β–hydroxybutyrate dehydrogenase with NAD^+ as the coenzyme. NADH is oxidized by NADH respiratory chain.

18. The high energy phosphate group of phosphoenolpyruvate is directly transferred to ADP by pyruvate kinase to generate two molecules of ATP per molecule of glucose oxidized. This is an example of substrate–level phosphorylation in glycolysis.

19. The uncoupling protein, which exists in brown adipose mitochondria, provides a proton channel through the inner mitochondrial membrane and reduces ATP synthesis.

20. Thyroid hormone induces the expression of Na^+, K^+–ATPase in cell membrane and accelerates the hydrolysis of ATP, which enhances oxidative phosphorylation. It also induces uncoupling protein gene expression, which causes metabolite oxidation to increase energy release and the rate of heat production.

21. Besides ATP, there is still another high–energy compound called phosphocreatine (PC) inside the

human muscle cells. When PC is broken down, energy is released for the resynthesis of ATP.

22. The physiological significance of special shuttle systems is that carries reducing equivalents from cytosolic NADH into mitochondria by an indirect route. The malate-aspartate shuttle transferring reducing equivalents from cytosolic NADH into the mitochondrial matrix for oxidation operates in liver, kidney, and heart.

23. The glycerol-3-phosphate shuttle transfers reducing equivalents from cytosolic NADH into the mitochondrion through the FAD prosthetic group of the membrane-bound enzyme to generate $FADH_2$, which is oxidizedby the FAD respiratory chain. The malate-aspartate shuttle passes reducing equivalent NADH in the cytosol into the mitochondrion matrix, the resulting NADH is oxidized by the NADH respiratory chain.

24. 2,4-Dinitrophenol is unique in that it disconnects the ordinarily tight coupling of electron transport and phosphorylation. In its presence, though electrons can still be transferred along the respiratory chain, the energy is released in the form of heat instead of ATP. Barbiturates, the antibiotic piericidin A, the fish poison rotenone, and cyanide all act by inhibiting the electron transport chain at some point.

25. Luft disease is a hypermetabolic disorder of striated muscle caused by an abnormal quantity and type of mitochondria producing excessive cellular respiration; it is characterized by profuse perspiration, asthenia, and progressive weakness. Evidence suggests that decoupling of electron transport chain occurs in the patients.

26. The standard reduction potential of individual electron carriers has been determined experimentally. Electrons should flow from carriers of lower E'° to higher E'°. The sequence of electronic carriers in the respiratory chain is: NADH ($FADH_2$)→Q→Cyt b→Cyt c_1→Cyt c→Cyt(aa_3)→O_2.

27. Complex II is also known as succinate dehydrogenase; it is the only FAD-linked enzyme of the TCA cycle.

28. Succinyl-CoA is converted to succinate by the enzyme succinate-CoA synthetase. This is the only example of substrate-level phosphorylation in the TCA cycle. Cleavage of the high-energy thioester bond of succinyl CoA is coupled to phosphorylation of GDP to GTP, whose energy content is the same as that of ATP.

29. Oxidation and decarboxylation of isocitrate in the TCA cycle is catalyzed by isocitrate dehydrogenase; it is associated with the release of the first NADH and the first CO_2. Isocitrate dehydrogenase is a NAD^+-dependent enzyme.

30. Succinate dehydrogenase is bound to the inner surface of the inner mitochondrial membrane. It is also called complex Ⅱ.

31. Complex IV, cytochrome c oxidase, transfers electrons from cytochrome c to O_2.

32. The sequence of electronic carriersin the electron transport chain to transfer electrons is NADH ($FADH_2$)→Q→Cyt b→Cyt c_1→Cyt c→Cyt(aa_3)→O_2. The last carrier to become reduced is cytochrome a_3.

33. Cytochrome c, as a freely water-soluble protein, is mobile on the outer surface of the inner membrane.

34. The phosphorylation of ADP to form ATP requires energy. The hydrolysis of ATP to ADP releases energy which is utilized in energy-requiring processes.

35. The oxidative phosphorylation is the main source of energy in the aerobic cell. The majority of ATP is generated in oxidative phosphorylation. Oxidative phosphorylation refers to the synthesis of ATP, which is energetically coupled to the oxidation of reducing equivalents to waterin the mitochondria.

36. Enzymes that catalyze glycolysis and palmitate synthesis are distributed in the cytoplasm. Enzymes

catalyzing β-oxidation of fatty acids are present in the mitochondria matrix. Mitochondria are described as cellular power plants. Nutrients ultimately release energy from mitochondria throughthe citric acid cycle and oxidative phosphorylation.

37. A: Flavin adenine dinucleotide contains vitamin B_2 and is a component of the respiratory chain. B: Nicotinamide adenine dinucleotide, an active form of vitamin PP in vivo, is the component of the respiratory chain. C: It contains vitamin B_6, but it is not a component of the respiratory chain. D: It's a component of the respiratory chain, but not vitamins B. E: It contains folic acid, but it is not a component of the respiratory chain.

38. Cytochromes are electron carriers of the respiratory chain, but not hydrogen carriers.

39. In iron-sulfur proteins, the iron is present in association with inorganic sulfur atoms, with the sulfur atoms of Cys residues, or both. Iron-sulfur centers are their prosthetic groups. Iron-sulfur proteins take part in the transferring of electron as components of complex I, complex II and complex III in the respiratory chain.

40. Hemoglobin, myoglobin, catalase and cytochrome all contain heme as their prosthetic groups, which allow for them to bind oxygen or participate in electron transfer. In iron-sulfur proteins, the iron is present not in heme but in association with inorganic sulfur atoms, with the sulfur atoms of Cys residues, or both.

41. Substrate level phosphorylation is a term used to describe the synthesis of ATP coupled to a dehydrogenation (or dehydration) of an organic substrate, independent of the respiratory chain. It involves chemical intermediates with the high energy bond, such as 1,3-bisphosphoglycerate, phosphoenol pyruvate or succinyl CoA.

42. High energy bondsrelease more than 25.0 kJ/mol of free energy when hydrolyzed. ATP, UTP, CTP, GTP, creatine phosphate and 1,3-bisphosphoglycerate contain high-energy phosphatebonds. Acetyl CoA contains high energy thioester bond.

43. According to the measurement of P/O ratio and calculation of free energy release, NADH→CoQ, CoQ→Cyt c and Cyt $aa_3 \rightarrow O_2$ can provide sufficient energy for ATP formation from ADP and phosphate.

44. 2,4-Dinitrophenol is a classic example of an uncoupler. It disconnects the ordinarily tightly coupling of electron transport and ADP phosphorylation. In its presence, electrons can transfer normally along the respiratory chain, but the energy released cannot be used to synthesize ATP.

45. The ATP/ADP ratio is an important regulatory factor for oxidative phosphorylation. When cells' energy demand is increased, the concentration of ADP increases. ADP enters mitochondria and accelerates oxidative phosphorylation. When the ATP/ADP ratio is normalized, the rate of oxidative phosphorylation also slows down. Thyroxine regulates Na^+, K^+-ATPase activity in the cell membrane and accelerates the hydrolysis of ATP, which enhances the oxidative phosphorylation.

46. A and C are two special shuttle systems that carry reducing equivalents from cytosolic NADH into mitochondria by an indirect route. B: Ammonia produced by muscle metabolism is transported to the liver. D: The lactate produced in muscle tissue is transported into the liver for gluconeogenesis. E: Acetyl CoA in the mitochondrial matrix is transported to the cytosol for the synthesis of fatty acids and cholesterol.

47. A: NADH is not permeable to the mitochondrial membrane. Two special shuttle systems carry reducing equivalents from cytosolic NADH into mitochondria by an indirect route. B: Phosphate translocase transports $H_2PO_4^-$ and H^+ into the mitochondrial matrix in the same direction. C: Adenine nucleotide translocase transports ADP into the mitochondrial matrix and newly synthesized ATP out to the cytosol. D: Acetyl CoA in the mitochondrial matrix is transported to the cytosol through citrate pyruvate cycle. E: Oxaloacetate enters and exits the inner mitochondrial membrane by converting to malate or aspartate.

48. The human microsomal cytochrome P450 monooxygenase incorporates only one atom of the oxygen molecule into the substrate while the other oxygen atom is reduced to water by hydrogen (derived from NADPH + H^+). Thus, it is named mixed-function oxidase or hydroxylase. It is involved in the generation of steroid hormones, bile acids and bile pigments. The microsomal cytochrome P450 monooxygenase is also involved in the biotransformation processes of drugs, toxins and products of endogenous metabolisms, such as bilirubin.

(***Zhang Weijuan***)

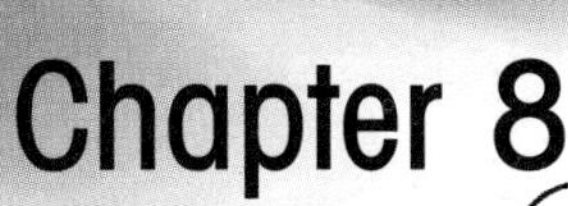

Metabolism of Lipids

Examination Syllabus

1. **The physiological function of lipids**: storage and supply of energy; components of biomembranes; regulatory function of lipid derivatives; essential fatty acids.

2. **Digestion and absorption of fats**: enzymes needed for fat emulsification and digestion; monoacylglycerol pathway of triglyceride synthesis and chylomicron.

3. **Anabolism of triacylglycerol**: synthetic site; building blocks; the basic pathway of synthesis.

4. **Synthesis of fatty acids**: synthetic site; building blocks; synthetic process.

5. **Fat catabolism**: fat mobilization; the basic process of fatty acid β-oxidation; ketone bodies.

6. **Metabolism of glycerophospholipids**: basic structure and classification of glycerophospholipids; site, materials and pathway of glycerophospholipid synthesis; degradation of glycerophospholipids.

7. **Cholesterol metabolism**: site, materials and key enzymes of cholesterol synthesis; regulation of cholesterol synthesis; conversion of cholesterol.

8. **Plasma lipoprotein metabolism**: plasma lipid and its composition; classification and function of plasma lipoprotein; plasma lipoprotein metabolism; hyperlipoproteinemia.

Ⅰ. The physiological function of lipids

Lipids are the general term for fats and lipoids.

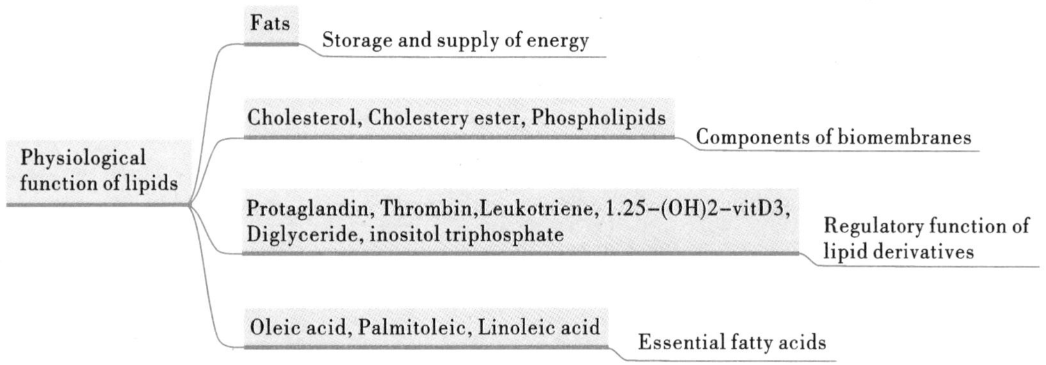

A. Storage and supply of energy

Fats, also called triglycerides, have the function:

1. Store energy and providing energy.
2. Maintain body temperature.
3. Protect the internal organs.
4. Assist the absorption of fat-soluble vitamins.

B. Components of biomembranes

1. Lipoids, including cholesterol, cholesteryl ester, phospholipids, and glycolipids, act as structural components of cell membranes.
2. Phospholipids are the major component of biomembranes, which can form lipid bilayer due to their amphiphilic characteristic.
3. Cholesterol plays an important role in maintaining biomembranefluidity.

C. The regulatory function of lipid derivatives

1. Some lipid derivatives, such as prostaglandin, thrombin and leukotriene, play an important role in the regulation of cell metabolism.
2. 1,25-$(OH)_2$-$VitD_3$, the product of cholesterol conversion, can regulate the metabolism of calcium and phosphorus.
3. Diglyceride and inositol triphosphate, the products of phosphatidylinositol degradation, are important second messengers.

D. Essential fatty acids

1. The main unsaturated fatty acids in the human body include oleic acid, palmitoleic acid, linoleic acid, α-linolenic acid and arachidonic acid.
2. The essential fatty acids must be supplied in the diet, mainly including **linoleic acid**, **α-linolenic acid** and **arachidonic acid**.
3. Linolenic acid and arachidonic acid can be derived from linoleic acid and **arachidonic acid** is the precursor of prostaglandin (PG), so the lack of the essential fatty acids can lead to the decrease of PG.

Ⅱ. Digestion and absorption of fats

A. Substances needed for fat emulsification and digestion

1. Bile salts act as an effective emulsifier and emulsify the fat into small micelles, thus increasing the surface area of fat, which allows the lipase to break down the fat more effectively.
2. The pancreatic juice contains pancreatic lipase, cholesteryl esterase, phospholipase A_2, and colipase, which are important digestive enzymes for attacking lipids (Table 8.1).

Table 8.1 Substances needed for fat emulsification and digestion

	Main function	End products or action
Bile salts	Acting as a fat emulsifier and making triglyceride and cholesteryl esters into small micelles	Finely emulsified micelles
Pancreatic lipase	Specifically hydrolyzing the ester bonds in 1 and 3 positions of triglyceride	2-monoglyceride+2 molecular fatty acids
Colipase	Co-factor of lipase and anchoring lipase to the surface of lipid micelles	Enhancing lipase activity
Phospholipase A_2	Hydrolyzing the ester bond at position 2 of phospholipids	Fatty acids+ Lysophospholipids
Cholesteryl esterase	Catalyzing the hydrolysis of cholesteryl esters	Fat acids + Cholesterol

B. Monoacylglycerol pathway of triglyceride synthesis and chylomicron

1. The products of lipid digestion together with bile acids form mixed micelles, which enter into the intestinal mucosal cells.

2. The short-chain, middle chain fatty acids and glycerol can directly enter the liver through the portal vein.

3. The majority of the digestive products are used to rebuild triglycerides and phospholipids in intestinal mucosal cells, where the triglycerides are then combined with a small number of phospholipids, cholesterols, cholesterol esters, and apolipoproteins B48 (ApoB48) to form **chylomicrons**.

4. **Chylomicrons** are a kind of lipoprotein that can pass fat-based substances into the blood circulation.

5. The process that lipid digestive products (monoacylglycerol and fatty acids) are used to synthesize triglyceride in small intestinal mucosa cells is called monoacylglycerol pathway.

[*Practices*]

[**A1 type**]

[1] If there is a chronic lack of vegetable oil, the substance that will decrease in the body is ____________

A. palmitoleic acid B. oleic acid C. arachidonic acid

D. cholesterol E. arachidic acid

[2] The material that can be used as the building block for the synthesis of prostaglandins is ____________ ______.

A. palmitic acid B. stearic acid C. arachidonic acid.

D. palmitoleic acid E. oleic acid

[3] Which one of the following arachidonic acid derivatives strongly promotes platelet aggregation, vasoconstriction, blood coagulation and thrombosis? __________

A. PGI_2 B. TXA_2 C. PGE_2

D. LTC_4 E. PGA_2

[4] In the intestinal tract, the non-enzymatic components that help the digestion and absorption of dietary lipids are ____________.

A. cholesterol B. bilirubin C. bile acid

D. biliverdin E. choline

[X Type]

[5] The factors that directly involved in the intestinal lipid digestion are

A. lipoprotein lipase B. bile salts C. the pancreatic lipase
D. colipase E. cholesteryl esterase

Ⅲ. Anabolism of triacylglycerol

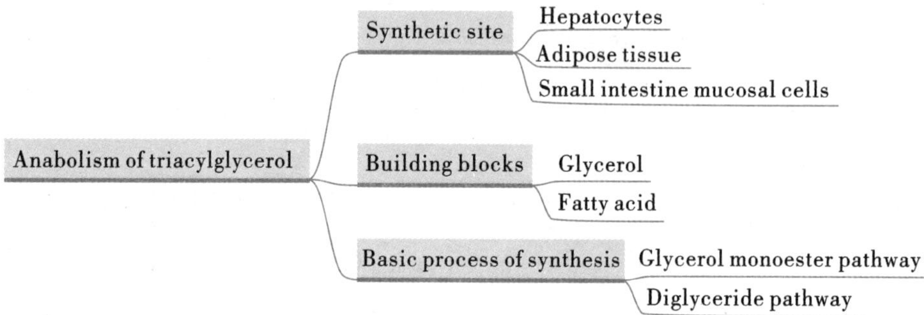

A. Synthetic site

1. **Liver**: Triacylglycerols synthesized in the live rmainly combines with ApoB100, phospholipids and cholesterol to produce very-low-density lipoprotein (VLDL), which is transported to extrahepatic tissues via the blood.

2. **Adipose tissue**: Adipocytes can both synthesize and store triacylglycerols, which are used as "energy storage" for fasting or starvation.

3. **Small intestine**: Small intestinal mucosal cells can use the digestive products of lipids in the diet to resynthesize triacylglycerols, which then enter the blood circulation through the lymphatic vessels in the form of chylomicrons.

B. Building blocks

Glycerol and fatty acid, in which 3-phosphate glycerol has two sources: mainly from glucose metabolism; from free glycerol.

C. The basic process of synthesis

1. **Glycerol monoester pathway**: mainly takes places in small intestinal mucosal cells.
2. **Diglyceride pathway**: mainly takes places in the liver and adipose tissues.

Ⅳ. Synthesis of fatty acids

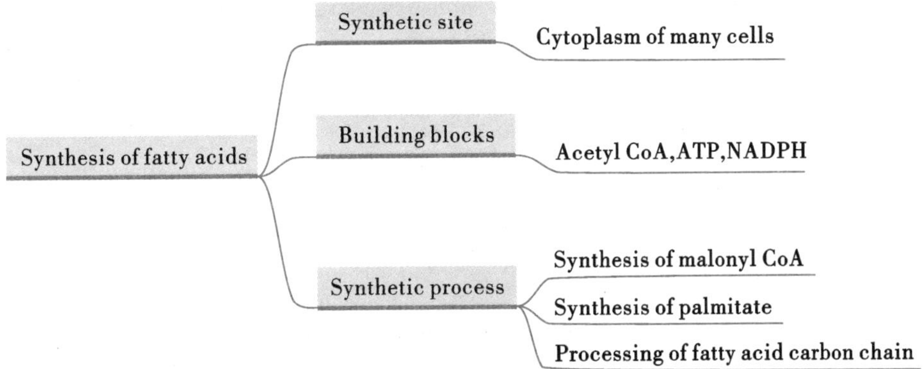

A. Synthetic site:

the cytosols of the liver (main site), kidney, brain, lung, breast, and adipose tissue

B. Building blocks

1. Acetyl CoA is mainly from glucose metabolism.

2. The acetyl CoA produced in the mitochondria must be transferred to the cytosol via **citrate-pyruvate cycle** to allow fatty acid synthesis.

3. NADPH is mainly from the pentose phosphate pathway.

C. Synthetic process

1. Synthesis of malonyl CoA

$$ATP + HCO_3^- + acetyl\ CoA \longrightarrow malonyl\ CoA + ADP + Pi$$

(1) This step is catalyzed by **acetyl CoA carboxylase**, which is the rate-limiting enzyme of fatty acid synthesis.

(2) Long-chain acyl CoA and glucagon inhibit acetyl CoA carboxylase activity, while citrate, isocitrate and insulin stimulate it.

2. Synthesis of palmitate

$$acetyl\ CoA + 7malonyl\ CoA + 14NADPH + H^+ \rightarrow palmitate$$

Palmitate is generated under the catalysis of fatty acid synthetase system through 7 repeated addition reactions including condensation, hydrogenation, dehydration and rehydrogenation.

3. Processing of fatty acid carbon chain

The production of fatty acids beyond 16-carbon is elongated on the basis of palmitate, which is carried out in the endoplasmic reticulum or mitochondria.

[*Practices*]

[**A1 type**]

[6] Which one of the following tissues has the strongest ability to synthesize triacylglycerols? ______

A. Adipose tissue B. Liver C. Small intestine

D. Kidney E. Muscle

[7] What is the main source of acetyl CoA for fatty acid synthesis? ____________

A. The catabolism of glucose B. The catabolism of fatty acids C. The catabolism of cholesterols

D. The catabolism of glucogenic amino acids E. The catabolism of ketogenic amino acids

[8] What is the pathway that transfers acetyl CoA from mitochondria to the cytosol to initiate fatty acid synthesis? ____________

A. Tricarboxylate cycle B. Lactate cycle C. Uronate cycle

D. Citrate-pyruvate cycle E. Alanine-glucose cycle

[9] Which is the carrier of the acyl in the synthesis of fatty acids? ____________

A. CoA B. Carnitine C. ACP

D. Malonyl CoA E. Oxaloacetate

V. Fat catabolism

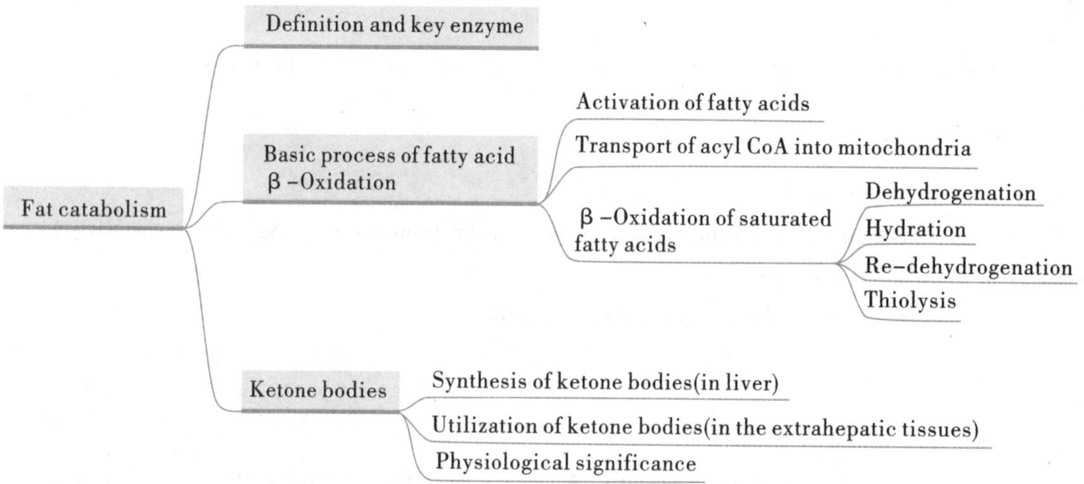

A. Fat mobilization

1. **Fat mobilization** is the process that triglyceride in adipocytes is hydrolyzed into free fatty acids and glycerol by a series of lipases, which are then released into the blood and transported to other tissues for oxidation.

2. **The key enzyme** of fat mobilization is hormone-sensitive triglyceride lipase (**HSL**), which is regulated by multiple hormones.

3. **Lipolytic hormone** promotes fat mobilization, such as epinephrine, glucagon, ACTH, TRH and so on.

4. **Anti-lipolytic hormone** inhibits fat mobilization, such as insulin, prostaglandin E2 and so on.

B. Basic process of fatty acid β-oxidation

1. **Activation of fatty acid** is catalyzed by **acyl CoA synthetase** in the cytosol, which consumes two molecules of ATP to form acyl CoA.

2. **Transport of acyl CoA into mitochondria**

(1) The acyl CoA produced in the cytosol is carried into the mitochondria by the carnitine with the aid

of the **carnitine acyltransferase** Ⅰ, Ⅱ and **carnitine - carnitine acyl translocation enzyme** in the mitochondrial membrane.

(2) This step is the major rate-limiting step and **carnitine acyltransferase I** is the rate-limiting enzyme in the β-oxidation of fatty acids.

3. **β-Oxidation of saturated fatty acids** consists of a sequence of four reactions involving the β-carbon that results in shortening the fatty acid chain by two carbons at the carboxyl terminal and releasing a molecule of acetyl CoA.

(1) **Dehydrogenation**: producing a molecule of $FADH_2$.

(2) **Hydration**

(3) **Re-dehydrogenation**: producing a molecule of NADH.

(4) **Thiolysis**: releasing a molecule of acetyl CoA and a molecule of acyl CoA shortened 2 carbon atoms.

4. **Energy generation**

(1) Fatty acids of even-numbered carbons can undergo ($n/2-1$) times of β-oxidation (where n is the number of carbons) to produce ($n/2-1$) molecules of $FADH_2$, ($n/2-1$) molecules of NADH and $n/2$ molecules of acetyl-CoA.

(2) The activation of fatty acids requires 2 ATP, therefore the total energy is $1.5 \times (n/2 - 1) + 2.5 \times (n/2 - 1) + 10 \times n/2 - 2 = (7n-6)$ ATPs.

(3) Take the oxidation of 16-carbon palmitate, for example, it requires 7 cycles of β-Oxidation:

Products	produced energy (ATP)
8 molecules of acetyl CoA	8×10
7 molecules of NADH + H^+	7×2.5
7 molecules of $FADH_2$	7×1.5
108 ATPs	

The activationof palmitateconsumes 2 ATPs, so net generated ATPs are 106 from a molecule of palmitate.

C. Ketone bodies

are the special intermediate products of fatty acid catabolism in the liver, including acetoacetate, β-hydroxybutyrate, and acetone.

1. **Synthesis of ketone bodies** take places in hepatic mitochondria and HMG-CoA synthase is the key enzyme, which mainly present in the liver.

2. **Utilization of ketone bodies** takes places in the extrahepatic tissue, such as heart, kidney, brain and skeletal muscle, which have highly active enzymes that can utilize ketone bodies to produce energy.

3. **Physiological significance**

(1) Ketone bodies are a kind of energy output from the liver and an important source of energy for muscles, especially brain when glucose is in short supply.

(2) The production of ketone bodies beyond the oxidative ability of extrahepatic tissues may result in **ketoacidosis**.

[*Practices*]

[A1 type]

[10] Which one of the following is the key enzyme of fat mobilization? ____________

A. HL B. HSL C. LPL

D. PLP E. LCAT

[11] Which one of the following is the product of fat mobilization? ________

A. Glycerol and fatty acids B. 3-Glycerophosphate C. 3-glyceraldehyde phosphate
D. 1,3-diphosphoglycerate E. 2,3-diphosphoglycerate

[12] Which one of the following is the place of fatty acid β-oxidation? ________

A. Mitochondrial matrix B. Mitochondrial inner membrane C. Cytoplasm
D. Nuclei E. Microsomes

[13] Which one of the following is the key enzyme for the β-oxidation of fatty acids? ________

A. Acetyl CoA carboxylase B. HMG-CoA synthase C. Argininosuccinate synthase
D. Carnitineacyltransferase I E. HMG-CoA reductase

[14] Which of the following enzymes activity is so low that the ketone body cannot be oxidized in the liver? ________

A. Acetoacetate CoA thiolase B. Acetoacetate lyase C. HMG-CoA reductase
D. HMG-CoA synthase E. Succinyl CoA thiophorase

[15] Which one of the following situations can make the brain mainly oxidizes ketone bodies for energy? ________

A. Prolonged starvation B. Controlled diabetes C. Fasting
D. Satiety E. Strenuous exercise

Ⅵ. Metabolism of glycerophospholipids

A. Basic structure and classification of glycerophospholipids

1. Glycerophospholipids are composed of glycerol, fatty acid, a phosphate group and/or substituents.
2. Glycerophospholipids are classified according to the substituents on phosphate.
3. Phosphatidylcholine (lecithin) and phosphatidylethanolamine (cephalin) are the most abundant in the human body.

B. Site, materials and pathway of glycerophospholipid synthesis

1. **Site**: in the endoplasmic reticulums, mainly in liver, kidney and intestine
2. **Building blocks**: glycerols, fatty acids, phosphates, choline, serine, inositol, S-adenosylmethionine, CTP and the like
3. **Pathways**: diacylglycerol pathway and CDP-diacylglycerol pathway

C. Glycerophospholipid degradation

Glycerophospholipid degradation is catalyzed by a series of phospholipases, including phospholipase A_1, A_2, B_1, B_2, C and D.

Ⅶ. Cholesterol metabolism

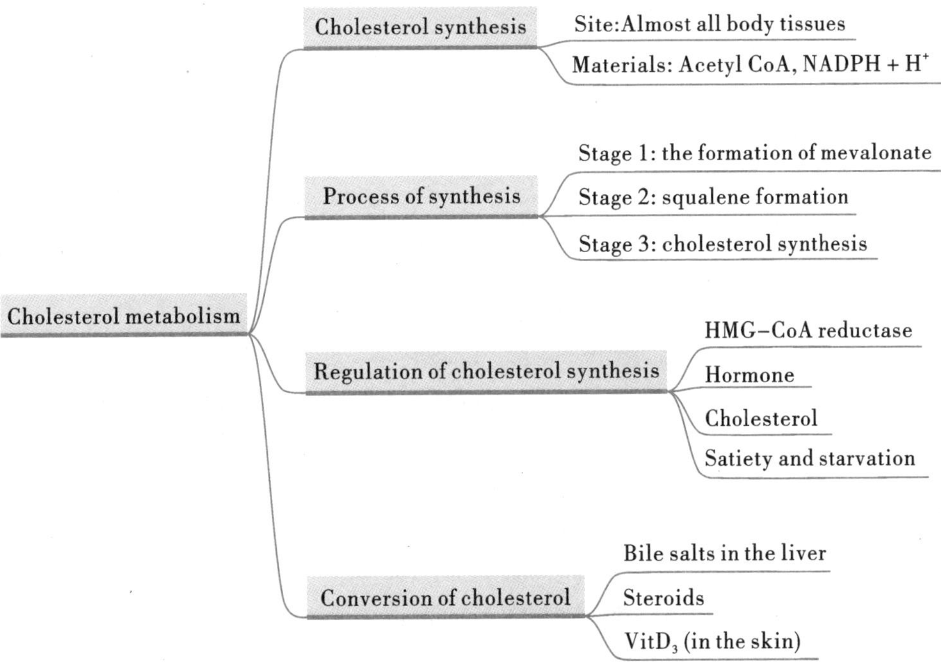

A. Cholesterol synthesis

1. **Synthetic site**

(1) Almost all human tissues can synthesize cholesterol (except for adult brain and mature red blood cells), among which liver is the main site.

(2) The synthesis is carried out in **cytosol**.

2. **Synthetic materials**: Acetyl CoA, NADPH + H^+ and ATP

The synthesis of 1 molecule of cholesterol requires 18 molecules of acetyl CoA, 16 molecules of NADPH + H^+ and 36 molecules of ATP.

3. **Key enzyme**: HMG CoA reductase

B. Process of synthesis

1. **Stage 1**: the formation of mevalonate
2. **Stage 2**: squalene formation
3. **Stage 3**: cholesterol synthesis

C. Regulation of cholesterol synthesis

1. **HMG-CoA reductase** is the rate-limiting enzyme in cholesterol synthesis, which is:

(1) the major control point;

(2) controlled by allosteric regulation, chemical modification regulation and enzyme amount regulation.

2. **Hormone**: Insulin and thyroid hormone can increase cholesterol synthesis by inducing the synthesis of HMG-CoA reductase, whereas glucagon and cortisol do the opposite.

3. **Cholesterol**: The increase of cellular cholesterol level can lower cholesterol synthesis by inhibiting

the synthesis of HMG-CoA reductase.

4. **Satiety and starvation**: Satiety increases the synthesis of HMG-CoA reductase, whereas starvation does the opposite.

D. Conversion of cholesterol

1. Bile salts in the liver.
2. Steroids.
3. $VitD_3$(in the skin).

[Practices]

[A1 type]

[16] The substance needed to synthesize cephalins is__________.

A. CDP-ethanolamine. B. CDP-choline. C. UDP-choline.
D. UDP-ethanolamine. E. GDP-ethanolamine.

[17] Which of the following phospholipases can hydrolyze 1 molecule of lecithin to form 1 molecule of diglyceride and 1 molecule of phosphorylcholine? __________

A. PLA_1. B. PLA_2. C. PLB_2.
D. PLC. E. PLD.

[18] Key enzymes in the biosynthesis of cholesterol in the body are__________.

A. HMG - CoA synthase B. HMG - CoA reductase C. HMG - CoA lyase
D. ALA synthase E. acetoacetate kinase.

[19] The enzyme that catalyzes the transfer of fatty acyl group to cholesterol to form cholesteryl ester in cells is__________.

A. ACAT B. LCAT C. phospholipase C
D. phospholipase D E. carnitine acyltransferase I

[20] The decrease of plasma cholesterol level in patients with liver disease is due to__________.

A. the increasein LDL activity
B. the decrease of LCAT
C. the increase in cholesterol esterase activity
D. the decrease in cholesterol esterase activity
E. the decrease in cholesterol synthesis

Ⅷ. Plasma lipoprotein metabolism

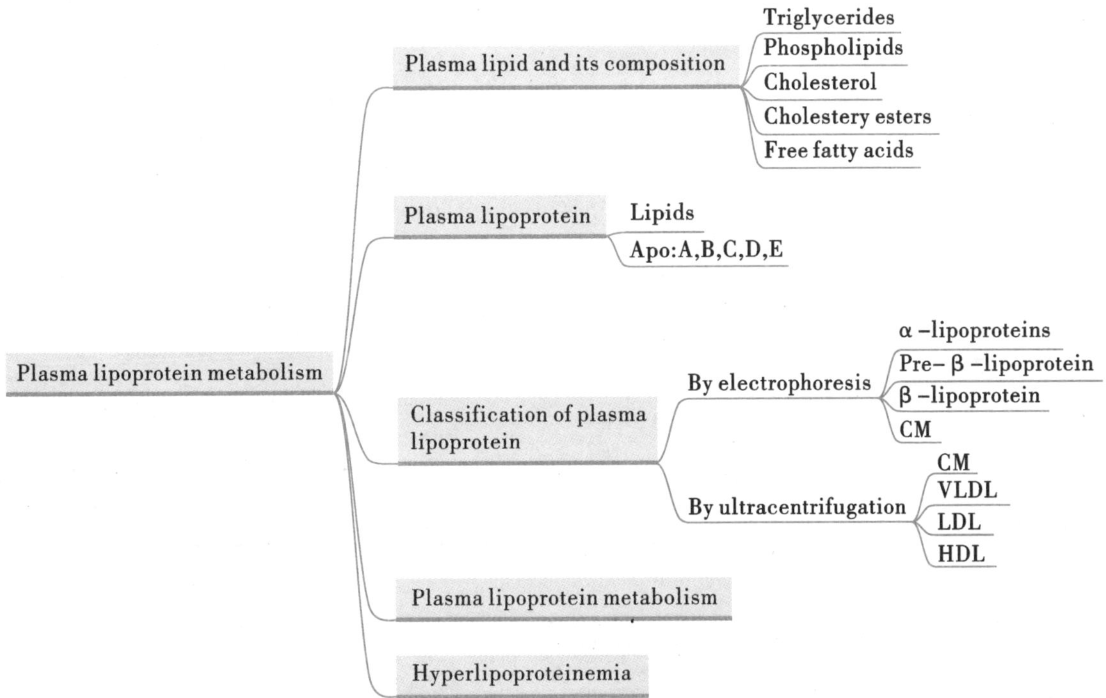

A. Plasma lipid and its composition

1. The lipids in plasma are called blood lipids, including triglycerides, phospholipids, cholesterol, cholesteryl esters and free fatty acids.

2. The sources of plasma lipids can either be exogenous or endogenous. The former comes from diet and the latter is synthesized and released by the liver or adipose tissue.

B. Plasma lipoprotein is a complex of lipids and proteins (Table 8.2).

1. The proteins in plasma lipoprotein are called **apolipoproteins** (Apo), which are mainly divided into five categories: A, B, C, D and E.

2. The distribution and content of apolipoprotein in different lipoproteins are different.

C. Classification of plasma lipoprotein

1. **By electrophoresis**: α-lipoproteins, pre-β-lipoprotein, β-lipoprotein and CM

2. **By ultracentrifugation**: chylomicron (CM), very low-density lipoprotein (VLDL), low-density lipoprotein (LDL) and high-density lipoprotein (HDL).

Table 8.2 Classification, composition and function of plasma lipoprotein

	CM	VLDL	LDL	HDL
Name	Chylomicron	Very low-density lipoprotein	Low-density lipoprotein	High-density lipoprotein
Density	CM<VLDL<LDL<HDL			
Electrophoresis corresponding protein	Chylomicron	Pre-β-lipoprotein	β-lipoprotein	α-lipoprotein
Protein content	1%	5% ~10%	20% ~25%	Most, 50 %
TG content	80% ~90%	50% ~70%	10%	5%
Cholesterol content	1% ~4%	15%	45% ~50%	20%
Characteristic apo	apo B48	apoE	apoB100	$apoA_1$, A_2
Synthetic site	Small intestinal mucosal cells	Liver cells	From VLDL in plasma	Liver, intestine, plasma
Function	Transport exogenous TG and cholesterol	Transport endogenous TG and cholesterol	Transport endogenous cholesterol	Transport cholesterol reversely

D. Plasma lipoprotein metabolism

1. CM

(1) **CM** is synthesized in the small intestine and its characteristic apolipoprotein is apoB48.

(2) **Metabolism**: nascent CM generated in small intestinal mucosal cells →into the blood via lymphatic circulation → mature CM →hydrolysis of triglycerides by LPL→ CM remnants → liver.

(3) **apoC Ⅱ** is the activator of **lipoprotein lipase** (LPL) in peripheral tissues.

2. VLDL

(1) VLDL is synthesized in the liver and its characteristic lipoprotein is apoB100.

(2) **Metabolism**: VLDL produced in the liver →into the blood → hydrolysis of triglycerides by LPL→ VLDL remnants (IDL).

3. LDL

(1) **LDL** is derived from VLDL in the blood and its characteristic apolipoprotein is apoB100.

(2) **Metabolism**: VLDL → VLDL remnants (IDL) → LDL → 2/3 of LDL enters the target cells (such as liver, renal cortex, gonads) via the LDL receptor pathway and 1/3 of LDL is cleared by mononuclear-phagocyte.

(3) The rise of intracellular free cholesterol level has a negative-feedback effect on cholesterol synthesis through the following mechanism:

- inhibition of HMG-CoA reductase activity.
- activation of ACAT activity to accelerate cholesterol esterification.
- inhibition of LDL receptor synthesis.

4. HDL

(1) HDL is synthesized in the liver or small intestine and its characteristic apolipoprotein is apoA Ⅰ.

(2) **Metabolism**: nascent HDL (discoid, no core) in the liver or small intestine→into the blood→taking in unesterified cholesterol→ HDL3 → HDL2 → liver.

(3) **Lecithin**: cholesteryl–acyltransferase (**LCAT**) catalyzes the esterification of cholesterol and transfers cholesteryl ester to the HDL core, increasing its particle size progressively, which is activated by apoA Ⅰ.

E. Hyperlipoproteinemia

1. **Concept**: It refers to the abnormally elevated levels of any or all lipids or lipoproteins in the blood.

2. **Classification**: six types based on the pattern of lipoproteins on electrophoresis or ultracentrifugation (Table 8.3).

Table 8.3 Classification of hyperlipoproteinemia

Type	Changes of lipoproteins	Changes of triglycerides	Changes of cholesterol
Ⅰ	CM ↑	↑ ↑ ↑	↑
Ⅱa	LDL ↑		↑ ↑
Ⅱb	LDL ↑ +VLDL ↑	↑ ↑	↑ ↑
Ⅲ	IDL ↑	↑ ↑	↑ ↑
Ⅳ	VLDL ↑	↑ ↑	
Ⅴ	VLDL ↑ +CM ↑	↑ ↑ ↑	↑

[*Practices*]

[**A1 type**]

[21] Which of the following is not contained in plasma lipoprotein? ____________

A. Cholesterol　B. Triglyceride　C. Phospholipid

D. Free fatty acid　E. Cholesteryl ester

[22] After the agarose electrophoresis, which of the following is the proper order from cathode to anode? ____________

A. CM, VLDL, LDL, HDL　B. CM, LDL, VLDL, HDL　C. HDL, VLDL, LDL, CM

D. HDL, LDL, VLDL, CM　E. CM, HDL, VLDL, LDL

[23] Which of the following plasma lipoprotein mainly contains apoB48? ____________

A. VLDL　B. LDL　C. HDL

D. IDL　E. CM

[24] Which of the following lipoprotein formation disorders is closely related to the fatty liver? ____________

A. CM　B. VLDL　C. LDL

D. HDL　E. IDL

[25] Which of the following enzyme in plasma catalyzes the production of cholesteryl esters and stores them in HDL? ____________

A. Carnitine acyltransferase　B. LCAT　C. ACAT

D. Acyl CoA synthase　E. HSL

Answers

1	2	3	4	5	6	7	8	9	10
C	C	B	C	BCDE	B	A	D	C	B
11	12	13	14	15	16	17	18	19	20
A	A	D	E	A	A	D	B	A	E
21	22	23	24	25					
D	B	E	B	B					

Brief Explanations for Practices

1-3. Linolenic acid, linoleic acid and arachidonic acid (20 carbon polyunsaturated fatty acid) are essential fatty acids, which cannot be synthesized in the body and must be supplied in the diet. Prostaglandins are derivatives of unsaturated fatty acids. When cells are stimulated by external factors, phospholipase A_2 in the cell membrane will be activated and then hydrolyzes the phospholipids to release arachidonic acids, which are utilized to synthesize prostaglandin (PG), thromboxane (TX) and leukotriene (LT) by a series of enzymes. All of these substances are called arachidonic acid derivatives. Therefore, when the essential fatty acids are in short supply, the deficiency of phospholipids can lead to decreased synthesis of PG, LT and TX. TXA_2, also named thromboxane, strongly promotes platelet aggregation, vasoconstriction, blood coagulation and thrombosis.

4. Bile acids are converted from cholesterols in the liver and are strongly amphiphilic. They are secreted into the intestine with bile, as an emulsifier to help disperse food lipids and to promote their digestion and absorption.

5. The digestion of lipids in the diet is mainly carried out in the small intestine. In the combined action of pancreatic lipase, colipase, cholesterol esterase, phospholipase and bile salts, the fat is hydrolyzed to glycerol, fatty acids and monoacylglycerols. Phospholipids are hydrolyzed to lysophospholipids and fatty acids, while cholesteryl esters are degraded into free cholesterol and fatty acids.

6. Omitting.

7. Acetyl CoA is the main building block for fatty acid synthesis, which is produced from various metabolic processes, but mainly from glucose metabolism.

8. The synthesis of fatty acid occurs in the cytosol, so the acetyl CoA in the mitochondria can initiate the synthesis only after it is transported to the cytosol via citrate-pyruvate cycle.

9. The acyl carrier of the β-oxidation is CoA, and the acyl carrier of fatty acid synthesis is ACP.

10. The key enzyme in fat mobilization is HSL.

11. The products of fat mobilization are glycerol and fatty acids.

12. The β-oxidation of fatty acids takes place in the mitochondrial matrix.

13. A is the key enzyme of fatty acid synthesis; B is the key enzyme of ketone body synthesis; C is the key enzyme of urea synthesis, and E is the key enzyme of cholesterol synthesis.

14. Omitting.

15. When the patient is in prolonged starvation, he can't use glucose for energy, and fat mobilization is strengthened, which produces a large number of ketone bodies in the liver.

16. CTP is particularly important in phospholipid synthesis because it is necessary for the synthesis of activated intermediates such as CDP-ethanolamine, CDP-choline, and CDP-diglycerides. CDP-ethanola-

mine is required for the synthesis of cephalin; CDP-choline is required for the synthesis of lecithin, and CDP-phosphatidylglycerol is required for the synthesis of cardiolipin.

17. Four different types of phospholipases hydrolyze the ester bonds of glycerophospholipid at different sites, resulting in the corresponding products. PLA hydrolyzes the ester bonds at 1 or 2 position and the resulting products are free fatty acids and lysophospholipids. PLC hydrolyzes the 3-position ester bond between the hydroxyl group and the phosphate to produce diglycerides and phosphorylated substances. PLD hydrolyzes the ester bond between phosphate and the substituent, resulting in phosphatidate and free substrates. The substrate of PLB is lysophospholipid and the product includes 1 molecule of free fatty acid and 1 molecule of incompletely hydrolyzed substance without a hydrophobic tail.

18. The synthetic material of cholesterol is mainly acetyl CoA and the key enzyme is HMG-CoA reductase.

19. The pathway for the synthesis of cholesteryl ester in the cell is different from that in plasma: acyl CoA: cholesterol acyltransferase (ACAT) is present in the endoplasmic reticulum of various cells and catalyzes the esterification of cholesterol to cholesteryl ester. Under the catalysis of lecithin cholesterol acyltransferase (LCAT) in plasma, the acyl at position 2 of lecithin is transferred to hydroxyl at position 3 of cholesteryl to form lysolecithin and cholesteryl ester. LCAT is synthesized in liver parenchyma cells, secreted into the blood and plays a role in the plasma.

20. The main site of cholesterol synthesis is in the liver, and liver disease can result in a decrease of cholesterol synthesis.

21. Omitting.

22. α-Lipoprotein and β-lipoprotein correspond to HDL and LDL, respectively. VLDL corresponds to the pre-β-lipoprotein, which is in front of the β-band (close to the anode).

23. The characteristic apolipoprotein of CM is apoB48.

24. VLDL is mainly synthesized by liver cells, and its main function is to transport endogenous triglycerides out of the liver. When hepatocytes cannot generate VLDL, the triglycerides synthesized accumulate in the liver to form the fatty liver.

25. LCAT catalyzes the esterification of cholesterol and transfers cholesteryl ester to the HDL core.

[*ComprehensivePractices*]

[A1 type]

[1] What is the main synthetic site of endogenous triglycerides? ____________

A. Fatty tissue B. Liver C. Muscle

D. Kidney E. Small intestine

[2] When massive fat is mobilized, what does the acetyl CoA produced in the liver mainly convert to? ____________

A. Glucose B. Cholesterol C. Fatty acid

D. Ketone body E. Malonyl CoA

[3] Which one is the enzyme involved in the actaivation of fatty acid? ____________

A. Thiolase B. Acyl CoA synthetase C. HMG-CoA synthase

D. Lipoprotein lipase E. Triglyceride lipase

[4] Which of the following statements is not correct about hormone-sensitive lipase? ____________

A. Catalyzing the hydrolysis of triglyceride stored in adipose.

B. Glucagon can activate it by phosphorylation.

C. Insulin inhibits it by dephosphorylation.

D. The reaction it catalyzed is the rate-limiting step of triglyceride hydrolysis.

E. It belongs to lipoprotein lipases.

[5] Which one is not required for the oxidation of fatty acids?____________

A. Carnitine B. NAD^+ C. $NADP^+$

D. FAD E. CoA

[6] What are the key enzymes for biosynthesis of fatty acids and cholesterol, respectively? __________

A. Acetyl CoA carboxylase, HMG-CoA reductase

B. HMG-CoA synthase, HMG-CoA reductase

C. Acetyl CoA carboxylase, HMG-CoA lyase

D. Acyl CoA synthetase, HMG-CoA synthase

E. Carnitineacyltransferase I, Lecithin: cholesterol acyltransferase

[7] Which is the common intermediate product in fat and glycerophospholipid synthesis?__________

A. Diacylglycerol B. Phosphatidate C. Monoacylglycerol

D. Acetyl CoA E. CDP-diacylglycerol

[8] Which is the common intermediate product of the fatty acid β-oxidation, ketone body formation and cholesterol synthesis?___________

A. AcetoacetylCoA B. Mevalonic acid C. HMG-CoA

D. Acetoacetate E. β-ketoacyl CoA

[9] What are the primary metabolic disorders of homozygous patients with familial hypercholesterolemia? ____________.

A. Increased LDL produced by VLDL

B. Cell membrane LDL receptor dysfunction

C. Increased activity of HMG-CoA reductase in liver

D. Lack of apolipoprotein B

E. Decreased activity of ACAT

[10] Which of the following functions does apolipoprotein **not** have?____________

A. Activating extrahepatic LPL

B. Activating LCAT

C. Activating lipase in adipose

D. Acting as the ligand of the lipoprotein receptor

E. Stabilizing lipoprotein structure

[11] What kind of nucleotide is needed in addition to ATP to provide energy when synthesizing glycerophospholipids? ___________

A. UTP B. cAMP C. CTP

D. cGMP E. GTP

[12] Which process occurs completely within the mitochondria? ___________

A. Synthesis of palmitic acid B. Protein synthesis C. Gluconeogenesis

D. Glucose aerobic oxidation E. Ketone body formation

[13] Which of the following hormones can inhibit fat mobilization? ___________

A. Epinephrine B. Prostaglandin E_2 C. Glucagon

D. Adrenocorticotropic hormone E. Thyrotropin

[14] Which of the following hormones can inhibit fat mobilization? ___________

A. Adrenaline B. Norepinephrine C. Insulin

D. Glucagon E. ACTH

[15] How many molecules of ATP will be released when a molecule of stearic acid (18 carbon saturated fatty acid) is completely decomposed into CO_2 and H_2O by β-oxidation? ___________

A. 30 B. 32 C. 106

D. 120 E. 124

[16] When fat is mobilized, what is the transport form of fatty acids in blood? ___________

A. Combination with albumin B. Combination with β-globulin C. Combination with Y protein

D. CM E. VLDL

[17] Which one of the following apolipoproteins is the most abundant in HDL? ___________

A. apoA Ⅰ B. apoA Ⅱ C. apoC

D. apoD E. apoE

[18] Which of the following statements about the apolipoprotein function is **incorrect**? ___________

A. LPL is activated by apoC Ⅰ.

B. The main apolipoprotein of LDL is apoB100.

C. LCAT is activated by apoA Ⅰ.

D. Both apoB and apoE participate in the recognition of the LDL receptor.

E. apoA Ⅱ can activate HL.

[19] Which of the following apolipoproteins is not elevated in all types of hyperlipoproteinemias? ___________

A. HDL B. IDL C. CM

D. VLDL E. LDL

[20] Which of the following changes stand for blood lipid results of a type Ⅱ a hyperlipoproteinemia patient? ___________

A. TG↑↑↑, cholesterol↑ B. Cholesterol↑↑, TG↑↑ C. Cholesterol↑↑

D. TG↑↑ E. Cholesterol↑↑, TG↑↑ and the electrophoresis showed broadband.

[B1 type]

No. 21 ~ 23 shared the following suggested answers

A. bile acids B. LCAT C. PLA_2

D. ACAT E. LPL

[21] The enzyme that catalyzes the esterification of cholesterol in plasma is___________.

[22] The enzyme that decomposes the membrane of the red cell and produces lysophosphatidic is___________.

[23] The enzyme that can catalyze the esterification of cholesterol in cells is___________.

No. 24 ~ 27 shared the following suggested answers

A. Lipase in tissue B. LPL C. Hepatic lipase

D. Pancreatic lipase E. HSL

[24] The enzyme that decomposes dietary triglycerides is___________.

[25] The enzyme that decomposes triglycerides in adipocytes is___________.

[26] The enzyme that decomposes triglycerides in lysosomes is___________.

[27] The enzyme that decomposes triglycerides in CM and VLDL is___________.

No. 28 ~ 32 shared the following suggested answers

A. Cytoplasm and endoplasmic reticulum

B. Mitochondrion

C. Cytoplasm

D. Cytoplasm and microsomes

E. Endoplasmic reticulum

[28] The site of cholesterol synthesis is____________.

[29] The site of bile acid synthesis is____________.

[30] The site of fatty acids synthesis is____________.

[31] The site of ketone bodies synthesis is____________.

[32] The site of phospholipid synthesis is____________.

No. 33 ~ 34 shared the following suggested answers

A. lysosomes B. endoplasmic reticulum C. mitochondrion

D. cytosol E. golgi apparatus

[33] The common metabolic site of gluconeogenesis and the tricarboxylic acid cycle is____________.

[34] The common metabolic site for cholesterol synthesis and phospholipid synthesis is__________.

[X type]

[35] Fatty acids can't be used to produce sugar because____________.

A. Acetyl CoA can't be converted to pyruvate

B. The reaction catalyzed by pyruvate dehydrogenase complex is irreversible

C. The fatty acids are oxidized in the mitochondria, while gluconeogenesis is performed in cytoplasm

D. The products of oxidation of fatty acids are CO_2, H_2O and ATP

E. The energy produced by oxidation of fatty acids is not sufficient to convert its metabolites to gluconeogenesis

[36] Which of the following compounds are involved in fatty acid β-oxidation? ____________

A. NAD^+ B. $NADP^+$ C. CoA-SH

D. FAD E. FMN

[37] The esterification of cholesterol in plasma needs____________.

A. fatty acyl CoA B. acetyl CoA C. phospholipids

D. LCAT E. ACAT

[38] The transfer of fatty acyl from the cytoplasm into the mitochondria needs____________.

A. carnitine

B. carnitine acyl transferase I

C. carnitine acyl transferase II

D. carnitine acyl carnitinetranslocase

E. coenzyme A

[39] The mechanism that free cholesterol regulates the level of cholesterol in cells include____________.

A. the inhibition of LACT

B. the activation of ACAT

C. the inhibition of HMG-CoA reductase

D. the inhibition of LDL receptor expression

E. the activation of acetyl CoA carboxylase

[40] Plasma components that are proportional to the incidence of atherosclerosis include__________.

A. LDL B. ox-LDL C. HDL

D. Triglycerides E. apoA I

Answers

1	2	3	4	5	6	7	8	9	10
B	D	B	E	C	A	B	A	B	C
11	12	13	14	15	16	17	18	19	20
C	E	B	C	D	A	A	A	A	C
21	22	23	24	25	26	27	28	29	30
B	C	D	D	E	A	B	A	D	C
31	32	33	34	35	36	37	38	39	40
B	E	C	B	AB	ACD	CD	ABCDE	BCD	ABD

Brief Explanations forcomprehensive practices

1. Although adipose tissue is the main triglyceride storage site of the body, which can also synthesize triglyceride, the liver is the most active organ of endogenous triglyceride synthesis.

2. When glucose is insufficient, massive fats are mobilized in the body increases, which result in a large amount of acetyl CoA produced by the oxidation of fatty acids in the liver. Acetyl CoA partially enters into the citric acid cycle and the rest is converted to the ketone bodies for extrahepatic tissue utilization.

3. The activation of fatty acids refers that fatty acids combine with CoA to form fatty acyl CoA catalyzed by fatty acyl-CoA synthase.

4. HSL, the key enzyme of fat mobilization, is located in adipose tissue and regulated by several hormones, such as glucagon and epinephrine, which can activate it by phosphorylation to accelerate fat mobilization. In contrast, insulin promotes the dephosphorylation of HSL and inhibits its activity. Lipoprotein lipase catalyzes the hydrolysis of triglyceride in CM and VLDL.

5. Two hydrogens produced by dehydrogenations of fatty acid β-oxidation are accepted by FAD and NAD^+, respectively, where no $NADP^+$ involved. Carnitine participates in the transport of fatty acyl-CoA to mitochondria and CoA participates in the activation of fatty acids before oxidation.

6. There are three enzymes involved in HMG-CoA formation: HMG-CoA synthase, HMG-CoA lyase and HMG-CoA reductase. Among them, HMG-CoA lyase is not involved in cholesterol synthesis and HMG-CoA reductase is not involved in ketone body synthesis. Acyl-CoA synthase is involved in the activation of fatty acids, and the resulting acyl-CoA can either enter the synthetic pathway or enter the decomposition pathway. Carnitine acyltransferase Ⅰ is a key enzyme for fatty acid oxidation.

7. Phosphatidate is produced during the synthesis of fat and glycerophospholipids.

8. Acetoacetyl-CoA is the common intermediate of fatty acid β-oxidation, ketone body formation and cholesterol synthesis. Mevalonate is an intermediate of cholesterol synthesis. HMG-COA is the common intermediate of ketone body and cholesterol synthesis. Acetoacetate is one member of the ketone bodies.

9. The primary metabolic disorder in the homozygote of familial hypercholesterolemia is that the LDL receptor deficiency leads to the abnormal metabolism of LDL, which causes the increase of cholesterol in the blood.

10. Apolipoproteins have three major functions: combining lipids, composing and stabilizing lipoproteins; regulating the activity of key enzymes in lipoprotein metabolism (eg: apoA Ⅰ activates LCAT and apoC Ⅱ activates LPL.); acting as a ligand of the lipoprotein receptor to mediate its cellular uptake, (eg:

apoB is a ligand for LDL receptors, apoA Ⅰ is a ligand for HDL receptors, and apoE is a ligand for LDL and VLDL remnant receptors). For Answer C, the lipase of adipose tissue is hormone-sensitive lipase, which is mainly regulated by hormones, not apolipoprotein.

11. Omitting

12. Both A and B are performed in the cytoplasm; C and D are carried out both in cytoplasm and mitochondria.

13. There are many kinds of lipolysis hormone, but anti-lipolysis hormone is only founded in two kinds: insulin and prostaglandin E2.

14. Insulin and prostaglandin E2 can inhibit fat mobilization, belonging to anti-lipolysis hormone. Epinephrine, norepinephrine and glucagon can promote fat mobilization, belonging to lipolysis hormone. Adrenocorticotropic hormone can promote fat mobilization and lipolysis indirectly because of promoting epinephrine secretion.

15. The number of ATP generated by the decomposition of saturated fatty acids of even-numbered carbons is 7n-6 (where n is the number of carbons). When $n = 18$, the result of the calculation is 120.

16. Free fatty acids are transported by albumins in the blood. C: The Y protein is a carrier protein involved in the entry of bilirubin into hepatocytes. D and E are the transport forms of exogenous and endogenous triglycerides in plasma, respectively.

17. The apolipoproteins of HDL are predominantly apoA, including apoA Ⅰ (about 65%-70%) and apoA Ⅱ (about 20%-25% of), as well as a small amount of apoC, apoD, and apoE.

18. The lipoprotein lipase is activated by apoC Ⅱ rather than apoC Ⅰ.

19-20. Referring to table 8.3.

21-23. Omitting.

24-27. Question 24-27 is related to the metabolism of triglycerides. Dietary lipid digestion is mainly carried out in the small intestine. Under the combined actions of pancreatic lipase, co-lipase, cholesterol esterase, phospholipase and bile salts, the fat in the diet is hydrolyzed to glycerol, fatty acids and monoglycerides; the phospholipids are hydrolyzed to lysophospholipids and fatty acids, and the cholesteryl ester is degraded to free cholesterol and fatty acids. The fats in adipocytes are gradually hydrolyzed by a series of lipases to release free fatty acids and glycerol into the bloodstream for other tissue utilization, in which triglyceride lipase (hormone-sensitive lipase, HSL) is hormone-regulated and is the rate-limiting enzyme for fat mobilization. Triglycerides in the lysosomes are metabolized by tissue lipases. LPL catalyzes the hydrolysis of triglycerides in CM and VLDL to generate glycerol and fatty acids.

28-32. Question 28-32 is related to lipid metabolic sites. Cholesterol is mainly synthesized in the cytoplasm and endoplasmic reticulum of hepatocytes, while bile acids are produced in liver microsomes and cytoplasm and intestine. Fatty acids are synthesized in the cytoplasm, whereas ketone bodies are generated in liver mitochondria and phospholipids in the endoplasmic reticulum.

33-34. Omitting.

35. The inability of fatty acids to synthesize glucose is mainly due to the irreversible reaction of the pyruvate dehydrogenase system, and acetyl CoA cannot be converted to pyruvate.

36. FAD participates in dehydrogenation and NAD^+ (not $NADP^+$) is involved in re-dehydrogenation. CoA-SH participates in the thiolysis.

37. The esterification of cholesterol in plasma is catalyzed by LCAT and the fatty acid residue is provided by lecithin.

38. The carnitine shuttle system involved in fatty acyl CoA transport consists of three enzymes (car-

nitineacyltransferase Ⅰ, carnitineacyltransferase Ⅱ, and carnitine/acyl carnitine translocase) and two carriers (carnitine and coenzyme A).

39. A: LACT is involved in the metabolism of plasma HDL and is not related to intracellular cholesterol levels. E: Acetyl CoA carboxylase is a key enzyme for fatty acid synthesis and has nothing to do with cholesterol.

40. HDL performs the reverse transport cholesterol and its plasma level is inversely proportional to the incidence of AS. E: apoA Ⅰ is unique to HDL, and its plasma level is also inversely proportional to the incidence of AS.

(***Zhang Haifeng***)

Chapter 9

Metabolism of Amino acid

1. **The physiological function and nutritive value of protein**: functions of protein and amino acids; The concept and the types of nutrition essential amino acid; nitrogen balance

2. **Digestion, absorption and putrefaction of protein in the intestinal tract**: role of protease in protein digestion; absorption of the amino acids; putrefaction of the protein

3. **General metabolism of amino acids**: transamination; deamination of the amino acid; metabolism of the α-keto acid

4. **Metabolism of the ammonia**: resource of the ammonia; transport of the ammonia; the outlet of the ammonia

5. **Metabolisms of individual amino acid**: decarboxylation of amino acid; concepts, sources, carriers and significance of one carbon unit; methionine cycle, SAM, PAPS; Metabolism of phenylalanine, histidine and tyrosine

Major points and Practices

Ⅰ. The physiological function and nutritive value of protein

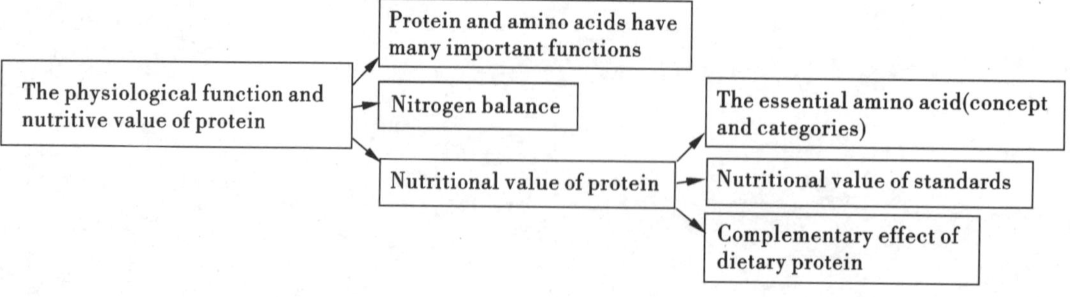

A. Theimportant functions of protein *in vivo*

1. Proteins can maintain cell tissue growth, renewal and repair.
2. Proteins are involved in a variety of important physiological activities.
3. Oxidation and energy supply

B. Nitrogen balance

1. **Concept**: a balanced relationship between nitrogen intake and discharge of nitrogen.
2. **Categories**: total nitrogen balance, positive nitrogen balance, negative nitrogen balance.

C. Nutritional value of protein

1. **Nutritional essential amino acid**

(1) **Definition**: the amino acids are needed *in vivo* and cannot be synthesized by the organism, must be provided by food.

(2) **Categories**: leucine, isoleucine, threonine, valine, lysine, methionine, phenylalanine, histidine and tryptophan.

2. **Nutritional value of standards**: the content of essential amino acids is high, assortment, in appropriate proportions of essential amino acids

3. **Complementary effect of dietary protein**: the nutritional value of the protein is improved by the combination of the lower nutritive proteins, and the essential amino acids can be supplemented mutually.

Ⅱ. Digestion, absorption and putrefaction of the protein in the intestinal tract.

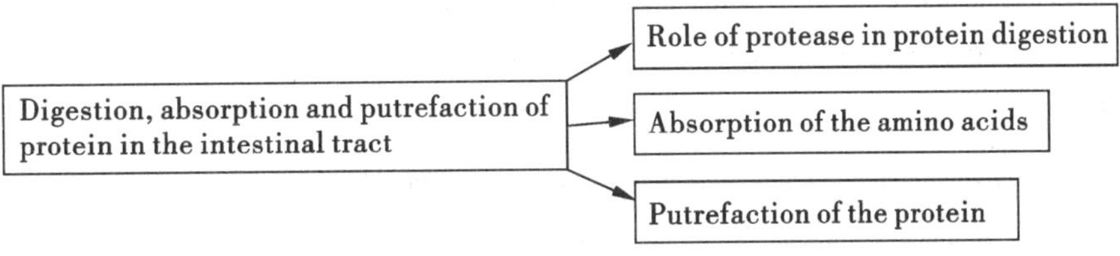

A. Role of protease in protein digestion

1. The digestion of food proteins begins in the stomach, but mainly in the small intestine.

2. The food proteins are gradually hydrolyzed into polypeptides and amino acids by protease. Pepsin and trypsin are the main enzymes of protein digestion.

3. Proteases are secreted in the form of zymogens, which need to be activated by an intestinal kinase or active protease.

B. Absorption of the amino acids

1. The absorption of amino acid/small peptide is mainly in the small intestine, which is an energy-consuming and positive transport process.

2. There are 3 kinds of absorption mechanisms: carrier protein, γ-glutamyl cycle, dipeptide/tripeptide transport system

C. Putrefaction of the protein

1. **Definition**: undigested proteins and undigested amino acids from the food protein are degraded by *E. coli* in the lower part of the colon.

2. **Product**: most of the products are harmful, including amines, ammonia and other harmful substances.

Ⅲ. General metabolism of amino acids

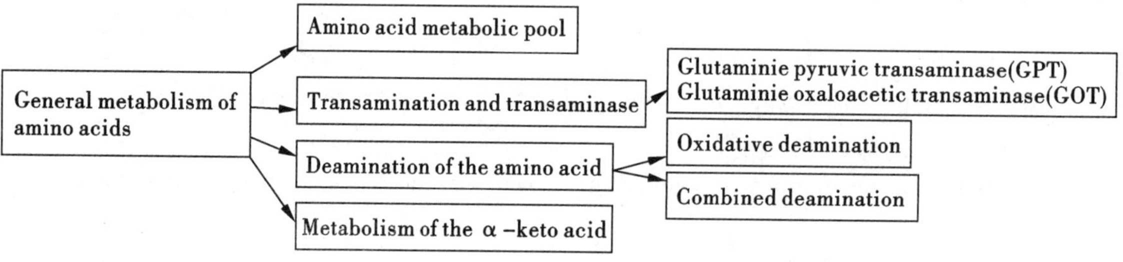

A. The pathway of protein degradation

1. **Non-dependent ATP-degrading pathway**: the **lysosomal** protein degradation pathway.

2. **Degradation pathway dependent on ATP and ubiquitin**: need ubiquitin participation to degrade the abnormal protein and short life protein, consuming ATP.

B. Metabolic pool of amino acids

Food protein by digestion and absorption of amino acids (exogenous amino acids) and *in vivo* tissue protein degradation produced by amino acids and *in vivo* synthesis of non-essential amino acids (endogenous amino acids) are distributed throughout the body, participate in metabolism, called **the amino acid metabolic pool** (Figure 9.1).

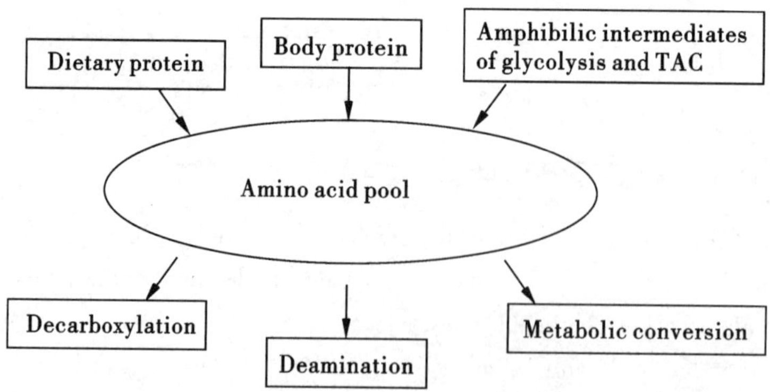

Figure 9.1 Amino acid metabolic pool

C. Deamination of amino acids

1. Oxidative deamination

The oxidation of amino acids is the process which amino acid is the first dehydrogenation and then hydrolyzed to free ammonia. *L*–glutamate is the only amino acid that can oxidize and remove ammonia at a relatively high rate in mammalian tissue cells.

2. Transamination

(1) The transamination refers to the reaction that transfer α–amino group of amino acids to the ketone base of an α–ketone acid catalyzed by the aminotransferase or transaminase, and the original amino acid is transformed into a new α–keto acid.

(2) Transamination is a reversible reaction, the coenzyme of transaminase is phosphopyridoxal, the most common transaminase *in vivo* are glutamic pyruvic transaminase (GPT or ALT) and glutamic oxaloacetictransaminase (GOT or AST).

3. Combined deamination

The combined deamination is the major deamination

(1) The coupling **between transamination and deamination.**

The amino acid was first formed with α–ketog lutarate to produceα–keto acid and glutamate, and then glutamate released free ammonia by oxidative deamination under the action of *L*–glutamic dehydrogenase.

The combination of transaminase and *L*–glutamic dehydrogenase was mainly performed in the liver and kidney tissues, and the inverse of the reaction was the main way to synthesize non–essential amino acids in vivo.

(2) The association of deamination with the purine **nucleotide cycle.**

The purine nucleotide cycle composed by the transaminase and adenosine deaminase was mainly performed in muscle tissue.

D. α–keto acid metabolism

1. Non–essential amino acids are produced by amination.

2. Fat and carbohydrates are produced from the amino acids.

The amino acids that can be converted into sugars in vivo are called glucogenic amino acid, which can be transformed into ketone bodies are called ketogenic amino acids, both of them are called glucogenic and ketogenic amino acids (Table 9.1).

Table 9.1 Classification of glucogenic amino acid and ketogenic amino acids

Types	Amino acids
glucogenic amino acid:	Glycine, serine, valine, histidine, arginine, cysteine, proline, alanine, glutamic acid, glutamine, aspartate, aspartic acid, aspartate, *L*–glutamine, *L*–aspartic acid, aspartic acid
ketogenic amino acid:	Leucine, lysine
gluco–and ketogenic amino acid:	*L*–leucine, phenylalanine, tyrosine, threonine, tryptophan

3. Oxidation and energy supply

[Practices]

[A1 type]

[1] The following options, which conform to the characteristics of protease-degrading proteins is ____________.

A. no need for ubiquitin participation

B. degrade exogenous proteins mainly

C. consuming ATP

D. the main way to degrade the protein of prokaryotic organism

E. degradation in the lysozyme body

[2] The main way of amino acid deamination in muscle is____________

A. purine nucleotide cycle

B. glutamate oxidative deamination

C. transamination

D. combination of the transamination and glutamate oxidative deamination

E. hydration deamination

[3] Coenzyme of transaminase is____________

A. phosphopyridoxal   B. thiamine pyrophosphate   C. biotin

D. FH_4 E. pantothenate

Ⅳ. Metabolism of ammonia

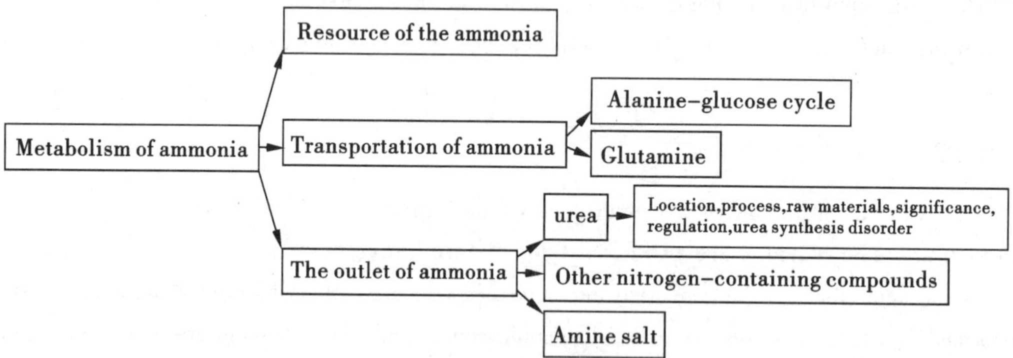

A. Resource of the ammonia

1. Amino acid deamination and amine decomposition produce ammonia.
2. Intestinal bacterial putrefaction produces ammonia.
3. The ammonia secreted by renal tubular epithelial cells is mainly from glutamine.

B. Transportation of ammonia

1. The alanine-glucose cycle

(1) The ammonia in the muscle is transported by the blood to the liver in the form of alanine (non-toxic), while the liver provides pyruvate for the muscle to produce glucose (Figure 9.2).

(2) Physiological significance: the realization of toxic ammonia transport; provides energy for muscular activity.

2. Ammonia-transporting effect of glutamine

(1) Glutamine is from the synthesis of ammonia and glutamic acid in the brain, muscle, and other tis-

sues, glutamine is transported by the blood to liver or kidney metabolism. Glutamine plays an important role in the transfer of ammonia, especially in the brain.

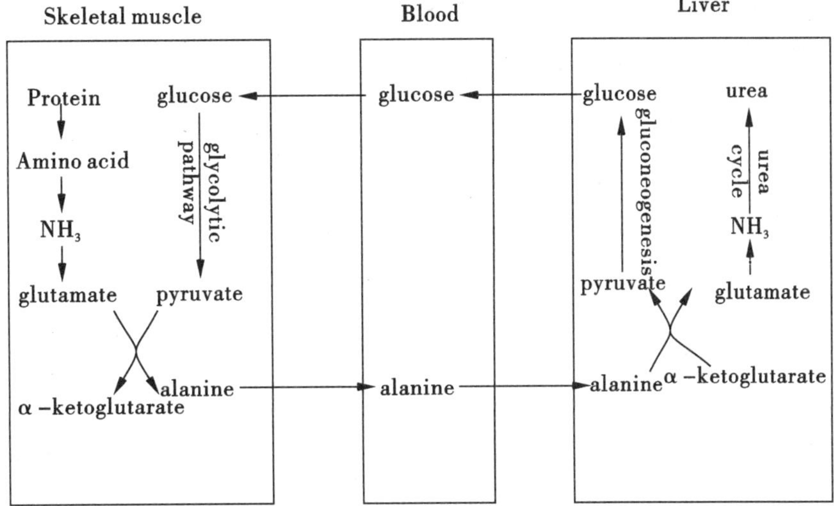

Figure 9.2 Alanine-glucose cycle

(2) Physiological Significance: Glutamine is the detoxification, storage and transport form of ammonia. The synthesis and decomposition reactions of glutamine are as follows:

$$\underset{\text{glutamate}}{\begin{array}{c} COOH \\ | \\ (CH_2)_2 \\ | \\ CHNH_2 \\ | \\ COOH \end{array}} \underset{\text{glutamate: } NH_3 \leftarrow \quad \rightarrow H_2O}{\overset{NH_3+ATP \rightarrow \text{glutamine synthetase} \rightarrow ADP+Pi}{\rightleftharpoons}} \underset{\text{Glutamite}}{\begin{array}{c} CONH_2 \\ | \\ (CH_2)_2 \\ | \\ CHNH_2 \\ | \\ COOH \end{array}}$$

C. The fate of ammonia

1. Synthesis of urea (main path).
2. Synthesis of non-essential amino acids and other nitrogen-containing compounds
3. Discharge in the form of amine salt

D. The generation of urea

1. **Location**: liver (mitochondria, cytosol)

2. **Process**: ornithine cycle, including 4 reaction steps (Figure 9.3):

(1) Synthesis of carbamyl phosphate: in the mitochondria, consume 2 ATPs, carbamoylphosphate synthetase (CPS-Ⅰ) is the key regulation enzyme.

(2) Synthesis of the citrulline: cytosol.

(3) Synthesis of the argininosuccinate: in the cytoplasm, consume 1 ATP (2 high-energy phosphate bonds), arginino succinate synthetase (ASS) is the key enzyme.

(4) Arginine produces urea: cytoplasm.

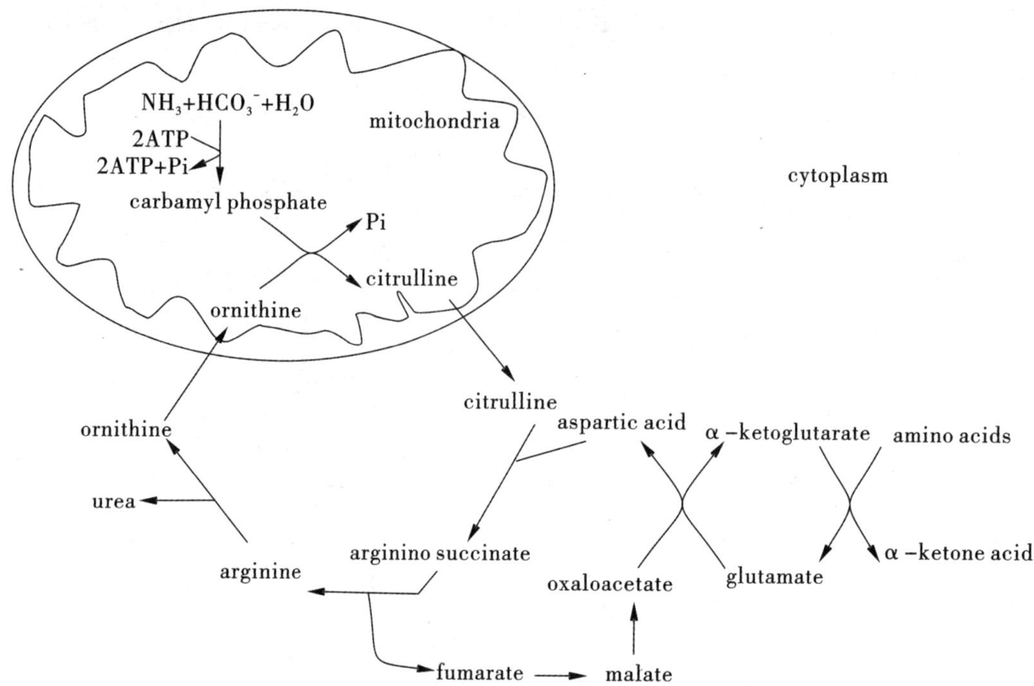

Figure 9.3 The intermediate steps and cell location of urea cycle

3. **Raw materials**: $2NH_3$ (from free ammonia and aspartic acid, respectively), CO_2.
4. **Energy consumption**: 3 ATPs, 4 high energy phosphate bonds.
5. **Excretion**: kidney.
6. **Significance**: relieving ammonia toxin.
7. **Regulation of urea synthesis**: regulated by food, CPS-Ⅰ and urea-producing enzyme system.
8. **Urea synthesis disorder**: may cause the hyperammonemia and the ammonia intoxication, the main reason is the liver dysfunction.

[*Practices*]

[**A1 type**]

[4] Ammonia in the blood is transported mainly based on which form of the following________.

A. glutamate B. aspartic acid C. glutamine
D. aspartic amide E. glycine

[5] The mechanism that the ammonia is transported through the blood from muscle tissue to the liver is________.

A. tri-carboxylic acid cycle B. ornithine cycle C. alanine-glucose cycle
D. methionine cycle E. cori cycle

[6] The main outlet path to ammonia in the brain is________.

A. diffusion into blood B. synthesis of urea C. synthesis of alanine
D. synthesis of amino acid E. synthesis of glutamine

[7] The ammonia from the deamination of amino acids in the body can be involved in the synthesis of________.

A. uric acid B. creatine C. tryptophan
D. glutamine E. polyamine

[8] In the ornithine cycle, which substance is the intermediate product to generate urea directly?________.

A. Arginine B. Citrulline C. Ornithine
D. Argininosuccinate E. Carbamyl phosphate

[9] Which organ is the main organ for the synthesis of urea in the body?____________

A. Heart B. Skeletal muscle C. Kidney
D. Brain E. Liver

[10] Which description is correct for the ornithine cycle? ____________.

A. Ornithine cycle starts from ornithine and ammonia to form glycine directly
B. Ornithine cycling starts with the synthesis of carbamyl phosphate
C. 3 molecule ammonia consumed per cycle of ornithine
D. 2 molecule ATPs consumed per cycle of ornithine
E. Ornithine cycle is mainly in the liver

V. Metabolisms of individual amino acid

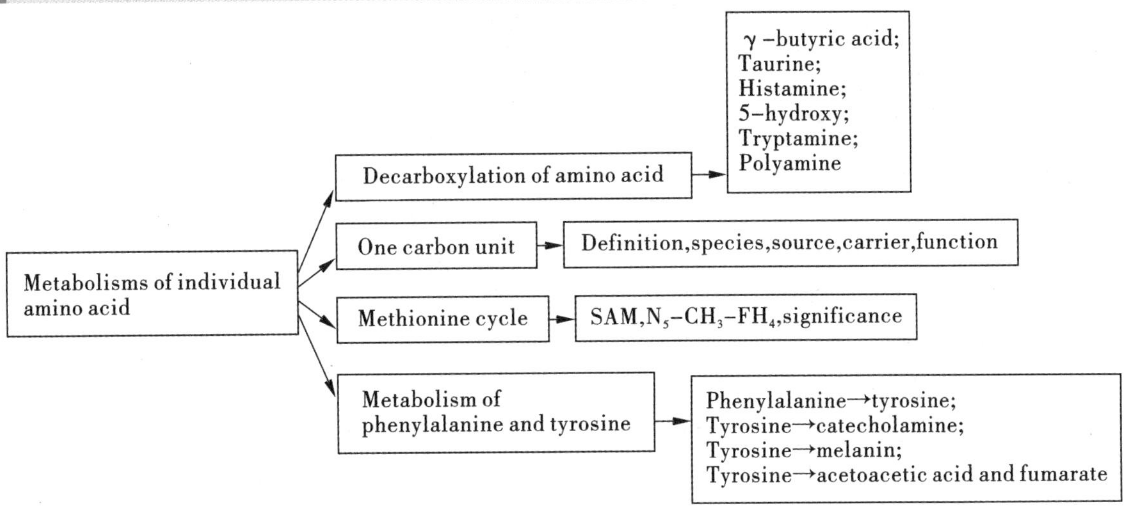

A. Decarboxylation

Some amino acids can bedecarboxylated to form corresponding amines (decarboxylation). The enzymes that catalyze these reactions are decarboxylase, the coenzyme of amino acid decarboxylase is the phosphopyridoxal.

1. **γ-aminobutyric acid (GABA)**: as an inhibitory neurotransmitter, formed by glutamic acid decarboxylation

2. **Taurine**: the composition of conjugated bile acid, formed by cysteine oxidation

3. **Histamine**: vasodilation, formed by histidine decarboxylation

4. **5-hydroxy tryptamine (5-HT)**: inhibitory neurotransmitter, formed by hydroxylation and decarboxylation of tryptophan

5. **Polyamine** (putrescine, spermidine, and spermine): regulation cell growth, formed by ornithine decarboxylation

B. Metabolism of one carbon unit

1. **Definition**: some amino acids in the process of catabolism can generate a unit containing a carbon atom called onecarbon unit.

2. **Species**: methyl ($—CH_3$), methyleneethylene ($—CH_2—$), methenyl ($—CH =$), formyl ($—CHO$) and formimino ($—CH =NH$)

3. **Source**: serine, glycine, histidine and tryptophan

4. **Carrier**: tetrahydrofalic acid (FH_4)

5. **Function**: participate in the synthesis of purine and pyrimidine

C. Metabolism of the methionine (Figure 9.4)

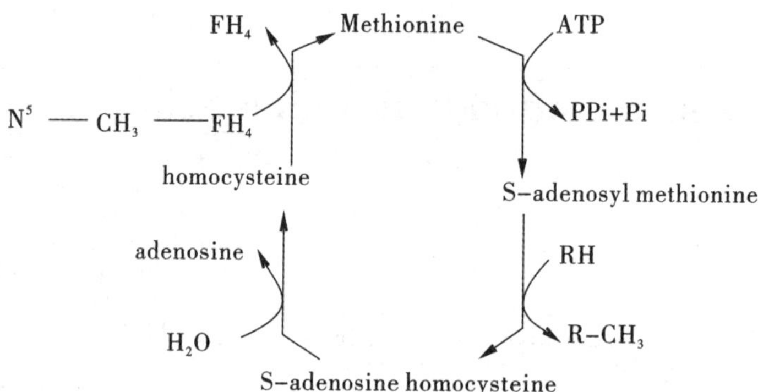

Figure 9.4 Methionine cycle

1. **S-adenosine methionine (SAM)**: direct donor of methyl group *in vivo*

2. **N^5-CH_3-FH_4**: indirect donor of the methyl group *in vivo*.

3. **Physiological significance**: provide methyl group for methylation reaction, to produce the containing methyl group compounds, such as epinephrine, choline, carnitine, creatine and so on.

D. The metabolism of phenylalanine and tyrosine (Figure 9.5)

1. **Phenylalanine→tyrosine**: the key enzyme is **phenylalanine hydroxylase**, a lack of enzyme that leads to phenylketonuria (**PKU**).

2. **Tyrosine→catecholamine**: mainly in the nervous tissue and adrenal gland.

3. **Tyrosine→melanin** (melanin cell): the key enzyme is **tyrosinase**, the lack of it caused the **albinism**.

4. **Tyrosine decomposition** can produce acetoacetic acid and fumarate, if homogentisicacid oxidase is lack, can cause alcaptonuria.

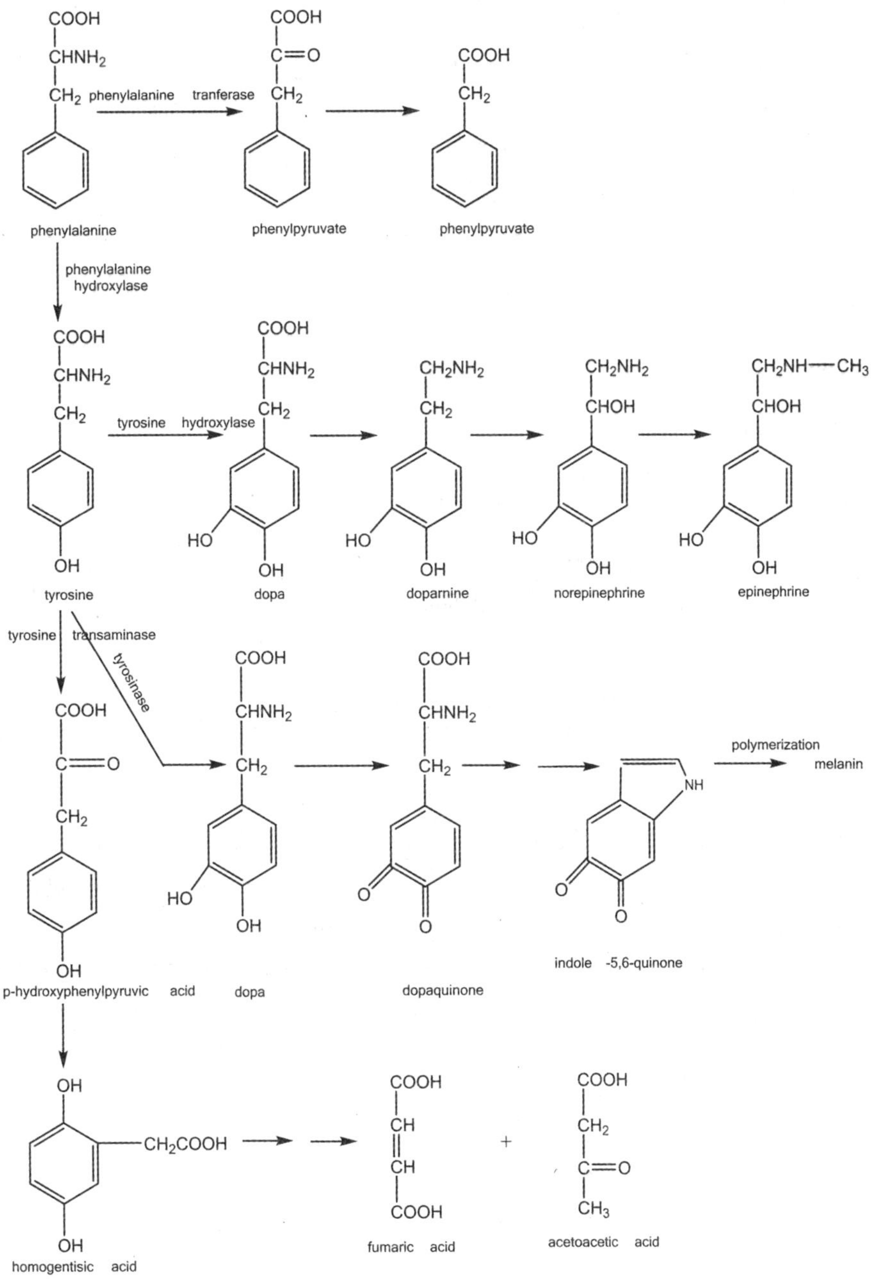

Figure 9.5 The metabolism of phenylalanine and tyrosine

[Practices]

[A1 type]

[11] Which amino acid is produced to γ-aminobutyric acid after decarboxylation?________

A. Tyrosine B. Histidine C. Aspartic acid

D. Glutamate E. Glutamine

[12] Which amino acid can be metabolized to produce taurine?____________

A. Cysteine B. Methionine C. Threonine

D. Glutamate E. Valine

[13] The carrier of one carbon unit transported in vivo is____________

A. folic acid B. biotin C. vitamin B_{12}

D. FH_4 E. SAM

[14] In the human body, tyrosine cannot be converted to____________

A. adrenaline B. melanin C. fumaric acid

D. phenylalanine E. acetylacetic acid

[15] Which one of the following amino acids can be converted to catecholamine?____________

A. Aspartate B. Tryptophan C. Tyrosine

D. Cysteine E. Methionine

[X type]

[16] Tyrosine in the body can be converted into____________

A. thyroxine B. adrenaline C. fumaric acid

D. acetylacetic acid E. melanin

Answers

1	2	3	4	5	6	7	8	9	10
E	A	A	C	C	E	D	A	E	E
11	12	13	14	15	16				
D	A	D	D	C	ABCDE				

Brief Explanations for Practices

1. Eukaryotic cells degrade intracellular proteins mainly through the following two pathways: the first one is the lysosomal protein degradation pathway, and the other is the pathway needing ubiquitin participation to degrade the abnormal protein and short life protein, consuming ATP.

2. The activity of L-glutamic dehydrogenase in muscle tissue is low, and amino acid is very difficult to remove the amino group by the combination of transamination and glutamic acid oxidation, but mainly through the purine nucleotide cycle to remove the amino group.

3. Coenzyme of the transaminase is phosphopyridoxal.

4. Ammonia is toxic, glutamine an alanine are the forms of transporting ammonia in the blood.

5. Ammonia is transported from skeletal muscle to the liver through an alanine-glucose cycle. Muscle protein degradation produces amino acids, the amino group are transferred to pyruvate to produce alanine through the transamination, and alanine is transported to the liver by blood cycle. In the liver, alanine released ammonia and pyruvate through the combined deamination, ammonia is used to synthesize urea, and pyruvic acid can be converted into the glucose by the gluconeogenesis. Glucose is transported from blood to the muscle tissue, it is converted into pyruvate via glycolysis, then receives the amino group and generate alanine. Alanine and glucose are repeatedly converted between the muscles and the liver, the ammonia produced in the muscle is continuously transported to the liver for synthesis of urea, so this pathway is known as the "alanine-glucose cycle". Through this cycle, the ammonia in the muscle is transported by the blood to the liver in the form of Non-toxic——alanine, while the liver provides pyruvate for the muscle to produce glucose.

6. In the brain and muscle, glutamine synthase catalyzed ammonia and glutamate to form glutamine, glutamine was transported to the liver or kidney through the blood, then it is hydrolyzed into glutamate and ammonia by glutaminase. Ammonia is synthesized into the urea in the liver and discharged through the renal excretion.

7. The amino acid in vivo produced ammonia in the body has 3 metabolic outlet pathways: ① synthesis

of urea (mainly); ②synthesis of non-essential amino acids and other nitrogenous compounds; ③ discharged in the form of ammonium salts.

8. The final step of the urea cycle is to produce urea from arginine hydrolysis catalyzed by arginase. Therefore the intermediate product of the direct generation of urea is arginine.

9. The liver is the main organ of urea synthesis, the kidney is the main organs of excretion of urea

10. The liver is the main organ of urea synthesis, ornithine cycle is a synthetic way of urea, including four steps: ①Synthesis of carbamyl phosphate; ②Synthesis of the citrulline; ③Synthesis of the argininosuccinate; ④Arginine produces urea. Synthesis of the urea is a process consuming ATP, one urea needs 3 ATPs, 4 high energy phosphate bonds. And the nitrogen of urea is from free ammonia and aspartic acid (from other amino acids).

11. γ-butyric acid: As an inhibitory neurotransmitter, formed by glutamic acid decarboxylation.

12. Taurine: the composition part of conjugated bile acid, formed by cysteine oxidation.

13. Some amino acids in the process of catabolism can generate a unit containing a carbon atom called a carbon unit. One carbon unit cannot exist alone and must be combined with a carrier to transport and participate in metabolism. Tetrahydrofalic acid (FH4) is the carrier of one-carbon unit. One carbon unit is usually combined with the N^5 and N^{10} position of the FH_4.

14. Phenylalanine is hydroxylated to produce tyrosine catalyzed by phenylalanine hydroxylase, the catalytic reaction is irreversible, so tyrosine cannot become phenylalanine back.

15. Dopamine, norepinephrine, and epinephrine are collectively called catecholamine. In the adrenal medulla, tyrosine is converted into the 3,4-dihydroxyphenylalanine (dopamine) by the hydroxylase. dopamine is catalyzed by dopamine beta-hydroxylase, produced norepinephrine, which is further methylated into epinephrine.

16. Dopamine, norepinephrine, and epinephrine are collectively called catecholamine (Dopamine, norepinephrine, and epinephrine). Tyrosine can synthesize thyroxine in the thyroid gland. In the melanin cells, tyrosine is produced to melanin. Tyrosine is produced toacetoacetic acid and fumaric acid through the decomposition.

[*Comprehensive Practices*]

[A1 type]

[1] The main way of amino acid deamination in the organism is______

A. combined deamination　　B. reductive deamination　　C. transamination

D. deamination directly　　E. oxidative deamination

[2] Which one is the major deamination to get the α-keto acids in the organism? ______

A. Reductive deamination　　B. Oxidative deamination　　C. Combined deamination

D. Transamination　　E. Non-oxidative deamination

[3] The activity of GPT in serum is abnormally elevated, indicates which organ's cell is in damage?______

A. Myocardial cell　　B. Hepatic cell　　C. Kidney cells

D. Pneumocyte　　E. Gastric cells

[4] The main reason for the elevated blood ammonia is______

A. excessive ingestion of diet protein

B. renal dysfunction

C. hepatic dysfunction

D. increased ammonia uptake from the intestinal absorption______

E. increased ammonia intake of renal tubules

[5] The key enzyme of urea synthesis is____________

A. carbamoyl phosphate synthetase- I

B. ornithine carbamoyl transferase

C. arginase

D. arginino succinate synthetase

E. arginino succinate lyase

[6] The amino acid after decarboxylation which can be used as the raw material for the production of spermine and spermidine is____________

A. arginine   B. leucine   C. ornithine

D. histidine   E. glycine

[7] FH_4 is not a carrier of which one of the following groups or compounds? ____________

A. —CHO   B. CO_2 C. $—CH_3$

D. —CH=   E. —CH=NH

[8] Albinism is caused by a lack of which enzymes in the body that leads to melanin synthesis disorders?____________

A. Tyrosine transaminase   B. Tyrosinase   C. Phenylalanine hydroxylase

D. DOPA decarboxylase   E. Tyrosine hydroxylase

[9] Which of the following compounds cannot be synthesized by tyrosine? ____________

A. Adrenaline   B. Thyroxine   C. Phenylalanine

D. Dopamine   E. Noradrenaline

[B1 type]

Question 10 ~ 12 sharethe following suggested answers.

A. γ-glutamyl cycle   B. ornithine cycle   C. methionine cycle

D. purine nucleotide cycle   E. tri-carboxylic acid cycle

[10] Involved in the synthesis of urea is____________

[11] Involved in the deamination is____________

[12] Participating in the generation of SAM and providing methyl group is____________

Question 13 ~ 15 sharethe following suggested answers.

A. transaminase

B. tyrosinase

C. phenylalanine hydroxylase

D. homogentisic acid oxidase

E. tyrosine hydroxylase

[13] The lack of phenylketonuria is____________

[14] Alcaptonuria is lack of____________

[15] Albinism is lack of____________

Question 16 ~ 18 share the following suggested answers.

A. Glycine   B. Arginine   C. Tyrosine

D. Serine   E. Alanine

[16] Which one can be transformed into choline in the body?____________

[17] Which one can be produced in the process of urea synthesis?____________

[18] Which substance can be converted into melanin in vivo?____________

[X type]

[19] The enzymes that can catalyze the deamination reaction are including ____________

A. *L*–amino acid oxidase B. transaminase C. glutamine oxidase

D. *L*–glutamate dehydrogenase E. *L*–glutamate oxidase

[20] The cell organelles involved in the synthesis of urea include____________

A. cytoplasm B. mitochondria C. Golgi complex

D. lysosome E. microsome

[21] The amino acids that can produce onecarbon unit during catabolism are____________

A. Gly B. Ser C. Tyr

D. Trp E. Gln

Answers

1	2	3	4	5	6	7	8	9	10
A	D	B	C	D	C	B	B	C	B
11	12	13	14	15	16	17	18	19	20
D	C	C	D	B	D	B	C	BD	AB
21									
ABD									

(***Li Jianning***)

Chapter 10

Nucleotide Metabolism

Examination Syllabus

1. **Nucleotide metabolism**: raw materials of *de novo* and salvage synthesis pathway of purine and pyrimidine nucleotides; end products of the degradation of purine and pyrimidine nucleotides

2. **Regulation of nucleotide metabolism**: key enzymes of synthesis pathway of purine and pyrimidine nucleotides; biochemical mechanism of nucleotide antimetabolites.

Ⅰ. Functions of nucleotides

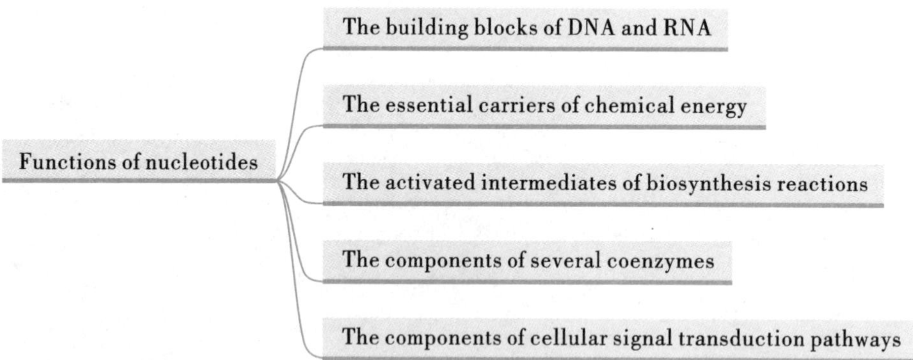

1. Nucleotides serve as the building blocks of DNA and RNA.

2. Nucleoside triphosphates, especially ATP, are essential carriers of chemical energy.

3. Some nucleotides derivatives, such as UDPG, SAM, PAPS, etc. are activated intermediates of biosynthesis reactions.

4. Some nucleotides are components of several coenzymes (CoA, NAD^+ and FAD).

5. Some nucleotides such as cAMP and cGMP are essential components of cellular signal transduction pathways.

Ⅱ. Synthesis and degradation of purine nucleotides

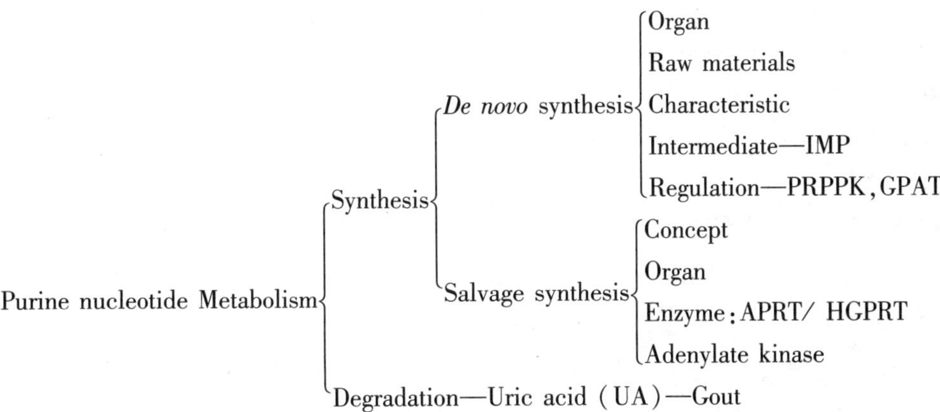

A. Two types of biosynthesis pathways of purine nucleotides

1. *De novo* synthesis pathway of purine nucleotides

(1) **Organ**: liver (main), small intestine mucosa, thymus

(2) **Raw materials**: glutamine, glycine, aspartic acid, one carbon unit, CO_2, and ribose-5-phosphate (R-5-P)—from pentose phosphate pathway.

①The carbon atoms are built into the purine ring by the following methods:

● Incorporation of glycine skeleton into the ring system (C_4, C_5);

● Addition of formyl group by transformylase with one carbon unit as donor (C_2, C_8);

● Addition of carboxyl group by carboxylase with CO_2 as donor (C_6).

②The nitrogen atoms are added to the purine ring by the following ways:

● Incorporation of glycine skeleton into the ring system (N_7);

● Addition of amino group by amidotransferase with glutamine as donor (N_3, N_9);

● Addition of amino group by a two-step mechanism with aspartic acid as donor (N_1).

(3) **Characteristic**: R-5-P is used as a **starting substrate** to build the purine ring system in an eleven-step process and the first intermediate with a **complete purine ring** is inosine-5′-monophosphate (**IMP**).

(4) **The first two key step**:

● R-5-P+ATP→**PRPP**(5′-phosphoribosyl-1′-pyrophosphate)

Enzyme: PRPP Synthefase.

● **PRPP** + glutamine(Gln)→**PRA**

Enzyme: glutamine PRPP amidotransferase (**GPAT**)

(5) **IMP →AMP or GMP**

● AMP is generated by using aspartic acid as amino group donor and using GTP as energy source.

● GMP is formed by using glutamine as amino group donor and using ATP as energy source.

(6) **Regulation**: **Key enzymes**: PRPPK and GPAT

● **Positive regulation**: substrates

PRPPK: R-5-P and ATP; GPAT: PRPP.

● **Negative regulation**: feedback inhibition

PRPPK: IMP, ADP, GDP and some other metabolites.

GPAT: IMP, AMP and GMP.

2. **Salvage synthesis pathway of purine nucleotides**

(1) **Concept**: Free purine bases or nucleosidescan be salvaged and recycled to produce nucleotides.

(2) **Organ and tissue**: brain, marrow—**salvage synthesis only**

(3) **Two kinds of salvage enzymes**:

① **The major kind**: adenosine phosphoribosyl transferase (APRT) and hypoxanthine-guanine phosphoribosyl transferase (HGPRT).

Theycatalyze the following reactions and also undergo feedback inhibition by their products, respectively.

$$A+PRPP \xrightarrow{APRT} AMP+PPi$$

$$H/G+PRPP \xrightarrow{HGPRT} IMP/GMP+PPi$$

② **Another kind**: adenylate kinase, catalyze adenosine phosphorylation.

$$Adenosine+ATP \xrightarrow{\text{adenylate kinase}} AMP+ADP$$

(4) A **genetic lack of HGPRT activity** blocks salvage synthesis pathway of purine nucleotides leads to a bizarre set of symptoms called **Lesch-Nyhan syndrome**.

B. Degradation of purines nucleotides

1. Nucleotides are usually degraded in the liver, intestine and kidneys.
2. Nucleotides are firstly hydrolyzed to nucleosides and phosphate by nucleotidase.
3. Nucleosides are cleaved again to free bases and ribose-1-phosphate bynucleoside phosphorylase.
4. Some purine bases and nucleosides can be recycled to form purine nucleotides by salvage synthesis pathway. Others are oxidized to the excreted end products, uric acid, whichstill has an intact purine ring.
5. **Uric acid** is the end product of purine degradation in primates, birds and some other animals. Elevated concentration of serum uric acid induces **gout**, a disease in which salts of uric acid crystallize and damage joints and kidneys.

Ⅲ. Synthesis and degradation of pyrimidine nucleotides

A. Two types of biosynthesis pathways of pyrimidine nucleotides

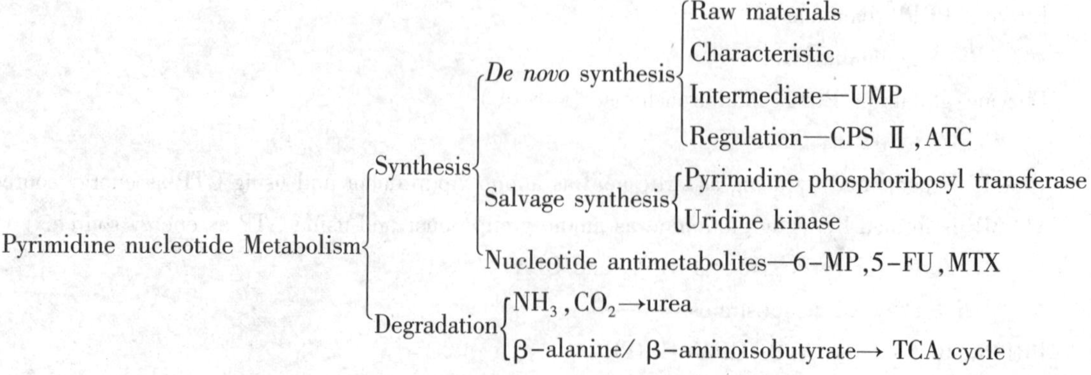

1. ***De novo* synthesis pathway of pyrimidine nucleotides**

(1) **Raw materials**: CO_2, aspartate, glutamine and R-5-P. Biosynthesis of dTMP also needs one carbon unit.

(2) **Characteristic**: unlike the case of purine nucleotide, **the pyrimidine ring is synthesized first and then reacts with PRPP to form pyrimidine nucleotide** (Table 10.1).

Table 10.1 The difference between purine and pyrimidine nucleotide synthesis

	Purine nucleotide	Pyrimidine nucleotide
Major feature	PRPP is used as a starting substrate to build the purine ring system	The pyrimidine ring is synthesized first and then reacts with PRPP
Substrates	Glutamine, Asparate, CO_2, Glycine, R-5-P and one carbon unit	Glutamine, Asparate, CO_2 and R-5-P *
First reaction	R-5-P+ATP→PRPP	CO_2+Gln→carbamoyl phosphate
Key enzymes	PRPPK and GPAT	PRPPK, CPS Ⅱ and ATC
Major intermediate	IMP	OMP, UMP
End products	AMP, GMP	UMP, CTP and dTMP

* Biosynthesis of dTMP also needs one carbon unit.

(3) The first synthesized common pyrimidine nucleotide is **UMP**, and then UMP can be converted into other pyrimidine nucleotides.

(4) **Synthesis of UMP**

● The formation of carbamoyl phosphate from CO_2 and glutamine, which is catalyzed by carbamoyl phosphate synthetase Ⅱ (**CPS Ⅱ**).

● Aspartate transcarbamoylase (**ATC**) transfers the carbamoyl group to aspartate and results in the formation of *N*-carbamoylaspartate.

● Aftercyclization and dehydrogenation, orotate, which has an intact **pyrimidine ring**, is formed.

● Orotate can react with **PRPP** to yield OMP, and then OMP is decarboxylated to form UMP.

(5) **Synthes is of CTP**

● UMP →UDP→UTP

● **UTP→CTP**: CTP synthetase (**CTPS**), -1ATP

(6) **Synthesis of dTMP**

● **Formation** of dUMP: dUDP→dUMP; dCMP→dUMP (main)

● **dUMP→dTMP** (dTMP synthase, methyl group donor: N^5, N^{10}-CH_2-FH_4)

(7) **Regulation: key enzymes**: CPS Ⅱ, ATC and PRPPK.

● Positive regulation: ATP, as a substrate, can activate the key enzyme PRPPK and CPS Ⅱ. And enzyme OPRT also can be activated by its substrates PRPP.

● Negative regulation: CPS Ⅱ is inhibited by UTP, ATC and CTPS is inhibited by CTP.

● PRPP is a common substrate and activator of both purine and pyrimidine nucleotides *de novo* synthesis pathway. And PRPPK also can be feedback inhibited by the end products of both pathways.

2. **Salvage synthesis pathway of pyrimidine nucleotides**

There are also two kinds of salvage enzymes for this pathway.

(1)**Pyrimidine phosphoribosyl transferase** can reconnect the phosphoribosyl group of PRPP to pyrimidines such as U,T and even O,but not C to form UMP,TMP and OMP,respectively.

$$\text{U/T/O+PRPP} \xrightarrow{\text{OPRT}} \text{UMP/TMP/OMP+PPi}$$

(2)**Uridine kinase**

$$\text{Uridine+ATP} \xrightarrow{\text{uridine kinase}}$$

3. NMP→ NTP

$$\text{AMP+ATP} \xrightleftharpoons{\text{adenylate kinase}} \text{ADP+ADP}$$

$$\text{NMP+ATP} \rightleftharpoons \text{NDP+ADP}$$

$$\text{XDP+YTP} \rightleftharpoons \text{XTP+YDP}$$

4. Synthesis of deoxyribonucleotides: at NDPs(N = A, G, C, U) level (Figure 10. 1); **dUMP→dTMP**

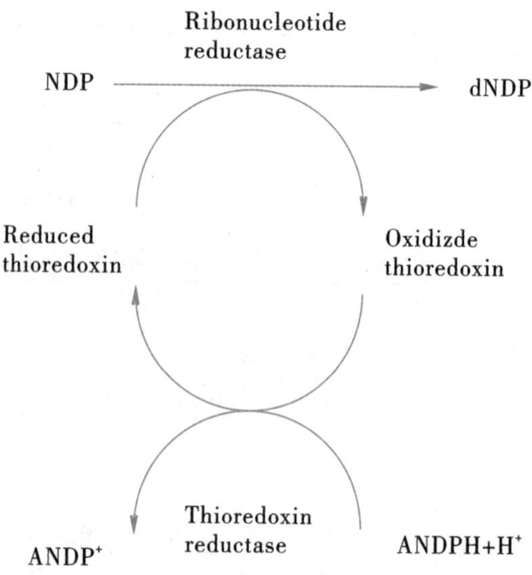

Figure 10.1 Synthesis of deoxyribonucleotides

5. Nucleotide antimetabolites

(1)Nucleotide antimetabolites are **analogs** of the normal metabolites involved in the nucleotide biosynthesis pathway.

●They can **inhibit and block** the nucleotide **biosynthesis** pathway,and then halt the DNA production and cell growth.

●Some of these nucleotide antimetabolites are not only used to search a novel metabolic pathway in scientific research but also can be used as drugs to treat disease such as cancer.

(2)Nucleotide antimetabolites include several types of analogs:

●analogs of purine and pyrimidine metabolites:6-mercaptopurine (6-MP),5-fluoruracil (5-FU)

●analogs of glutamine:azaserine (AS) and diazonorleucine (DAL)

●analogs of folate:aminopterine (AP) and methotrexate (MTX)

●analogs of nucleosides:arabinosyl cytosine and cyclocytidine

B. Degradation of pyrimidine nucleotides

1. The feature of degradation of pyrimidines is different from that of purines (Table 10.2). The end

product of purines, uric acid still has an intact purine ring, whereas the pyrimidine ring is broken during the degradation process.

2. In usual, the degradation of pyrimidines leads to ammonia production and then urea synthesis.

Table 10.2 The difference between the degradation of purine and pyrimidine

	Purine nucleotide	Pyrimidine nucleotide
Major feature	The purine ring is intact	The pyrimidine ring is broken
Substrates	A, G	U, C and T
End products	Uric acid	NH_3, CO_2 → urea β-alanine/β-aminoisobutyrate

[*Practices*]

[A1 type]

[1] Which one of the following contributes nitrogen to both purine and pyrimidine rings?__________.

A. Glutamate B. Carbamoyl phosphate C. Carbon dioxide
D. Glutamine E. Tetrahydrofolate

[2] The principal nitrogenous urinary excretion product in humans resulting from the catabolism of AMP is __________.

A. creatinine B. urea C. uric acid
D. thiamine E. thymine

[3] Which of the following is the common material for the synthesis of pyrimidine and purine nucleotide?__________.

A. Fumarate B. Glutamine C. Formic Acid
D. Asparagines E. Glycine

[4] Which of the following is not a physiological role of nucleotides?__________.

A. Carriers of chemical energy
B. Intermediates for biosynthetic processes
C. Components of many proteins
D. Components of the coenzymes FAD and CoA
E. Intracellular signaling molecules

[5] Which of the following substance directly links nucleotide synthesis and glucose metabolism?__________.

A. Glucose B. Glucose-6-phosphate C. Glucose-1-phosphate
D. Glucose-1,6-diphosphate E. Ribose-5-phosphate

[6] Which one of the following does not occurs in the degradation pathway of AMP?__________.

A. Adenine is converted to hypoxanthine
B. The end product is urea
C. AMP is converted to adenosine
D. Adenosine is converted to adenine
E. None of the above

[7] One of the properties in *de novo* synthesis of purine nucleotides is that__________.

A. the first step is the production of PRPP

B. one carbon unit is provided by S–adenosylmethionine

C. orotate phosphate is an intermediate

D. glycine does not incorporate in the pathway

E. asparagine is the direct nitrogen donor

[8] Which of the following is not the direct material of *de novo* synthesis of purine nucleotide?___________.

A. Glycine B. Aspartate C. Glutamate

D. CO_2 E. One carbon unit

[9] Which one of the following amino acid is involved in the synthesis of nucleotide?____________.

A. Asparagine B. Glutamine C. Alanine

D. Lysine E. Methionine

[10] Glutamine and glycine can be co–involved in the synthesis of which substance?____________.

A. Coenzyme A B. Purine nucleotide C. Pyrimidine nucleotide

D. Chlorophyll E. SAM

[11] The direct precursor of dTMP synthesis is____________.

A. dUMP B. TMP C. TDP D. dUDP E. dCMP

[12] Which vitamin deficiency can interfere with DNA synthesis?____________.

A. Vitamin B_1 B. Vitamin B_2 C. Vitamin B_6

D. Vitamin PP E. Folate

[13] Which one of the following enzymes is the common key enzyme for the synthesis of pyrimidine and purine nucleotide?____________.

A. Glutamine PRPP amidotransferase (GPAT)

B. Adenylosuccinate synthetase (ASS)

C. Carbamoyl phosphate synthetase Ⅱ (CPS Ⅱ)

D. PRPP Synthetase

E. Aspartate transcarbamoylase (ATC)

[14] Which one of the following compounds does not participate in the synthesis of pyrimidine nucleotide?____________.

A. R–5–P B. Aspartate C. Glycine

D. Glutamine E. CO_2

[15] Which one of the following enzymes deficiency can lead to Lesch–Nyhan syndrome?____________.

A. Glutamine PRPP amidotransferase (GPAT)

B. Adenylosuccinate synthetase (ASS)

C. Adenosine phosphoribosyl transferase (APRT)

D. PRPP kinase (PRPPK)

E. Hypoxanthine–guanine phosphoribosyl transferase (HGPRT).

[16] Which one of the following nucleotide is the first nucleotide with a complete purine ring in *de novo* synthesis pathway of purine nucleotide?____________.

A. AMP B. UMP C. GMP D. IMP E. XMP

[17] Which of the following is the end product of purine nucleotide catabolic metabolism in humans?____________.

A. Urea B. Creatine C. Creatinine

D. Uric acid E. β–Alanine

[18] Which one of the following nucleotide is the first nucleotide with a complete pyrimidine ring in *de novo* synthesis pathway of pyrimidine nucleotide?____________.

A. IMP B. UMP C. TMP D. CMP E. CTP

[19] The synthesis of deoxynucleotides takes place at____________level.

A. ribose B. nucleoside C. NMP

D. NDP E. NTP

[20] Which one of the following compounds is one of the products of pyrimidine degradation in humans?____________.

A. β-alanine B. Creatine C. Creatinine

D. Uric acid E. α-Alanine

Answers

1	2	3	4	5	6	7	8	9	10
D	C	B	C	E	B	A	C	B	B
11	12	13	14	15	16	17	18	19	20
A	E	D	C	E	D	D	B	D	A

Brief Explanations

1. Glutamine is a common nitrogen source of both purine and pyrimidine rings.

2. Uric acid is the end product of purine degradation in primates, birds and some other animals.

3. CO_2, aspartate, glutamine and R-5-P are the common substrate of *de novo* synthesis pathways of purine and pyrimidine nucleotides.

4. Nucleotide is not a component of protein.

5. The ribose-5-phosphate (R-5-P), formed by the pentose phosphate pathway, is used as a common substrate of both biosynthesis pathways of purine and pyrimidine nucleotides

6. The end product of purine degradation is uric acid, not urea.

7. The first step of *de novo* synthesis of purine nucleotides is the production of PRPP.

8. Glutamate is not the direct material of *de novo* synthesis of purine nucleotide.

9. CO_2, aspartate, glutamine and R-5-P are the common substrate of *de novo* synthesis pathways of purine and pyrimidine nucleotides.

10. Glutamine and glycine are the substrates of *de novo* synthesis pathways of purine nucleotides.

11. The direct precursor of dTMP synthesis is dUMP.

12. The derivative of folate, tetrahydrofolate, is the carrier of one carbon unit.

13. The common key enzyme for the synthesis of pyrimidine and purine nucleotide is PRPP Synthetase.

14. Glycine is one of the raw materials of the synthesis of purine nucleotide, but not pyrimidine nucleotide.

15. Hypoxanthine-guanine phosphoribosyl transferase (HGPRT) deficiency can lead to Lesch-Nyhan syndrome.

16. IMP is the first nucleotide with a complete purine ring in *de novo* synthesis pathway of purine nucleotide.

17. Uric acid is the end product of purine nucleotide catabolic metabolism in humans.

18. UMP is the first nucleotide with a complete pyrimidine ring in *de novo* synthesis pathway of pyrimi-

dine nucleotide.

19. The synthesis of deoxynucleotides takes place at NDP level.

20. β-alanine is one of the products of pyrimidine degradation in humans.

[*Clinical Cases*]

[21] Pregnant women frequently suffer from folate deficiencies. A deficiency of folate would decrease the production of.

A. Creatine phosphate from creatine

B. All of the pyrimidines required for RNA synthesis

C. The thymine nucleotide required for DNA synthesis

D. Phosphatidyl choline from diacylglycerol and CDP-choline

E. None of above

[22] After several weeks of chemotherapy in the form of methotrexate, a cancer patient's tumor begins to show signs of resistance to treatment. Which of the following mechanisms is most likely to explain the tumor's methotrexate resistance?

A. Overproduction of dihydrofolate reductase

B. Overproduction of xanthine oxidase

C. Deficiency of PRPP synthase

D. Deficiency of thymidine kinase

E. Deficiency of thymidylate synthase

Answers and Brief Explanations

21. C. Biosynthesis of dTMP also needs one carbon unit. The derivative of folate, tetrahydrofolate, is the carrier of one carbon unit.

22. A. Methotrexate is an analog and antimetabolite of dihydrofolate, it can halt DNA synthesis via inhibiting the activity of dihydrofolate reductase.

(***Wang Xuejun***)

Chapter 11

Blood Biochemistry

Examination Syllabus

1. Chemical composition of blood: plasma proteins; non-protein nitrogen; nitrogen-free organic compound

2. Plasma proteins: classification, source, function

3. Metabolism of red blood cell: material, site and key enzyme of heme synthesis; metabolic characteristics of mature erythrocytes

[Major points and Practices]

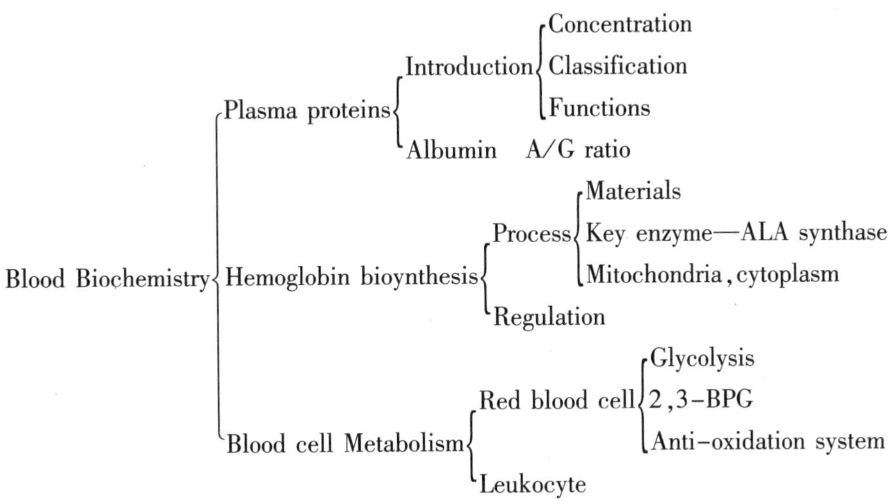

Blood is made up of plasma and blood cells (red blood cells, RBC; white blood cells, WBC, and platelets).

Plasma consists of water (92%) and soluble substances (inorganic compounds, organic compounds: plasma proteins, non-protein nitrogen).

Non-protein nitrogen: the nitrogen-containing small molecular weight metabolites, such as urea, creatine, creatinine, uric acid, bilirubin, and ammonia, except protein.

Polyacrylamide gel electrophoresis: a kind of electrophoresis technique using polyacrylamide gel as the support media. It can separate biological macromolecules, usually proteins or nucleic acids.

Acute phase protein: a class of proteins whose plasma concentrations increase (positive acute-phase proteins) or decrease (negative acute-phase proteins) in response to inflammation.

C-reactive protein, CRP: a pentameric protein with ring-shape, a kind of acute-phase protein whose levels rise in response to inflammation.

Colloid osmotic pressure: or oncotic pressure, is a form of osmotic pressure exerted by proteins, notably albumin, in a blood vessel's plasma (blood/liquid) that usually tends to pull water into the circulatory system.

Immune globulin: also known as antibody, is a large, Y-shaped protein produced mainly by plasma cells that is used by the immune system to neutralize pathogens such as pathogenic bacteria and viruses.

Hemoglobulin: iron-containing oxygen-transport metalloprotein

ALA synthase: also known as aminolevulinic acid synthase, is an enzyme (EC 2.3.1.37) that catalyzes the synthesis of δ-aminolevulinic acid.

Prophyria: a group of diseases in which substances called porphyrins build up, negatively affecting the skin or nervous system.

Erythropoietin: also known as hematopoietin or hemopoietin, is a glycoprotein cytokine secreted by the kidney in response to cellular hypoxia stimulating red blood cell production (erythropoiesis) in the bone marrow.

2,3-BPG shunt: also called the Luebering-Rapoport pathway, or the Luebering-Rapoport shunt, is a metabolic pathway in mature erythrocytes involving the formation of 2,3-bisphosphoglycerate (2,3-BPG), which regulates oxygen release from hemoglobin and delivery to tissues.

Ⅰ. Plasma proteins

A. Introduction

Plasma proteins are the most abundant solid substances in plasma.

1. The total concentration of plasma proteins is about 70 ~ 75 g/L.

(1) Plasma proteins can be separated into three major groups fibrinogen, albumin, and globulins by salting out methods.

(2) Plasma proteins can be separated into five bands in cellulose acetate film electrophoresis (barbital buffer, pH 8.6), designated albumin, α_1 globulin, α_2 globulin, β globulin and γ globulin from anode to cathode.

2. **Classification**: eight categories according to their physiological function.

Table 11.1 Classification of human plasma proteins

Category	Plasma protein
(1) Binding protein or carrier	Albumin, apoprotein, transferrin, copper-protein
(2) Immune defense	IgG, IgM, IgA, IgD, IgE and complement C1 ~ C9
(3) Coagulation and anti-coagulation	Blood coagulation factor Ⅶ, Ⅷ, prothrombin, plasminogen and so on
(4) Enzymes	Lecithin-cholesterol acyltransferase and so on
(5) Protease inhibitors	α_1 Antitrypsin, α_2 macroglobulin and so on
(6) Hormones	Erythropoietin, insulin and so on

Continue to Table 11.1

Category	Plasma protein
(7) Proteins participating in inflammatory response	C- reactive protein, acid glycoprotein and so on
(8) Oncofetal	α_1-Fetoprotein (AFP)

3. Functions:

(1) Maintenance of plasma colloid osmotic pressure

(2) Maintenance of plasma pH

(3) Transportation

(4) Immunization

(5) Catalysis

(6) Trophism

(7) Blood clotting, anticoagulation and fibrinolysis

B. Albumin

Albumin is the most abundant protein in plasma.

1. The mature albumin is a single strand peptide containing 585 amino acids, with 17 disulfide bonds.
2. The normal of albumin to globulin (A/G) ratio is 1.5 ~ 2.5.
3. Albumin is the most important transporting protein in the blood.

Ⅱ. Synthesis of hemoglobin

A. Process of heme synthesis

1. **The materials**: glycine, succinyl CoA, and Fe^{2+}.

2. **The key enzyme**: δ-aminolevulinate (ALA) synthase, catalyzes the condensation of glycine and succinyl-CoA to form δ-aminolevulinate.

3. Kinds of unsaturated intermediate compounds are produced during the heme synthesis pathway and responsible for the light sensitivity of heme synthesis obstruction.

4. The initial and final steps are taken place in **mitochondria** and the others are in the **cytoplasm.**

B. Regulation of heme synthesis

1. ALA synthase is a rate-limiting enzyme of heme synthesis, feedback inhibited by heme.

2. Erythropoietin (EPO) promotes proliferation, differentiation, mature of erythroblast, synthesis of heme and Hb.

Ⅲ. Metabolism of red blood cell and leukocyte

A. Metabolism of red blood cell

1. Glycolysis and pentose phosphate pathway are the main metabolisms of red blood cell glucose metabolism

2. 2,3-BPG is essential for regulating the ability of Hb to transport oxygen

3. NADPH, produced by pentose phosphate pathway, together with NADH, produced by glycolysis, consist the anti-oxidation system of red blood cell, to avoid the oxidation of red blood cell membrane, protein and Fe^{2+}.

4. Mature erythrocyte can not undertake *de nove* synthesis of fatty acids, but continuously exchanges its lipids with plasma lipoproteins by active participation or passive exchange to maintain its normal lipid composition, structure and function.

B. Metabolism of leukocyte

1. Glycolysis is the primary carbohydrate metabolism pathway.

2. NADPH oxidase transmitting-electron system plays an important role in maintaining the function of granulocyte and mononuclear phagocyte.

3. Under the catalysis of lipid oxidase, granulocytes and mononuclear phagocytes can change arachidonic acid into leukotriene, which is the slow-reacting substance of anaphylaxis produced in type Ⅰ hypersensitivity reaction.

4. Lots of histamines exist in granulocytes and it will release to participate allergy after white blood cell is activated.

[***Practices***]

[**A1 type**]

[1] What is the normal ratio of albumin to globulin in plasma?____________.

A. 0.5 ~ 1　　B. 1 ~ 1.5　　C. 1.5 ~ 2.5　　D. 2.5 ~ 3.5　　E. >3.5

[2] The main source of NPN is ____________.

A. urea nitrogen　　B. creatine nitrogen　　C. uric acid nitrogen

D. amino acid nitrogen　　E. peptide nitrogen

[3] About albumin, which one is not correct?____________.

A. It is a proteoglycan

B. It is the most abundant protein in human plasma

C. It can maintain the plasma colloid osmotic pressure

D. It can serve as a transporter for liposoluble substances

E. It is the smallest protein in plasma

[4] Using cellulose acetate film electrophoresis, serum proteins can be separated into____________(from anode to cathode).

A. fibrinogen, albumin and globulin

B. albumin, α_1 globulin, α_2 globulin, β globulin and γ globulin

C. albumin, α_2 globulin, α_1 globulin, β globulin and γ globulin

D. albumin, α_2 globulin, α_1 globulin, γ globulin and β globulin

E. fibrinogen, α_2 globulin, α_1 globulin, β globulin and γ globulin

[5] Which protein level will obviously decrease because of the liver damage?____________.

A. Albumin B. Fibrinogen C. EPO

D. ALA synthase E. Ceruloplasmin

[6] The main energy resource of the mature red blood cell is____________.

A. pentose phosphate pathway B. glycolysis C. fatty acid oxidation

D. carbohydrate aerobic oxidation E. ketone body oxidation

[7] Compared to plasma, there is no in serum____________.

A. Carbohydrates B. Vitamins C. Metabolites

D. Fibrinogen E. Electrolyte

[8] Respiratory burst means the rapid release of reactive oxygen species. It is a crucial reaction occurs in neutrophile granulocytes to degrade internalized particles and bacteria. Which following will be created when NADPH react with oxygen in respiratory burst of neutrophile granulocytes?____________.

A. O_2^- B. H_2O_2 C. $O_2^{\cdot}$ D. Peroxide E. H_2

[9] Plasma colloid osmotic pressure is mainly dependent on the amount of ____________.

A. inorganic acid anion B. organic metal ion C. globulin

D. albumin E. glucose

[10] The main storage form of iron in the body is ____________.

A. transferrin B. ferroprotein C. ceruloplasmin

D. heme E. catalase

[11] The coenzyme of ALA synthase is the active form of ____________.

A. Vit B_{12} B. Vit B_1 C. Vit B_2 D. Vit B_6 E. Vit PP

[12] Which following can feedback inhibit ALA synthase____________?

A. Hb B. ALA C. Linear tetrapyrrole

D. Heme E. Uroporphyrinogen III

[B1 type]

No. 13 ~ 14 share the following suggested answers.

A. ALA dehydratase B. Uroporphyrinogen decarboxylase C. ALA synthase

D. Porphobilinogen deaminase E. Uroporphyrinogen I synthase

[13] The enzyme which is rate-limitedin heme biosynthesis?____________.

[14] The enzyme which is very sensitive to heavy metal ion?____________.

No. 15 ~ 18 share the following suggested answers.

A. AFP B. CRP C. LCAT D. IgG E. apoprotein

[15] Which one is Oncofetal protein____________?

[16] Which one is Acute-phase protein ____________?

[17] Which one is Conjugated protein ____________?

[18] Which one is Plasma functional enzyme____________?

No. 19 ~ 20 share the following suggested answers.

A. main energy resource

B. participate inthe fatty acid synthesis

C. maintain the reduction of the cell membrane

D. reduce GSSH to GSH

E. react with oxygen to generate superoxide anion (O_2^-)

[19] The function of NADPH in red blood cell is ____________.

[20] The function of NADPH in neutrophile granulocytes is ____________.

[X type]

[21] Which one belong to hemoprotein ____________?

A. Hemoglobin B. Myoglobin C. Cytochrome c
D. Peroxidase E. Catalase

[22] Under ____________ condition, blood NPN will increase?

A. Enhanced protein catabolism
B. Renal dysfunction
C. Enhanced nucleic acid catabolism
D. Hypohepatia
E. Enhanced glucose catabolism

[23] To separate blood protein preliminarily, we can use ____________.

A. ultracentrifugation B. salting out C. ultraviolet spectrophotometry
D. electrophoresis E. ion exchange

[24] Which of the following result of cellulose acetate film electrophoresis are corred ____________.

A. The fastest-moving protein is γ globulin
B. The deepest band is γ globulin
C. Usually, five bands can be identified
D. The lightest band is α_1 globulin
E. The slowest moving protein is albumin

[25] $Hb-Fe^{3+}$ can be reduced to $Hb-Fe^{2+}$ by ____________.

A. NADPH + H^+ B. GSH C. NADH + H^+
D. Vit C E. Vit PP

[26] Hemoglobulin oxygenationis influenced by ____________.

A. plasma protein concentration
B. blood partial pressure of carbon dioxide (PCO_2)
C. blood pH value
D. 2,3-BPG concentration of RBC
E. 1,3-BPG concentration of RBC

Answers

1	2	3	4	5	6	7	8	9	10
C	A	A	B	A	B	D	A	D	B
11	12	13	14	15	16	17	18	19	20
D	D	C	A	A	B	E	C	C	E
21	22	23	24	25	26				
ABCD	ABC	BD	CD	ABCD	BCD				

Brief Explanations for Practices

1. The normal ration of albumin to globulin in plasma is 1.5 ~ 2.5. If this ratio is less than 1.5, it

means there is a problem of biosynthesis of albumin, mostly because of the liver disease and in this case, the tissue edema always will be happened.

3. Albumin is the most abundant and the smallest protein in human plasma. So it is the main protein to maintain the plasma colloid osmotic pressure. It consists only peptide but no saccharides, so it is not a glycoprotein.

5. In these five proteins, albumin, fibrinogen, ALA synthase and ceruloplasmin are synthesized in the liver. So when the liver is injured, Albumin synthesis will be obviously decreased.

9. Colloid osmotic pressure, also called oncotic pressure, is a form of osmotic pressure exerted by proteins, notably albumin, in a blood vessel's plasma.

22. NPN means non-protein nitrogen. In our blood, NPN includes urea, creatinine, uric acid, bilirubin, and ammonia, except protein. Usually, enhanced protein ornucleic acid catabolism will result in the increased blood concentration of urea or uric acid and the damage of the kidney will influence the excretion of NPNs. Hypohepatia is related to a decreased NPN level.

24. Five bands can be separated in cellulose acetate film electrophoresis. From anode to cathode, they are albumin, α_1 globulin, α_2 globulin, β globulin and γ globulin. Albumin is the deepest band because it is the most abundant amount of protein in human blood and α_1-globulin is the lightest band because it is the least amount protein in human blood.

25. The reaction of $Hb-Fe^{3+}$ reduced to $Hb-Fe^{2+}$ needs reducing equivalent. NADH $+H^+$ or NADPH + H^+ can both provide reducing equivalent to this reaction. GSH and Vit C are reducing agent.

26. Increased PCO_2 leads to decreased pH value, let the oxyhemoglobin dissociation curve shift to the right, making more oxygen available at the tissues. The increased 2,3-BPG level will decrease the affinity of hemoglobulin and oxygen resulting in the release of oxygen.

[*Clinical Case*]

In the emergency room, a 40-year-old emotional woman complains of severe abdominal pain and muscle weakness. Regular laboratory examination shows nothing abnormal. Then gave her 40 mg phenobarbital (a drug belongs to barbital) to help her calm down and observed in the clinic for several hours. But her condition deteriorated rapidly, progressing to respiratory failure, general weakness and sensory disturbance. Further laboratory examination found the urine porphobilinogen level was high and at last the patient was diagnosed as acute intermittent porphyria. The main abnormal enzymatic activities of this disease are a high level of ALA synthase and reduced activity by 50% of porphobilinogen deaminase.

[27] What is the substrate of ALA synthase?

[28] It seems that phenobarbital fastened the progress of acute intermittent porphyria. Please analyze this conclusion reasonably.

Answers

27. Glycine and succinyl CoA are two substrates of ALA synthase and pyridoxal phosphate is the coenzyme of ALA synthase.

28. High level of ALA synthase is one of the main abnormal enzymatic activity of this disease and it will result in high ALA level. Phenobarbital can induce the ALA synthase and increase the ALA synthesis and then deteriorate the disease.

(***Lu Xiaoling, Wang Lianghua***)

Chapter 12

Liver Biochemistry

Examination Syllabus

1. **Hepatic Biotransformation**: definition, characters, types, enzymes, factors
2. **Metabolism of Bile Acids**: structure, metabolite, regulation
3. **Metabolism of the Bile Pigment**: properties of free bilirubin and conjugated bilirubin, metabolite, jaundice

Major points and Practices

I. The Role of Liver in Metabolism

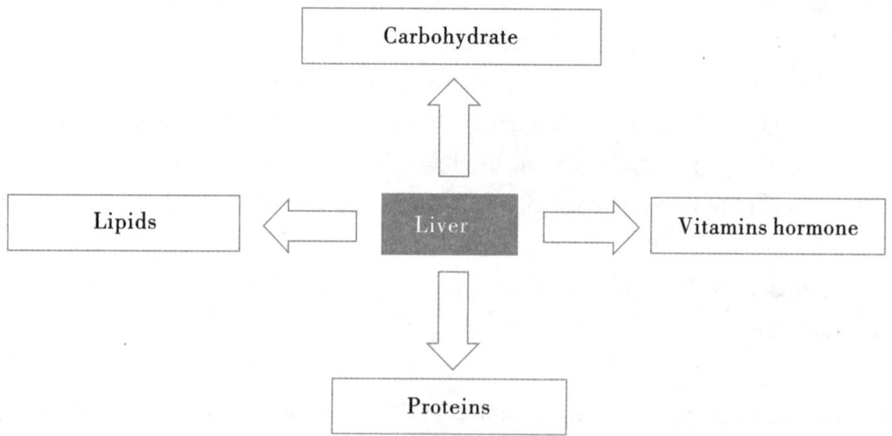

A. Carbohydrate metabolism

In carbohydrate metabolism, the main role of the liver is to maintain blood glucose relatively stable.

1. Glucose-6-phosphate is the hub of carbohydrate metabolism. In the fed state, the glucose in the liver is rapidly converted to its storage form of glycogen and the fat. In the fasting state, liver glycogen decomposes and produces glucose -6-phosphate to convert to glucose directly to supply the blood glucose under the glucose-6-phosphatase.

2. The liver is the active organ for gluconeogenesis that non-sugar compounds (amino acid, lactic acid, glycerin, etc) convert to glucose to supply the blood glucose, ensuring the blood glucose stable in the starvation state.

B. Lipid metabolism

The liver plays a very important role in digestion, absorption, catabolism, synthesis and transportation of lipids.

1. The liver is the hub of metabolism of triglyceride and fatty acid. In the starvation state, the liver decomposes fatty acid to produce a large amount acetyl CoA, whose outlet is to oxidation completely and to produce ketone body. In the fed state, the liver is the place to biosynthesis endogenous triglyceride, which effectively coordinates the oxidation of fatty acid and synthesis of triglyceride by esterification, and to change the length and transform the saturation degree of carbon chain for fatty acid.

2. The liver is not only the place of biosynthesis cholesterol but also takes part in the esterification of cholesterol.

C. Protein metabolism

The role of the liver in protein metabolism mainly includes synthesizing protein, decomposing amino acids, synthesizing urea, etc.

1. Most plasma proteins are synthesized in the liver, of which plasma albumin, prothrombin, and fibrinogen can only synthesize in the liver.

2. The ammonia produced by the decomposition of the amino acids is toxic to the body, which could excrete in the form of urea by ornithine cycle, where the liver is the sole organ to synthesis urea.

D. Vitamins

The liver plays a very important role in absorption, transportation, storage, and transformation, etc of vitamins.

1. The adequate intestinal absorption of the fat-soluble vitamins is critically dependent on adequate fatty acid micellization, which requires bile acids.

2. The liver is the reserve organ of vitamin A, E, K and B_{12}. It is responsible for converting some of these water-soluble vitamins to active coenzymes.

E. Hormone

1. Some hormone could bind to the receptor of the hepatocyte membrane and metabolite and transform in the hepatocytes through endocytosis.

2. Some steroid hormones enter into the hepatocytes through diffusion and conjugate with glucuronic acid or active sulfuric acid to inactivate.

[*Practices*]

[A1 type]

[1] Which one is incorrect to the characteristics of the metabolism of the liver? ____________

A. To synthesis urea using the ammonia from the decomposition of the amino acid

B. Important parts for the fatty acid oxidation

C. Finally to decompose glycogen into glucose

D. To synthesis and storage glycogen

E. The sole organ to gluconeogenesis

[2] Gluconeogenesis, the formation of ketone body and synthesis of urea all take part in ________.

A. brain B. muscle C. kidney D. heart E. liver

[3] Which one is the minimum storage amount in the liver? ________

A. Vitamin K B. Vitamin A C. Vitamin E

D. Vitamin B_{12} E. Vitamin D

[4] If the liver is impaired, which one is the main change of the proteins in the blood? ________.

A. The albumin content increased and the globulin content reduced

B. Albumin/globulin inversion

C. The albumin content increased

D. The globulin content reduced

E. Both the content of albumin and globulin reduced

[5] If the liver is impaired, the content of the urea is reduced and ________ is increased in the blood.

A. glucose B. fat C. ammonia D. cholesterol E. bilirubin

[6] The energy of the brain is mainly from ________ when people endure long starvation.

A. ketone body B. glucose C. amino acid

D. glycerol E. glycogen

[X type]

[7] ________ is the cross-over point of the carbohydrate and lipid metabolism.

A. Acetyl CoA

B. Fructose-6-phosphate

C. Dihydroxyacetone phosphate

D. 3-phosphoglyceraldehyde

E. Glucose-6-phosphate

[8] ________ lies in the mitochondrial system.

A. Uera cycle enzgne system

B. Fatty acid β-oxidase system

C. Various dehydrogenase of the respiratory chain

D. Citric acid circulating enzyme system

E. Glycolysis

Ⅱ. Hepatic Biotransformation

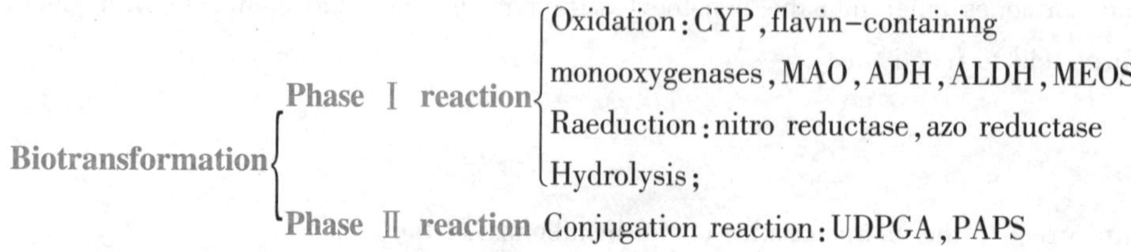

A. Concept

Biotransformation involves a series of enzyme-catalyzed processes, chemical alterations of the non nutritive substances such as exogenous and endogenous substances like nutrients, amino acids, toxins, drugs, et al, which make them reduce or lose the biological activity, reduce toxicity or detoxify, increase the water solubility and the polarity, more easily excreted with the urine or bile.

B. The major phase reactions

1. Phase I reactions

(1) involve oxidation, reduction and hydrolysis, while oxidation is the most common reaction.

(2) **Enzymes mainly take part in these reactions**: cytochrome P450 monooxygenases, monoamine oxidase (MAO), alcohol dehydrogenases, aldehyde dehydrogenases, nitroreductase and azoreductase, hydrolases.

2. Phase II reactions

(1) **Conjugation reaction**: conjugate with glucuronic acid, sulfuric acid, acetyl group, glutathione, glycine, methyl group, *etc*.

The conjugated reaction with glucuronic acid, sulfuric acid and acetyl group is most important, and the conjugated reaction with glucuronic acid is the most common.

(2) The active donor

- The active donor of glucuronic acid: Uridine diphosphate glucuronic acid (**UDPGA**).
- Thedonor of active sulfate: 3′-phosphoadenosine 5′-phosphosulfate (**PAPS**)
- The donor of an acetyl group: **Acetyl CoA**.
- The methyl donor: S-adenosylmethionine (**SAM**)

C. The characteristics of biotransformation

1. Continuity.
2. Diversity of the reaction.
3. The duality of detoxification and toxicity: The toxicity of some substances is enhanced after biotransformation, whose biotransformation has the characteristics of detoxification and toxicity in the liver.

D. The factors influencing hepatic biotransformation

1. Age, gender, genetic variability, nutrition, disease.
2. Exposure to other chemicals can inhibit or induce enzymes and dose levels.

[*Practices*]

[A1 type]

[9] The most important phase I reaction of the biotransformation is ______.

A. hydrolysis reaction　　B. synthetic reaction　　C. reduction reaction

D. decarboxylation reaction　　E. oxidation reaction

[10] The most relevant of the biotransformation process in liver is ______.

A. cytochrome P_{450}　　B. oxygen atom　　C. peroxidase

D. oxygen dehydrogenase　　E. catalase

[X type]

[11] Non-nutritional substances can ______ after biotransformation.

A. low down the toxicity

B. excrete from the bile acid or urine

C. low down the drug activity

D. increase the water solubility

E. increased lipid solubility

[12] The common donors of the conjugated substances are ______ in the phase II reaction of the liver.

A. UDPGA B. SAM C. acetyl CoA

D. PAPS E. UDPG

Ⅲ. Bile and the metabolism of bile acid

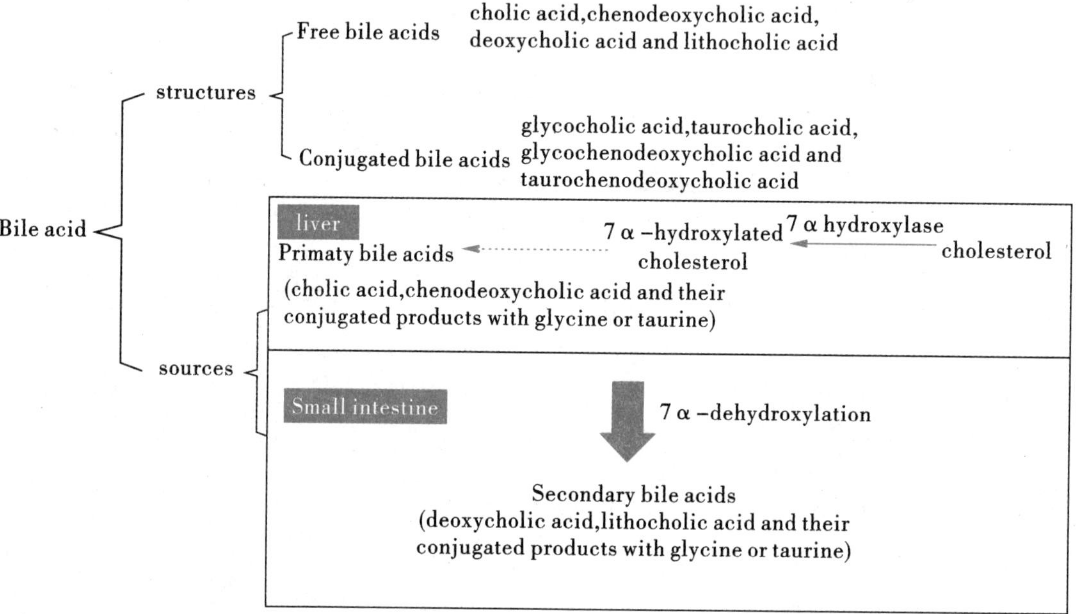

A. The classification of the bile acid

1. According to the structure

(1) **Free bile acids**: contain cholic acid, deoxycholic acid, chenodesoxycholic acid, lithocholic acid.

(2) **Conjugated bile acid**: the products of free bile acids separately combining with glycine or taurine.

2. According to the source

(1) **Primary bile acid directly**

● synthesized from cholesterol as a starting material.

● including cholic acid, chenodesoxycholic acid and the products of which separately combine with glycine or taurine.

(2) **Secondary bile acid**

● the products of the deoxidization of the primary bile acid by the intestinal bacteria

● mainly including deoxycholic acid, lithocholic acid and the products of which separately combine with glycine or taurine.

B. The physiologicfunction of bile acids

Bile acids are physiologic surfactant.

(1) Promote the digestion and absorption of lipids.

(2) Maintain the water-soluble of cholesterol to prevent forming the gallstones.

C. Bile acid metabolism and bile acid enterohepatic circulation

1. The biosynthetic pathway of the bile acids

(1) Synthetic site: the primary bile acid: liver; the secondary bile acid: intestine.

(2) The key enzyme: 7α-hydroxylase is negatively feedback regulated by the bile acids.

2. Bile acid enterohepatic circulation

The primary and secondary bile acids are absorbed almost exclusively in the ileum, returning to the liver by way of the portal circulation, where up to 90% of the bile acids are extracted by hepatocytes. After hepatic extraction, bile acids are recirculated into the biliary tree, completing the circuit. This is known as the enterohepatic circulation.

[*Practices*]

[A1 type]

[13] Which bile acid is the secondary bile acid?______

A. Cholic acid B. Glycine cholic acid C. Taurocholic acid

D. Lithocholic acid E. Chenodesoxycholic acid

[14] ______ is produced by the transformation of cholesterol in the liver cells.

A. Cholic acid B. Litocholic acid C. Bile pigment

D. Primary bile acid E. Secondary bile acid

[15] ______ is the most content solid component in bile acid.

A. Mineral salt B. Bile salt C. Mucin

D. Cholesterol E. Bile pigment

[16] Bile acid has the feedback inhibition of______.

A. HMG-CoA reductase B. Monooxygenase C. Cholesterol 7α hydroxylase

D. Acetyl-CoA carboxylase E. 12α-hydroxylase

[17] Which factor could make cholesterol precipitation? ______.

A. The concentration of bile salt

B. The concentration of lecithin

C. The concentration of cholesterol

D. Poorly soluble of cholesterol in water

E. The ratio of the bile salt/cholesterol and the lecithin/cholesterol

[X type]

[18] In the liver, ______ combine with the bile acid.

A. sulphuric acid B. taurine C. glucuronic acid

D. glycine E. nucleic acid

[19] Which one is the secondary bile acid?______

A. Lithocholic acid B. Taurolithocholic acid C. Chenodesoxycholic acid

D. Taurodeoxycholic acid E. Cholic acid

Ⅳ. Metabolism of bile pigment

A. Source

1. Bile pigment is the main decomposition products of the heme, including bilirubin, biliverdin, bilinogen and bilin. Bilirubin is the main pigment in the bile.

2. Bilirubin is produced by the catabolism of iron porphyrin compounds. The hemoglobin is released from the aging worn-out red blood cells, and 80% of total bilirubin derives mainly from hemoglobin. Heme degradation is catalyzed by HO^{-1} and some cofactors, including at least 3 molecules of oxygen and NADPH, and cytochrome P450, whose reaction is the speed limit step of bilirubin formation.

3. It has irreversible damage to the nervous system and needs to be transported to the liver by circulation to detoxify the liver in time and then excreted with the bile.

B. The transport of bilirubin in the blood

1. **Form**: **bilirubin-albumin** complex.

2. **Name**: The bilirubin which is transported by the albumin in the blood is called **free bilirubin**, **unconjugated bilirubin**, hemobilirubin, **indirect bilirubin**.

3. **Significance**:

(1) increase the water solubility of the bilirubin and the plasma transport capacity of bilirubin.

(2) restrict it to permeate various cell membranes and avoid toxic effects on tissues and cells.

C. Transformation of bilirubin in the liver

1. The bilirubin separate from the albumin before entering the hepatic cells, then quickly enter the hepatic cells, **combining with two carrier proteins** (**Y protein/Z protein**) separately in the cytoplasm to form complex compounds (**bilirubin-Y protein/ bilirubin-Z protein**).

2. The complex of bilirubin-Y protein/Z protein carries to the smooth endoplasmic reticulum, and most of the bilirubin is conjugated in the form of **bilirubin diglucuronide** under the catalysis of **UDP-glucuronyltransferase**, called conjugated bilirubin, which can be easily excreted from the bile (Table 12.1).

Table 12.1 The physical and chemical properties of two kinds of bilirubin

Properties	Unconjugated bilirubin (Indirect bilirubin)	Conjugated bilirubin (Direct bilirubin)
Water solubility	small	large
Fat solubility	large	small
Affinity with albumin	Large	small
Permeability and toxicity of cell membrane	Large	small
Pass through the glomerulus	No	Yes
Reactivity to diazonium reagents	Indirect positive	Direct positive

3. The conjugated bilirubin secreting from the hepatic cell enter to the bile duct, active transport of inverse concentration gradient, which is the rate-limiting step for hepatocytes to metabolize bilirubin.

D. The process of bilirubin in the intestinal tract and enterohepatic circulation

1. Process: As the conjugated bilirubin reaches the terminal ileum and the large intestine, the glucuronides are removed by specific bacterial enzymes (β-glucuronidases), and the pigment is subsequently reduced to bilinogens, further reduced to stercobilinogen.

2. In the terminal ileum and colon, a small fraction of the urobilinogens is reabsorbed and excreted again through the liver to constitute the enterohepatic circulation of urobilinogen.

E. Hyperbilirubinemia and jaundice

The concentration of total bilirubin in the serum of normal adults ranges from 3.4 ~ 17.1 μmol/L.

1. Hyperbilirubinemia

(1) When bilirubin in the blood exceeds 17.1 μmol/L, hyperbilirubinemia occurs.

(2) Hyperbilirubinemia may be due to the production of more bilirubin than the normal liver can excrete, or it may result from the failure of a damaged liver to excrete bilirubin produced in normal amounts.

2. Jaundice

(1) Bilirubin accumulates in the blood, and when it reaches a concentration of 34.2μmol/L (2mg/dL) or greater, it diffuses into the tissues, skin or sclera become yellow, called jaundice.

(2) Depending on the difference of episode mechanism, jaundice can be classified as **hemolytic jaundice, hepatic jaundice and extrahepatic jaundice**.

[*Practices*]

[A1 type]

[20] The primary source of the bilirubin is ______

A. aging red blood cell   B. myoglobin   C. peroxidase
D. catalase   E. cytochrome

[21] Bilirubin mainly combines with ______ in the blood.

A. fatty acid   B. amino acid   C. glucose
D. globulin   E. albumin

[22] ______ and bilirubin could competitively combine with the same carrier in the blood.

A. Fatty acid   B. Amino acid   C. Creatine
D. Fibrin   E. Uric acid

[23] Which is correct of the description of free bilirubin? ______

A. Bilirubin could combine with glucuronic acid
B. High water-soluble
C. Not easy to cross the biomembrane
D. Excretion of the kidneys with the urine
E. Indirect reaction with diazonium reagents

[24] The cause of the bilirubin in the urine is ______

A. Increase of the free bilirubin in the blood
B. Combine with albumin
C. Increase of the conjugated bilirubin in the blood

D. Combine with urobilinogen
E. Combine with Y protein

[X type]

[25] Which one can produce bile pigment?______
A. Myoglobin B. Hemoglobin C. Peroxidase
D. Cytochrome E. Cholesterol

[26] Unconjugated bilirubin is also called ______.
A. direct bilirubin B. indirect bilirubin C. free bilirubin
D. liver bilirubin E. hemobilirubin

[27] ______ can combine with the albumin.
A. Free bilirubin B. Conjugated bilirubin C. Fatty acid
D. Salicylic acid E. Glucose

[28] Which one has the enterohepatic circulation?____
A. Bile pigment B. Bile acid C. Glycine
D. PAPS E. SAM

Answers

1	2	3	4	5	6	7	8	9	10
E	E	E	B	C	A	AC	BCD	E	A
11	12	13	14	15	16	17	18	19	20
ABCD	ABCD	D	D	B	C	E	BD	ABD	A
21	22	23	24	25	26	27	28		
E	A	E	C	ABCD	BCE	ACD	AB		

Brief Explanations for Practices

1. Gluconeogenesis can take place in the liver and kidney.

2. Gluconeogenesis, the formation of ketone body and synthesis of urea all take part in the liver.

3. Vitamin D is the minimum storage amount in the liver.

4. The liver can synthesis most of the plasma proteins. The ratio of albumin/globulin is inverted when the liver is impaired.

5. The main path of the blood ammonia is to synthesis urea in the liver. If the liver is impaired, the content of the urea is reduced and ammonia is increased in the blood.

6. When people endure long starvation, ketone body provides the energy for the brain in order to save glucose utilization.

7. Acetyl CoA and dihydroxyacetone phosphate both take part in the carbohydrate and lipid metabolism.

8. Fatty acid β-oxidase system, various dehydrogenase of the respiratory chain and citric acid circulating enzyme system all lie in the mitochondrial system.

9. Oxidation reaction is the most important phase Ⅰ reaction of the biotransformation.

10. Cytochrome P_{450} is the most relevant of the biotransformation process in the liver.

11. In the biotransformation process, non-nutritional substances can be low down the toxicity and the drug activity, excreted from the bile acid or urine, increased the water solubility.

12. The common donor of the conjugated substances are UDPGA, SAM, Acetyl CoA, PAPS in the phase II reaction of the liver.

13. The secondary bile acid includes deoxycholic acid, lithocholic acid and the products of which separately combine with glycine or taurine.

14. Primary bile acid is produced by the transformation of cholesterol in the liver cells.

15. Bile salt is the most content solid component in bile acid.

16. Cholesterol could turn to bile acid in the liver. It has the feedback inhibition of cholesterol 7α hydroxylase.

17. In the liver, the ratio of the bile salt/ lecithin/cholesterol is fixed, which increase the water solubility of the cholesterol and decrease the cholesterol precipitation.

18. In the liver, the bile acids mainly take part in the conjugation reaction. The common conjugate reagent is taurine and glycine.

19. The secondary bile acid includes deoxycholic acid, lithocholic acid and the products of which separately combine with glycine or taurine.

20. 80% bilirubin comes from the aging red blood cell.

21. In the blood, bilirubin mainly combines with albumin.

22. In the blood, a fatty acid also combines with albumin.

23. Unconjugated bilirubin is called free bilirubin, which doesn't combine with the glucuronic acid. This bilirubin has the low water solubility, goes cross the biomembrane and eliminate from the body difficultly.

24. The free bilirubin turns into conjugated bilirubin in the liver. Then conjugated bilirubin eliminates from the body by the urine. The bilirubin is in the form of conjugated bilirubin in the urine.

25. Bile pigment comes from the myoglobin, hemoglobin, peroxidase, cytochrome.

26. Unconjugated bilirubin is also called free bilirubin, indirect bilirubin.

27. Free bilirubin and fatty acid combine with the albumin. Salicylic acid is a pharmaceutic preparation, it also can combine with the albumin.

28. Bile pigment and bile acid have the enterohepatic circulation

[*Comprehensive Practices*]

[A type]

[1] The following is about the function of the liver, which one is not right? ______

A. Store glycogen and vitamin B. Synthesis albumin C. Biotransformation

D. Synthesis digestive enzyme E. Synthesis urea

[2] Which one is peculiar to the liver?______

A. Glycogen decompose into glucose B. Glycogenesis C. Gluconeogenesis

D. Oxidation of carbohydrate E. Phosphopentose pathway

[3] The most important phase II reaction in the biotransformation is ______.

A. combine with glucuronic acid

B. combine with an acyl group

C. combine with sulphuric acid

D. combine with glutathione

E. combine witha methyl group

[4] Cholesterol couldn't convert to ______

A. vitamin D_3 B. bile acid C. bile pigment

D. estradiol E. testosterone

[5]The key enzyme of the synthetic of bile acid is ______.

A. bile acyl CoA synthetase B. hydroxyl cholesterol oxidase C. cholesterol 7α hydroxylase

D. bile acid synthetase E. HMG-CoA reductase

[6]The bile acid ______to regulate bile acid synthetic.

A. activate 7α hydroxylase B. activate 3α hydroxylase C. inhibit 7α hydroxylase

D. inhibit 3α hydroxylase E. None of the above

[7]Which is an incorrect description of direct bilirubin?______

A. Bilirubin glycuronic acid two ester

B. Large water solubility

C. Not easy to cross the biomembrane

D. Not pass through the glomerulus

E. Reactivity to diazonium reagents, direct reaction

[8]Which is an incorrect description of bile salts?______

A. Not inhibit the formation of the cholesterol stones

B. Cholesterol synthesis in the liver

C. Emulsifier absorbed by lipid

D. The metabolic product of the bile pigment

E. Reabsorbed by the enterohepatic circulation

[9]Which is the most important enzyme taking part in the oxidation reaction of the biotransformation? ______

A. Monooxygenase B. Dioxygenase C. Hydrolase

D. Amine oxidase E. Alcohol dehydrogenase

[10]Which one belongs to the biotransformation?______

A. Gluconeogenesis

B. Synthesis of the ketone body

C. Unconjugated bilirubins turn to conjugated bilirubins

D. Synthesis of protein

E. Protein decomposition

[11]The most significance of the biotransformation is to ______

A. inactivate bioactive substances

B. detoxicate

C. provide energy

D. increase the solubility of the non-nutritional substance and eliminate from the body

E. increase the bioactivity or the toxicity of the drugs

[12]Which bile pigment can easily go through the biomembrane?____

A. Bilirubin-albumin B. Free bilirubin C. Urobilinogen

D. Urobilin E. Conjugated bilirubin

[13]The toxicity of the bilirubin mainly show ______

A. hepatotoxicity B. nephrotoxicity C. gastrointestinal reaction

D. cardiotoxicity E. neurotoxicity

[B type]

No. 14 ~ 15 share the following suggested answers

A. primary bile acid B. secondary bile acid C. free bile acid
D. heme E. bilirubin

[14] Taurocholic acid is______.

[15] Bile pigment involves______.

No. 16 ~ 17 share the following suggested answers

A. Heme oxygenase B. Biliverdin reductase C. UDPGA transferase
D. 7α- hydroxylase E. Monooxygenase

[16]______ catalyze cholesterol to bile acid.

[17]______ catalyze heme to biliverdin.

No. 18 ~ 19 share the following suggested answers

A. UDPGA B. UDPG C. UDP D. GSH E. PAPS

[18] In the liver,______ combine with free bilirubin.

[19]______ is essential in the synthetic of the glucogen.

[X type]

[20] Which one is secondary bile acid?______

A. 7-deoxycholic acid B. Taurocholic acid C. Lithocholic acid
D. Chenodesoxycholic acid E. Cholic acid

[21] Which one can take part in the biotransformation?______

A. Nutritional substances
B. Bioactive substances
C. Some metabolite products
D. Exogenous substances like drugs, toxins, additives
E. Nucleic acid

[22] Which compound belongs to the bile pigment?______

A. Bilirubin B. Biliverdin C. Urobilinogen
D. Stercobilinogen E. Heme

Answers

1	2	3	4	5	6	7	8	9	10
D	A	A	C	C	C	D	D	A	C
11	12	13	14	15	16	17	18	19	20
D	B	E	A	E	D	A	A	B	AC
21	22								
BCD	ABCD								

(*Lu Xiaoling, Wang Lianghua*)

Chapter 13

Integration and Regulation of Metabolism

Examination Syllabus

1. The relationralation and integration of metabolism

2. Characteristic of metabolism in different organs

3. Regulation of metabolism: cellular regulation (allosteric regulation, chemical modification), hormone's regulation, the whole-body control of regulation.

[*Major points*]

Ⅰ. The relationralation and integration of metabolism

1. Metabolism of nutrients is correlated and controlled by each other to form integrity.

(1) Each metabolite has specific metabolic pathways;

(2) All the metabolic pathways are interrelated and controlled by each other to form integrity.

2. Each metabolite has its own metabolic pool

(1) In metabolic pool, the metabolites come from diverse sources, and will enter into different outcomes according to the demand of body.

(2) Generally, the income and outcome of a metabolic pool must maintain balance;

(3) Actually, the storage of each metabolite in its pool is not abundant, so that the metabolic pool must be rapidly rotated to maintain the demand of the body.

3. Three major nutrients are connected through intermediate metabolites

(1) The saccharide, fat and protein are completely oxidized to provide ATP through the common pathways such as tricarboxylic acid cycle and oxidative phosphorylation.

(2) From the perspective of energy, these three major nutriments could be replaced and complement each other.

4. Different nutrients are interconvertible through intermediate metabolites

(1) Most amino acids could be used for glyconeogenesis, metabolites of glucose are used for synthesis of amino acids.

(2) Glucose is the major material for biosynthesis of endogenous TG. Mobilization of TG produces glyc-

erol and fatty acid. Glycerol is the material of gluconeogenesis, however, fatty acid is oxidized to produce acetyl CoA, which is impossible to be converted to glucose.

(3) Amino acid could be converted to plenty of lipids, however, lipid has no possibility to be converted to amino acid.

5. Three major nutrients are used for synthesis of other biomolecules.

Nucleic acids are majorly obtained by endogenous synthesis with materials from metabolism of glucose and amino acid.

Ⅱ. The major manner of regulation of metabolism

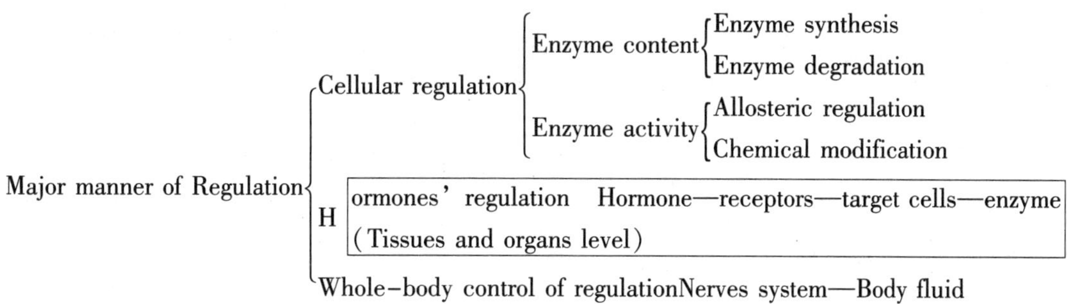

Regulation of metabolism is the essential feature of life. Regulation of metabolism may take place on different levels. The general model of metabolic regulation is shown in Figure 13.1.

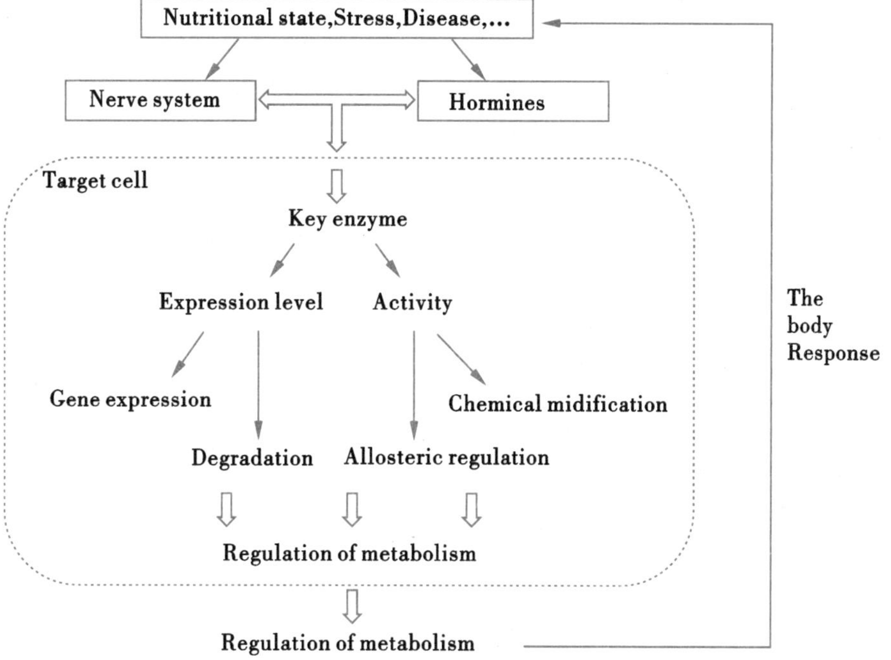

Figure 13.1 The general model of metabolic regulation

A. Cellular regulation

1. Key enzyme

Regulation of key enzymeis **the basis of metabolism regulation.**

(1) The general features of key enzyme

● catalyze the first reaction of a pathway, reaction on branch site, or reaction with the lowest rate.

● catalyze the unidirectional or nonequilibrium reaction, the activity of it decides direction of the entire metabolic pathway.

● besides the substrate or product, the activity of key enzyme is regulated by other metabolites or effectors.

(2) Changing the **amount** or **activity** of key enzyme is the **basic regulation mode** of metabolism. It is also the way how hormones regulate metabolism in cell level.

Some key enzymes in important metabolic pathway are listed in Table 13.1.

Table 13.1 Key enzymes of some metabolic pathways

Metabolic pathway	Key enzyme
Glycolysis	Hexokinase
	Phosphofructokinase-1
	Pyruvate kinase
Decarboxylation of pyruvate	Pyruvate dehydrogenase complex
Citric acid cycle	Citrate synthase
	Isocitrate dehydrogenase
	α-ketoglutaric acid dehydrogenase complex
Glycogenolysis	Phosphorylase
Glyconeogenesis	Glycogen synthase
Gluconeogenesis	Pyruvate carboxylase
	Phosphoenolpyruvatecarboxykinase
	Fructose-2,6-diphosphatase
	Glucose-6-phosphatase
Fatty acid synthesis	Acetyl CoA carboxylase
Fatty acid degradation	Carnitine acyl transferase
Cholesterol synthesis	HMG CoA reductase

2. Regulation of enzyme amount

The amount of enzyme is controlled by **regulating the expression or degradation of zymoprotein**.

(1) The chemical nature of enzymes in substance metabolism are proteins, which are also called, **zymoprotein**.

(2) Changing the velocity of **synthesis** or **degradation** of zymoprotein will control the amount and the entire activity of them. By this way, more ATP will be exhausted, and longer time needed, generally hours, even days, so that it is called slow regulation.

(3) The substrate, products, hormones or drugs may induce or repress the expression of enzyme.

3. Regulation of enzyme activity

Allosteric regulation is the **popular manner** for metabolic regulation in the living nature.

(1) Allosteric regulation.

● Allosteric regulation means small compound binds with the specific site of zymoprotein, changes the conformation and activity of enzyme.

● Allosteric regulator binds with enzyme at a site **out of** its **active center**.

● The allosteric **agents** might be the substrate, end product, or other small metabolites.

● The change of allosteric regulators sensitively reflects the strength of relevant metabolic pathway and the demand of the body.

● Some intermediate metabolites regulate key enzymes of different metabolic pathways to coordinate these pathways.

(2) **Chemical modification**

Chemical modification of enzyme is **rapid and popular**.

● Some amino acid residues on the side chain of a zymoprotein can be reversely modified by another enzyme, this chemical modification changes its activity.

● There are multiple forms of chemical modification of enzymes, such as phosphorylation and dephosphorylation, acetylation and deacetylation, methylation and demethylation, adenylation and deadenylation, and interconversion of sulfhydryl group SH and disulfide bond S–S, et al. **The most popular form is phosphorylation and dephosphorylation**.

● The chemical modification of enzyme is catalyzed by other enzymes, which are controlled by the upstream factors.

B. Hormones' regulation

Hormones are the key factors in metabolism regulation in the body

1. On the tissue and organs level, metabolisms in multiple cells are cooperated by hormones.

2. Hormones bind with the specific receptors on the target cells in different tissues and initiate intracellular signal pathways, change the activity or the number of enzymes in different metabolic pathways.

C. The whole-body control of regulation

The whole-body control of regulation is executed through **nerves** and **body fluid** system.

1. The entire control of regulation is executed through nerves and body fluid system.

2. Under the fed state, metabolism is related to the composition of diet.

3. Under the fasting state, the metabolism is characterized by glycogenolysis, gluconeogenesis, and moderate lipid mobilization.

Ⅲ. Characteristic of metabolism in different organs

1. Liver is the center of metabolism, and plays important roles in nutriments' and other substances' metabolism, such as vitamins, hormones, drugs, and poisons.

2. Each extrahepatic organ has its own characteristics of metabolism to fit for the special function of it (Table 13.2).

Table 13.2 Major metabolic features of organs

Organs or tissues	Major metabolic pathways	Fuels	Major metabolites
Brain	Aerobic oxidation Anaerobic oxidation Amino acid metabolism	Glucose Ketone bodies	Lactate CO_2, H_2O Neurotransmitter
Myocardium	β-oxidation and citric acid cycle	Fatty acid Ketone bodies lactate	CO_2, H_2O
Skeleton muscle	Aerobic oxidation Anaerobic oxidation β-oxidation and citric acid cycle	muscle glycogen fatty acid ketone bodies	Lactate, CO_2, H_2O
Adipose tissue	Lipogenesis lipolysis	Glucose	TG fatty acid, glycerol
Kidney	Gluconeogenesis Glycolysis, citric acid cycle, ketone bodies synthesis	Free fatty acid Ketone bodies	Glucose Lactate Ketone bodies
Erythrocyte	Anaerobic glycolysis Pentose phosphate pathway	Glucose	Lactate

Ⅳ. The abnormal metabolism in disease

1. Metabolic syndrome is clustered with a lot abnormal metabolic changes.
2. Metabolic reprogramming is one of the hallmarks of cancer.

[*Practices*]

[**A1 type**]

[1] Which is correct in the following descriptions?____________

A. Protein can completely turn into glucose

B. Materials for lipid synthesis could be provided by glucose

C. Lipid can turn into glucose

D. Protein is the primary material for gluconeogenesis

E. Lipid can turn into protein

[2] The common intermediate metabolite of glucose, lipid, and protein is____________

A. citric acid B. acetyl CoA C. lactic acid

D. pyruvate E. amino acid

[3] The following descriptions about allosteric regulation, which one is correct?____________

A. Most of allosteric regulation is feedback regulation

B. The allosteric agent firmly binds with the enzyme

C. The allosteric agent promotes the binding of substrate with the enzyme

D. The allosteric agent changes the conformation of enzyme

E. The allosteric agent binds with the active center of enzyme

[4] The chemical modification of enzyme means____________

A. The hydrolyzation of amino acids from zymoprotein

B. The binding of prothetic group with enzyme

C. Polymerization of the subunits of zymoprotein

D. The reversible modification of some amino acid residues by other enzymes

E. Small agent bind with enzyme and change its activity

[5] In short-time hungry, the primary material of gluconeogenesis is ____________

A. Lactic acid B. Glycerol C. Amino acid

D. Fatty acid E. Acetyl CoA

[6] Which compound is the major inhibitor of key enzyme in catabolism? ____________

A. ATP B. AMP C. ADP D. FMN E. FAD

[7] Which one is the major energy source of brain in long-term hungry? ____________

A. Glutamine B. Amino acid C. Glucose

D. Fatty acid E. Ketone bodies

[8] Which one of the following hormones has functional abnormity in obesity? ________

A. Glucagon B. Leptin C. Insulin

D. Growth hormones E. Adrenaline

[B1 type]

No. 9 ~ 11 share the following suggested answers

A. ATP B. pyruvate C. acetyl CoA

D. glycine E. CO_2

[9] The direct supplier of energy in the body is ____________.

[10] The product of decarboxylation includes ____________.

[11] The material for fatty acid biosynthesis is ____________.

No. 12 ~ 14 share the following suggested answers

A. hepatic glycogen B. glucose C. amino acid

D. fatty acid E. glycerol

[12] After fasting for 12 hours, the blood glucose is provided by ____________.

[13] After fasting for 36 hours, the primary fuel of body is ____________.

[14] The major materials for lipid synthesis are derived from ____________.

[X type]

[15] Which of the following are the outlets of amino acids in metabolic pool? ____________

A. Used as blocks for protein synthesis

B. Convert to other nitrogen-containing compounds

C. Degrade by deamination to form α-keto acids and amonia

D. Produce amine by decarboxylation

E. Passed out of the body in the urine

[16] In short-time fasting, the blood glucose is maintained through: ____________

A. hepatic glycogenolysis B. muscle glycogenolysis C. gluconeogenesis

D. lipid mobilization E. protein degradation

Answers

1	2	3	4	5	6	7	8		
B	B	D	D	C	A	E	C		
9	10	11	12	13	14	15	16		
A	E	C	A	D	B	ABCD	ACDE		

Brief Explanations for Practices

1. Metabolism of glucose provides glycerol and acetyl CoA for biosynthesis of fatty acid and TG.

2. The acetyl CoA is the common intermediate metabolites of glucose, protein, and lipid.

3. The allosteric regulation means small molecules bind with enzyme, change its conformation, and regulate its function.

4. Chemical modification means some amino acid residues in the enzyme are reversibly modified by other enzymes, such as phosphorylation and dephosphorylation.

5. In short-term hungry, the degradation of protein speed up and amino acids are used for biosynthesis of glucose.

6. Catabolism provides ATP, since the high concentration of ATP will inhibit the catabolism.

7. In long-term hungry, blood glucose decreases and can't provide enough ATP for the brain. Degradation of protein decreases, no extra amino acids are used for the brain. The brain can't utilize fatty acid. ketobodies are produced in liker, and are the main energy source of brain in long-term hungry.

8. Insulin resistance is a common mechanism of obesity.

9. ATP degradation provide energy for biological activation directly.

10. Decarboxylation of metabolite could produce CO_2, which is the major way of CO_2 production in the body.

11. In the above metabolites, only acetyl-CoA is used as direct material for fatty acid biosynthesis.

12. In short-time fasting, hepatic glycogen degrade to produce glucose, which is the major source of blood glucose in that time.

13. In long-time fasting, hepatic glycogen is exhausted, lipid mobilization enhances and fatty acid is the major fuel for the body.

14. Glucose is the major source for lipid biosynthesis in vivo, which enhanced in satiation.

15. The major outlet of free amino acids include the following: used as blocks for protein synthesis convert to other nitrogen-containing compounds, degrade by deamination to form α-keto-acids and amonia, produce amine by decarboxylation.

16. In short-time fasting, hepatic glycogen degrades firstly and provides glucose into blood; with decrease of hepatic glycogen, degradation of protein increases to provide amino acids for gluconeogenesis and helps to maintain blood glucose. lipid mobilization is intiated, and also plays a role for maintaining blood glucose.

[*Clinical Cases*]

One night, a 60-year-old insensible woman was sent to the emergency room. Regular laboratory examination showed that the blood glucose was 33. 3mmol/L, the blood ketone bodies was 4. 8 mmol/L. Her daughter told the doctor that her mother was diagnosed with diabetes 10 years ago, and she had a cold 3 days ago. Finally, women were diagnosed with diabetic ketoacidosis.

1. What is diabetic ketoacidosis?

2. Why do the blood ketone bodies increase in diabetes patient under-stimulation such as cold, stress, et al?

Answers:

1. Diabetic ketoacidosis (KDA) is a complication of diabetes, which happens most often in those with type 1 diabetes, but can also occur in those with other types of diabetes under certain circumstances, including infection, not taking insulin correctly, stroke, and certain medications such as steroids. DKA results from a shortage of insulin; in response the body switches to burning fatty acids with producing acidic ketone bodies. Diabetic ketoacidosis is typically diagnosed when testing finds high blood sugar, low blood pH, and α-ketoacids in either the blood or urine.

2. Under stimulation, the metabolic abnormity is enhanced in diabetic patient, blood glucose is increased, however, the glucose can't be utilized efficiently, which causes the energy supply is not enough in tissues, so the lipid mobilization speeds up, more ketone bodies are produced.

(***Yuan Ping***)

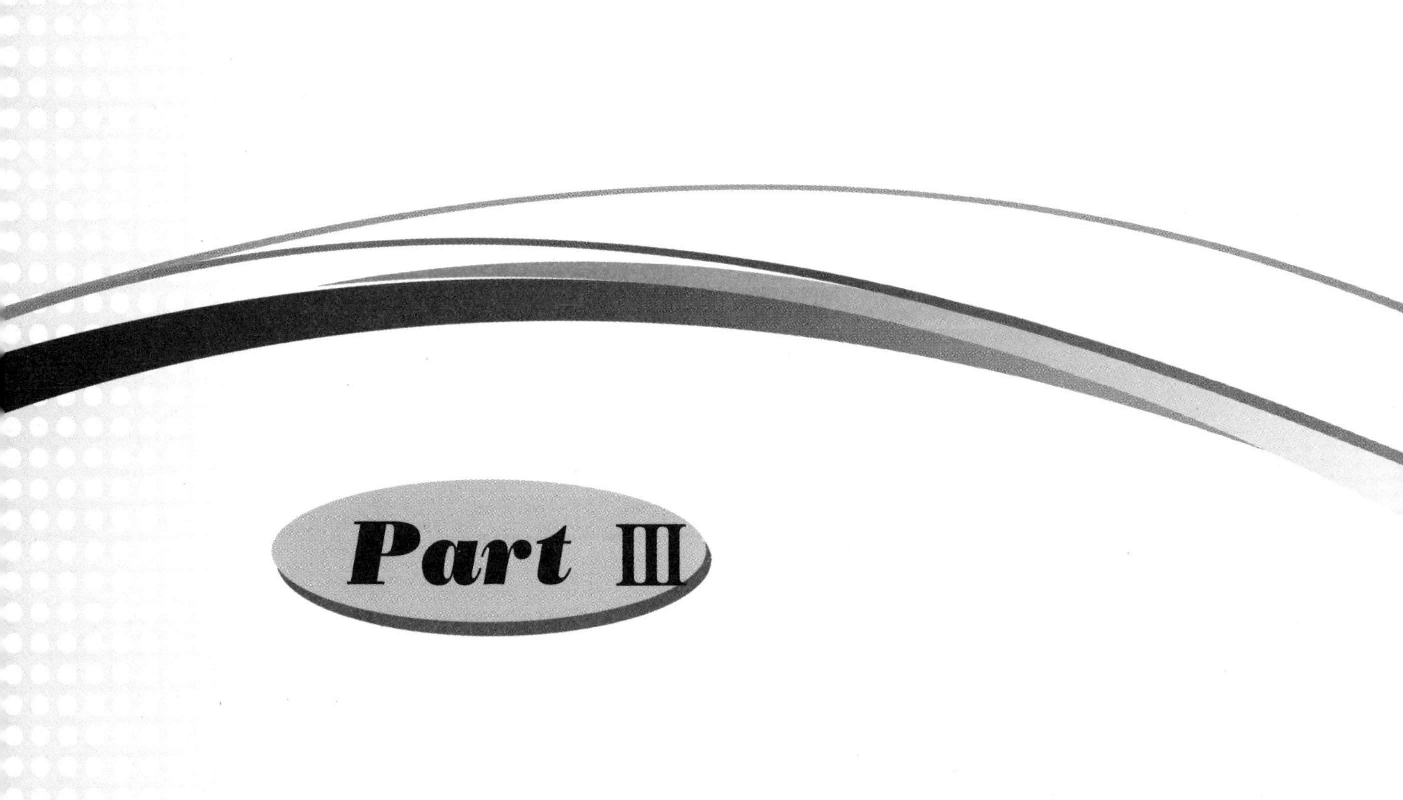

INFORMATION PATHWAYS

Chapter 14

DNA Synthesis

Examination Syllabus

1. The concept of DNA biosynthesis, the basic rules of DNA replication
2. The process of DNA replication
3. Reverse transcription

Ⅰ. Three basic rules of DNA replication

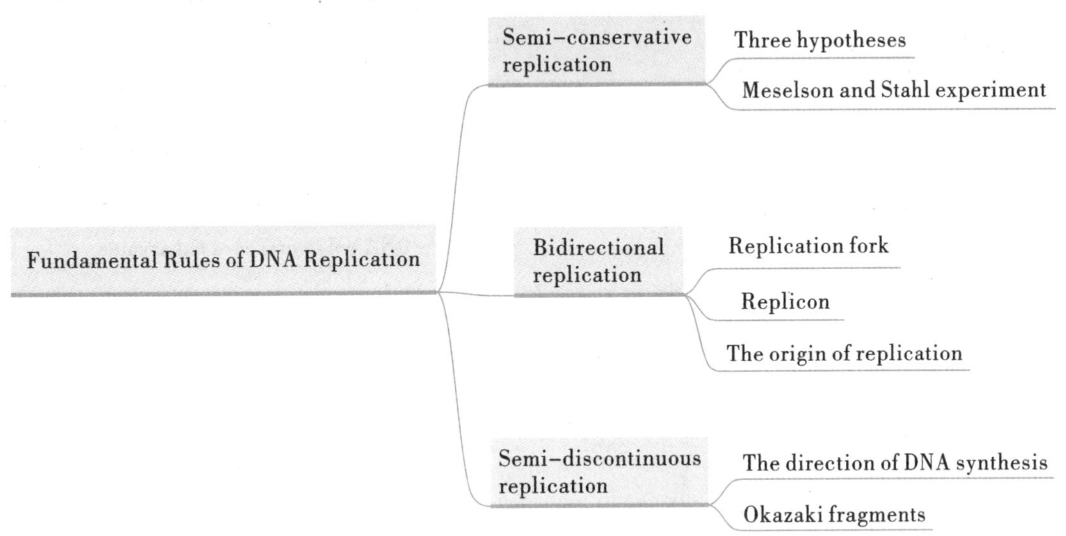

A. Semi-conservative

1. **The semi-reserved replication** means that the parent double-stranded DNA is untwisted into two single strands, each serving as a template, and the DNA double strands with complementary sequences are synthesized by base pairing.

2. **Three hypotheses** which are semiconservative, conservative and dispersive hypothesis had been initially proposed for the method of replication of DNA.

3. **Meselson and Stahl experiment** demonstrated that DNA is semi–conservative replication in natural.

B. Bidirectional

1. **Bidirectional replication** is a process in which DNA replication begins at an origin and usually proceeds with two replication forks moving in opposite directions.

2. **The replication fork** is the place where parental DNA is being unwound to separate the two helix strands to form a "Y" shape structure.

3. **The origin of replication** is a particular DNA sequence in a genome at which replication is initiated.

C. Semi–discontinuous

1. **The leading strand is synthesized continuously**, whose synthesized direction is in the same direction as the fork movement.

2. **The lagging strand** is synthesized discontinuously in the form of short fragments that are later connected covalently, whose synthesized direction is opposite to the direction of the fork movement.

3. **Okazaki fragments** which were first described in 1969 by Reijiand Tuneko Okazaki, are the short stretches produced during replication of lagging strand.

4. **Semidiscontinuous replication** is a mode in which the leading strand is synthesized continuously while the lagging strand is synthesized discontinuously.

Ⅱ. Various enzymes and proteins involved in DNA replication

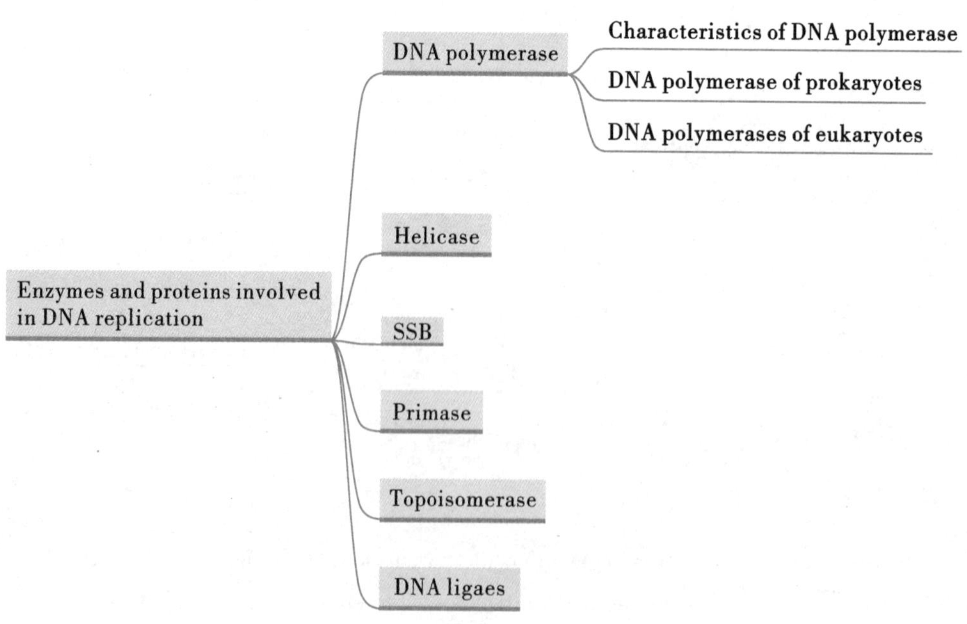

A. DNA polymerase:

An enzyme that can synthesize a new DNA strand on a template strand.

1. Characteristics of DNA polymerase

(1) All DNA polymerases require a template which can guide the polymerization reaction according to the base-pairing rules.

(2) DNA polymerase can extend a DNA chain by adding one nucleotide to a 3′-OH end at a time.

(3) Polymerase which cannot initiate DNA synthesis de novo requires a primer that can provide a free 3′-OH.

2. DNA polymerase of prokaryotes (Table 14.1)

Table 14.1 Comparison of DNA Polymerases of *E. coli*

Type	DNA polymerase		
	I	II	III
Structural gene	pol A	pol B	pol C
Subunits	1	7	10
molecular weight	103,000	88,000	791,500
3′→5′ Exonuclease	Yes	Yes	Yes
5′→3′ Exonuclease	Yes	No	No
Polymerization rate (nucleotides/s)	16-20	40	250-1,000
Processivity	3-200	1,500	500,000
Function	DNA repair; connection of Okazaki fragments	DNA repair	Major replication enzyme

3. DNA polymerases of Eukaryotes (Table 14.2)

Table 14.2 Comparison of DNA Polymerases of Eukaryotes

DNA polymerases	Function	Structure	Proofreading
Pol α	Primase; initiation of DNA replication	tetramer	No
Polβ	DNA repair	monomer	No
Polγ	Mitochondrial DNA replication	dimer	Yes
Pol δ	Replication of the lagging strand	tetramer	Yes
Pol ε	Replication of the leading strand	tetramer	Yes

B. Helicase is an enzyme which can break hydrogen bonds between complementary nucleotide bases, using the energy provided by ATP hydrolysis to separate the strands of a nucleic acid duplex.

C. Single-strand binding protein (SSB) attaches to single-stranded DNA with high affinity and in a sequence-independent manner, thereby preventing the DNA from forming a duplex.

D. Primase can synthesize short segments of RNA that will be used as primers for DNA replication.

E. Topoisomerases are enzymes that participate in the overwinding or underwinding of DNA by cutting the phosphate backbone of either one or both the DNA strands.

F. DNA ligase can catalyze the covalent joining of the adjacent Okazaki fragments into a long DNA strand.

Ⅲ. Prokaryotic DNA replication

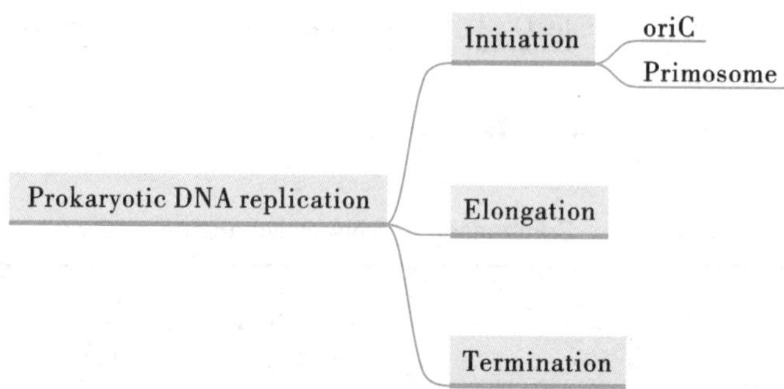

A. The initiation stage

Prokaryotes need to complete the recognition of the origin which is called oriC, the assembly of primosome and synthesis of primer.

1. oriC is composed of 245bp nucleotide sequences which are highly conserved among bacterial replication origins.

2. Primosome is formed which consists of DnaG primase, DnaB helicase, DnaC protein and primer at the replication fork.

B. The elongation stage

DNA polymerase Ⅲ binds to the RNA primer and initiate a nucleophilic attack onto the alpha phosphate of the deoxyribonucleotide, and then 3′→5′phosphodiester bonds are formed.

C. The termination stage

RNA primer need be removed and replaced with DNA sequences by DNA polymerase Ⅰ, and the remaining nick is sealed by DNA ligase.

Ⅳ. Eukaryotic DNA replication

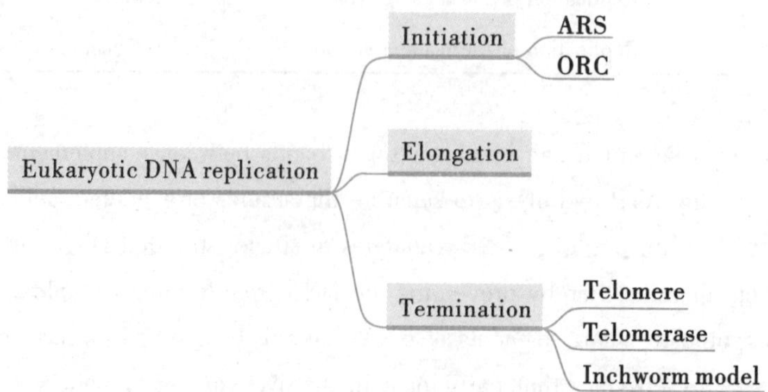

A. The initiation stage

Eukaryotes need to complete the recognition of the origin which is called autonomously replicating sequences(ARS), the assembly of the origin recognition complex (ORC).

1. **ARS** contains four regions (A, B1, B2, and B3), in which element A is highly conserved, consisting of the consensus sequence: 5′-T/ATTTAYRTTTT/A-3′.

2. **The ORC** is a protein complex which is associated with eukaryotic origins through the entire cell cycle.

B. The elongation stage

DNA polymerases α synthesize a short primer in two strands, whereas DNA polymerases ε and δ synthesize the leading strand and lagging strand respectively.

C. The termination stage

The requirement of DNA polymerases to have a primer causes a problem at the ends of linear templates, which is called telomere shortening, which can be resolved by telomere DNA and telomerase.

1. **A telomere** is a region of repetitive nucleotide sequences at each end of a chromosome, which protects the end of the chromosome from deterioration or from fusion with neighboring chromosomes.

2. **Telomerase** is a ribonucleoprotein which has both a polypeptide and an RNA component. The RNA (telomerase RNA, TERC) serves as a template to direct addition of nucleotides, and the polypeptide ((telomerase reverse transcriptase, TERT) acts as a reverse transcriptase to make a DNA copy of a hexanucleotide segment of the RNA.

3. **Inchworm model** solves the problem of telomere shortening, in which the single-stranded DNA structure can recruits telomerase which can add a hexanucleotide to the 3′ end of the parental DNA molecule using TERC as a template, then the enzyme shifts over and synthesizes another hexanucleotide. (Table 14.3)

Table 14.3 Comparisons between prokaryotic and eukaryotic DNA replication

type	Major points	Prokaryotic	Eukaryotic
the differences	location	cytoplasm	nucleus
	origin	1	thousands and tens of thousands
	the length of Okazaki fragment	1000-2000 nt	100-200 nt
	the speed of movement	fast	slow
	the end replication problem	no	yes
	the mechanism of DNA replication	simple	complex
the similarities	basic rules	semi-conservative replication, bidirectional replication, semi-discontinuous replication	
	the direction DNA was synthesized	in the 5' to 3'	
	reaction system	template, dNTP, primer, DNA polymerases	

V. Reverse Transcription and Other Models of DNA Replication

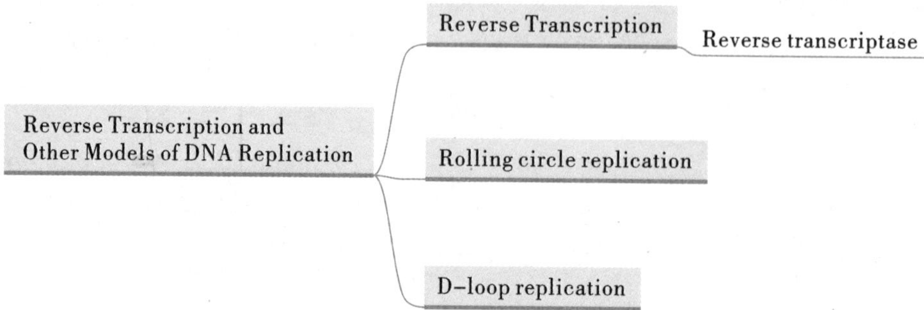

A. Reverse transcription (RT)

Reverse transcription is a process which generates complementary DNA (cDNA) from an RNA template under the action of reverse transcriptase.

Reverse transcriptase is an enzyme which has three enzyme activities:

1. RNA-dependent DNA polymerase activity which synthetizes the first DNA strand using retroviral RNA as atemplate;

2. Ribonuclease H which degrade the RNA strand of an RNA-DNA hybrid;

3. DNA-dependent DNA polymerase activity which synthetizes the second DNA strand using the first DNA strand as a template.

B. Rolling circle replication

Rolling circle replication is a process of unidirectional nucleic acid replication that can rapidly synthesize multiple copies of circular molecules of DNA or RNA, such as plasmids, bacteriophages, and some eukaryotic viruses.

C. D-loop replication

D-loop replication is a process of DNA replication in mitochondria and chloroplast DNA, in which there is a different origin between the leading strand and the lagging strand.

[*Practices*]

[A1 type]

[1] The accepted hypothesis for DNA replication is ____________.

A. conservative theory B. dispersive theory C. semi-conservative theory
D. evolutionary theory E. semi-dispersive theory

[2] DNA replication takes place in which direction? ____________

A. 3′ to 5′ B. 5′ to 3′ C. Randomly
D. Vary from organism to organism E. N terminus to C terminus

[3] Each replication bubbles consists of ____________.

A. 3 replication forks B. 2 replication forks C. 4 replication forks

D. 1 replication forks E. 5 replication forks

[4] DNA replication results in ____________.

A. two completely new DNA molecules

B. two DNA molecules such that each one contains a strand of the original

C. one new DNA molecules 1 old molecule

D. one new molecule of RNA

E. many new DNA molecules in one cell

[5] The Meselson-Stahl experiment demonstrated that DNA replication produces two DNA molecules each composed of ____________.

A. two old strands.

B. two new strands.

C. one old and one new strand.

D. two strands with variable proportions of new and old DNA.

E. a variable number of old and new strands.

[6] The leading strand of DNA is synthesized ____________.

A. discontinuously in a 5′ to 3′ direction

B. continuously in a 5′ to 3′ direction

C. discontinuously in a 3′ to 5′ direction

D. continuously in a 3′ to 5′ direction

E. in both directions

[7] Semiconservative replication of DNA means ____________.

A. only one strand is used as a template.

B. a double-stranded DNA is split into two single-stranded DNAs.

C. only half the genes are copied into the new cells.

D. each DNA strands contains one old strand and one new strand of DNA.

E. two strands with variable proportions of new and old DNA.

[8] DNA replication in bacteria begins at ____________.

A. a single origin and proceeds in one direction.

B. a single origin and proceeds in both directions.

C. two origins and proceeds in both directions.

D. many origins and proceeds in one direction.

E. many origins and proceeds in one direction.

[9] When DNA polymerase is in contact with guanine in the parental strand, what does it add to the growing daughter strand? ____________

A. Phosphate B. Cytosine C. Uracil

D. Guanine E. Hypoxanthine

[10] DNA polymerase can add new nucleotides to the free ____________ of a nucleotide strand.

A. 1′ end B. 3′ end C. 5′ end

D. A and B E. A, B and C

[11] Ligase forms a bond between ____________.

A. two segments of single-stranded DNA.

B. two molecules of double-stranded DNA.

C. two segments of single-stranded RNA.

D. two molecules of double-stranded RNA.

E. one segment of single-stranded DNA and one segment of single-stranded RNA.

[12] Which of the following reactions is required for proofreading during DNA replication by DNA polymerase Ⅲ? ____________

A. 3′→ 5′ exonuclease activity B. 5′→ 3′ exonuclease activity C. 3′→ 5′ endonuclease activity
D. 5′→ 3′ endonuclease activity E. 2′→ 5′ exonuclease activity

[13] Which of the following statements about DNA replication is TRUE? ____________

A. The leading strand is replicated continuously, while the lagging strand is replicated discontinuously
B. The leading strand is replicated discontinuously, while the lagging strand is replicated continuously
C. Both the leading and lagging strands are replicated continuously
D. Both the leading and lagging strands are replicated discontinuously
E. Each of the leading and lagging strands is replicated both continuously and discontinuously

[14] Arrange the following proteins in the proper order in which they participate in DNA replication. ____________

1 = Primase
2 = Helicase
3 = Single-strand binding proteins
4 = DNA polymerase

A. 1,2,3,4 B. 1,3,2,4 C. 2,3,1,4
D. 2,3,4,1 E. 2,4,3,1

[15] The short DNA segments formed on the discontinuously replicated strand are called____________.

A. Kawasaki fragments. B. Nagasaki fragments. C. Miyazaki fragments.
D. Okazaki fragments. E. Suzuki fragments.

[16] For the DNA replication in eukaryotes, the cell cycle consists of____________.

A. G_1, G_2 and M phases B. S, G_2 and M phases C. G_1, S, G_2 and M phases
D. G_2 and M phases E. S and M phases

[17] An important difference between eukaryotic and prokaryotic DNA replication is____________.

A. eukaryotic DNA polymerases are faster
B. more DNA polymerases are found in prokaryotes
C. multiple origins of replication in eukaryotes
D. RNA primers are not required in eukaryotes
E. RNA primers are not required in prokaryotes

[18] Length of Okazakifragments in Eukaryotes is____________.

A. 100-200 nucleotides B. 400-600 nucleotides C. 500-700 nucleotides
D. 1000 -2000 nucleotides E. 3000 -5000 nucleotides

[19] What feature of replication ensures that DNA is copied quickly in eukaryotes? ____________

A. The DNA molecule is replicated from both ends toward the center
B. Nucleotides are pre-assembled before replication begins
C. Replication occurs in thousands of places at one chromosome
D. DNA replicates during both interphase and mitosis
E. The length of Okazaki fragments of eukaryotes is longer than that of prokaryotes

[20] During which phase of the cell cycle is DNA replicated? ____________

A. G_1 phase. B. S phase. C. G_2 phase.

D. M phase. E. G_1 and S phase

[21] Which has primase activity in five polymerases of eukaryotes?____________

A. Pol α B. Pol β C. Pol γ

D. Pol δ E. Pol ε

[22] RNase H is a ____________.

A. exonuclease involved intheremoval of DNA primer

B. endonuclease involved inthe removal of DNA primer

C. exonuclease involved inthe removal of RNA primer

D. endonuclease involved inthe removal of RNA primer

E. endonuclease involved inthe removal of RNA prime and DNA primer

[23] In the rolling circle method of replication____________.

A. the 5′ tail of DNA is nicked

B. RNA is nicked

C. one strand of DNA in the circle is nicked

D. both strands of DNA in the circle are nicked

E. both strands of DNA in the circle are not nicked

[24] Retroviruses have____________.

A. one copy of single-stranded RNA

B. two copies of single-stranded RNA

C. one copy of double-stranded RNA

D. two copies of single-stranded DNA

E. one copy of single-stranded RNA and one copy of single-stranded DNA

[B1 type]

No. 25 ~ 26 share the following suggested answers

A. DNA polymerase Ⅰ B. DNA polymerase Ⅱ C. DNA polymerase Ⅲ

D. DNA polymerase Ⅳ E. DNA polymerase Ⅴ

[25] Klenow fragment is a proteolytic product of ____________in *E. coli.*

[26] The DNA polymerase in bacteria, mainly responsible for DNA synthesis is____________.

No. 27 ~ 30 share the following suggested answers.

A. Helicase B. Topoisomerase C. DNA binding proteins

D. Primase E. DNA polymerase

[27] The enzyme responsible for breaking the hydrogen bonds, and thus separating the DNA strands during DNA synthesis is ____________.

[28] Which of the following prevents supercoiling of the DNA strands ahead of the replication bubble?____________

[29] The enzyme responsible for proofreading base pairing is ____________.

[30] What enzyme associated with DNA replication produces an RNA primer?____________

[X type]

[31] Which of the following statements about the *E. coli* chromosome is correct? Please select all that apply. ____________

A. The *E. coli* chromosome is a single replicon.

B. Replication begins at oriC.

C. Replication can start at any point in the chromosome.

D. A single replication fork moves around the molecule until the chromosome is completely replicated.

E. Replication begins at ARS.

[32] Which of the follouing is True about leading strand ____________?

A. Continuous synthesis towards the replicative fork

B. Single RNA Primer is enough

C. The main enzyme is DNA polymerase Ⅲ

D. The main enzyme is DNA polymerase Ⅰ

E. The main enzyme is DNA polymerase Ⅱ

[33] Which of the following statements is correct? Please select all that apply____________.

A. Telomeres become shorter as cells age

B. Somatic cells have very little telomerase

C. Immortal cancer cell lines always have high levels of telomerase

D. Telomeres become longer as cells age

E. Somatic cells have very much telomerase

Answers

1	2	3	4	5	6	7	8	9	10
C	B	B	B	C	B	D	B	B	B
11	12	13	14	15	16	17	18	19	20
A	A	A	C	D	C	C	A	C	B
21	22	23	24	25	26	27	28	29	30
A	C	C	B	A	C	A	B	E	D
31	32	33							
AB	ABC	ABC							

Brief Explanations for Practices

1. Meselson and Stahl experiment demonstrated DNA is semi-conservative replication.

2. The nascent DNA strand must be synthesized in a 5′ to 3′ direction in the same direction as the replication fork moves.

3. DNA replication begins at an origin and usually proceeds with two replication forks moving in opposite directions.

4. During DNA replication the two strands separate from one another and each strand has a new complementary strand built onto it.

5. During DNA replication the two strands separate from one another and each strand has a new complementary strand built onto it.

6. The leading strand of DNA must be synthesized continuously in a 5′ to 3′ direction in the same direction as the replication fork moves.

7. Semiconservative replication of DNA means that each new DNA molecule is composed of one conserved strand from the original molecule and one new strand.

8. DNA replication in bacteria begins ata single-origin and proceeds in both directions.

9. Guanine is complementary to cytosine, and adenine is complementary to thymine in DNA replica-

tion.

10. All DNA polymerase can only extend a DNA chain by adding one nucleotide to a 3′-OH end at a time.

11. The Okazaki fragments which were left in lagging strand can then be sealed by DNA ligase, which catalyzes the covalent joiningof the adjacent fragments into a long DNA strand.

12. The proofreading activity possessed by many DNA polymerases is an exonuclease activity that degrades mismatched bases that have been wrongly incorporated into the growing chain. This is, therefore, an exonuclease activity (exonucleases digest from the end of a DNA chain) and it operates ′ backward′ from the 3′ growing end, i. e. 3′ – 5′.

13. DNA polymerase III uses one set of its core subunits to synthesize the leading strand continuously, while the other set of core subunits cycles from one Okazaki fragment to the next on the looped lagging strand.

14. First, a helicase binds to the unwound double helix, and unwinds the DNA bidirectionally, creating two potential replication forks. Second, many SSB molecules bind cooperatively to single-stranded DNA, stabilizing the separated DNA strands and preventing renaturation. Then the RNA primer which is a short RNA sequence at the replication origin is synthesized by primase which is activated by helicase and SSB. Finally, DNA polymerase Ⅲ binds to the RNA primer and initiate DNA replication.

15. The lagging strand is synthesized discontinuously as a series of short DNA fragments. These short segments were first described in 1969 by Reiji and Tuneko Okazaki, and are thus called "Okazaki fragments."

16. The cell cycle consists of four distinct phases: G_1 phase, S phase (DNA synthesis), G_2 phase and M phase.

17. DNA replication begins at an origin in Prokaryotes, whereas there are many origins of replication in eukaryotic.

18. Okazaki fragments are between 1000 and 2000 nucleotides long in prokaryotes and are roughly 100 to 200 nucleotides long in eukaryotes.

19. There are many origins of replication inthe eukaryotic linear chromosome, which enable it to replicate DNA quickly.

20. Eukaryotic DNA replication takes place during the S phase because it is strictly controlled within the context of the cell cycle.

21. DNA polymerase α has no proofreading 3′→5′ exonuclease activity, is believed to function only in the synthesis of short primers because it is unsuitable for high-fidelity DNA replication.

22. RNase H is responsible for removing RNA primer from the newly synthesized strand.

23. Rolling circle replication is initiated by an initiator protein which can nick one strand of the double -stranded.

24. A retrovirus which targets a host cell has two single-stranded positive-sense RNA strand(+ssRNA) with a DNA intermediate.

25. DNA polymerase I can be enzymatically cleaved into two fragments by the protease subtilisin inwhich the large protein fragment is called Klenow fragment.

26. DNA polymerase III is the principal replication enzyme in the prokaryotes.

27. DNA double helix must be separated by helicases using the energy from ATP hydrolysis, which is a process characterized by the breaking of hydrogen bonds between complementary nucleotide bases

28. During DNA replication, DNA becomes overwound ahead of a replication fork due to the inter-

twined nature of its double-helical structure. In order to prevent supercoiling of the DNA strands ahead of the replication bubble caused by the double helix, topoisomerases bind to DNA and cut the phosphate backbone of either one or both the DNA strands.

29. All DNA polymerases have 3′→5′ exonuclease activity which is responsible for proofreading base pairing.

30. The RNA primers which initiate synthesis of a chain of DNA de novo has been synthesized by primase.

31. DNA replication begins at an origin which is called oriC in *E. coli* and usually proceeds with two replication forks moving in opposite directions.

32. Leading strand begins with the synthesis of a short RNA primer at the replication origin. Leading strand synthesis proceeds continuously, keeping pace with the unwinding of DNA at the replication fork because the template strand of DNA runs in the 3′ to 5′ direction toward the fork. Inaddition, DNA polymerase III is the principal replication enzyme in the prokaryotes.

33. Telomerase is the enzyme that synthesizes chromosome ends, or telomeres. Telomerase occurs in germ cells which are continuously dividing. However, most cell lines need only to go through a certain number of cell divisions in their lifetime and the stock of telomerase is used up as they age. Most differentiated somatic cells have little or no telomerase and their telomeres get shorter as they divide. Many (but not all) cancer cell lines have high levels of telomerase, which allows them to carry on dividing unchecked.

(***Li Chongqi***)

Chapter 15

DNA Damage and Repair

Examination Syllabus

1. **DNA damage**: internal factors and external factors that cause DNA damage; types of DNA damage.
2. **DNA repair**: direct repair, excision repair, recombinational repair and SOS repair.
3. **Significance of DNA damage and repair**: double effect and the association with diseases.

Ⅰ. DNA damage

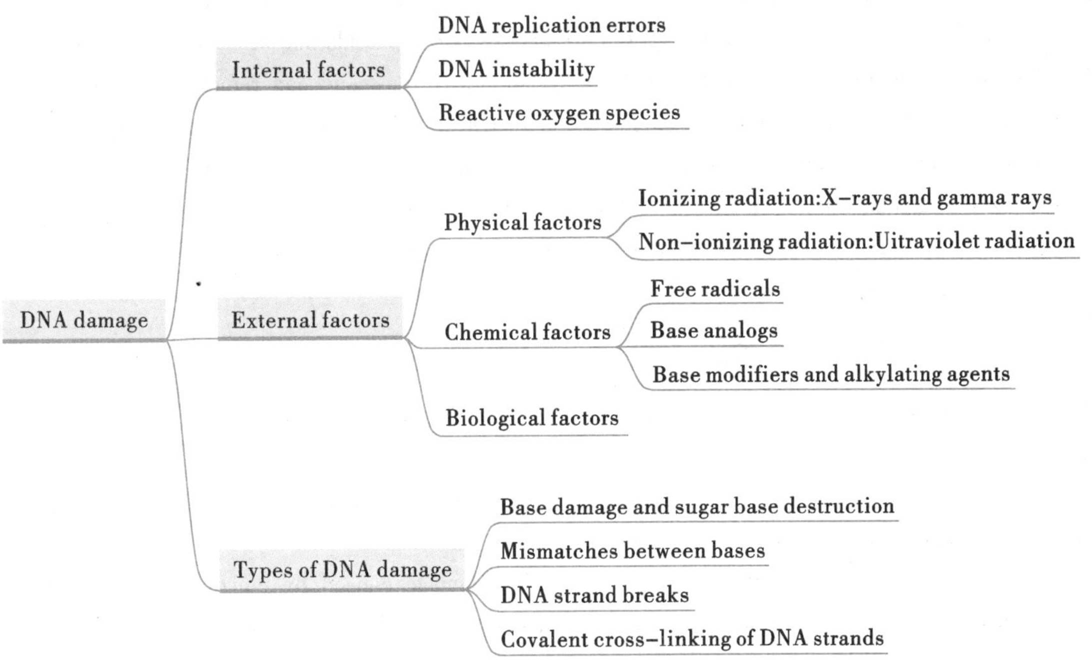

DNA damage is an alteration in the chemical structure of DNA, such as a break in a strand of DNA, a base deletion from the backbone of DNA, or a chemically changed base. DNA damage can occur naturally or via environmental factors and cause DNA mutation.

DNA damage is caused by both internal and external factors and results in many types of DNA damage.

A. Internal factors

1. **DNA replication errors** lead to base mismatches and missing or inserted fragments.

2. **The unstable nature of the DNA** molecule is the major causative factor of DNA damage caused by heating, pH changces or spontaneous deamination.

3. **Reactive oxygen species (ROS)** production during metabolism negatively affects DNA structure and function.

B. External factors

1. **Physical factors** act directly on DNA through breaking chemical bonds and destroying molecular structures. The most common physical factor is electromagnetic radiation divided into:

(1) **Ionizing radiation**, including alpha particles, beta particles, X-rays, and gamma rays, which directly destroying the molecular structure of DNA through breaking its chemical bonds, and also enhance the production of free radicals in cells, exerting indirect destructive effects to DNA integrity.

(2) **Non-ionizing radiation** is mainly UV, which enables DNA forming a **thymine dimer structure (TT)**.

2. **Chemical factors**

(1) **Free radicals** are atoms, groups, or molecules that can exist independently with unpaired electrons in their outer orbitals. 'OH' radicals have strong oxidizing properties, while 'H' have strong reducing properties.

(2) **Base analogs** are synthetic compounds that resemble the normal base structure of DNA and are commonly used as DNA mutagens or as anti-cancer drugs. 5-BU is an analogue of thymine.

(3) **Base modifiers and alkylating agents** are classes of compounds that alter DNA structure by modifying the chemical groups of nucleotides in the DNA strand. Nitrous acid can remove the amino group of adenines. Hypoxanthine cannot pair with thymine and instead pairs with cytosine. Nitrogen mustard, sulfur mustard and diethylnitrosamine can cause the alkylation of nitrogen atoms in DNA bases. Ethidium bromide and acridine orange can directly insert into DNA base pairs.

3. **Biological factors** mainly refer to viruses and fungi, whose protein or metabolic products can damage DNA, such as aflatoxin.

C. Types of DNA Damage

1. **Base damage and sugar base destruction** include deamination, formation of pyrimidine dimers, etc.

2. **Mismatches** between bases are caused by incorporation of base analogs and the action of base modifying agents, resulting in incorrect base-pairing.

3. **DNA strand breaks** are the **main form** of DNA damage caused by ionizing radiation. They are results from the cleavage of phosphodiester bonds, the destruction of deoxypentose, and the detachment of bases. **AP site (apurinic/apyrimidinic site)** located in DNA that has neither a purine or pyrimidine base can also cause DNA strand breaks.

4. **Covalent cross-linking** includes DNA intrastrand cross-linking, DNA interstrand cross-linking and DNA-protein cross-linking.

Ⅱ. DNA repair

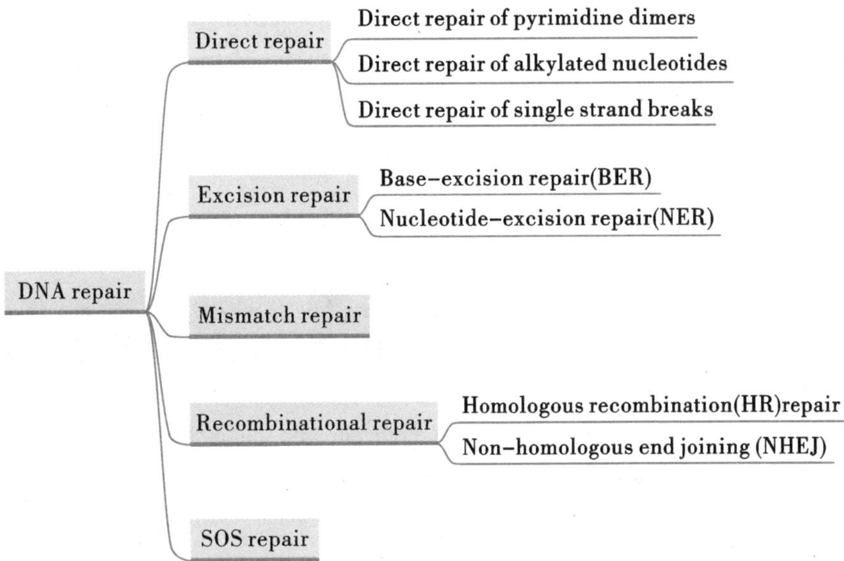

DNA repair is a collection of processes by which a cell identifies and corrects damage to the DNA molecules that encode its genome, and restore the DNA to its original state.

A. Direct repair

Direct repair does not remove a base or nucleotide, nevertheless, the repair enzymes act directly on damaged DNA and restore it to its original state. It is the simplest form of DNA repair and the most energy efficient method.

1. **Direct repair of pyrimidine dimers**: DNA photolyase directly reverses pyrimidine dimers dependent on energy derived from absorbed light.

2. **Direct repair of Alkylated nucleotides**: O^6-Alkylguanine-DNA alkyltransferase directly reverse the methylation of guanine bases.

3. **Direct repair of single-strand breaks**: DNA ligases directly seal the nick of 5′-phosphate group and the 3′-hydroxyl group of another nucleotide.

B. Excision repair

Excision repair is **the most common form** of DNA repair. It corrects a DNA lesion by removal of the damaged DNA segment and its replacement with a new segment using the undamaged strand as a template.

1. **Base excision repair (BER)** removes an altered base by a DNA glycosylase, followed by excision of the resulting sugar phosphate. The processes include DNA glycosylases recognition and hydrolysis, AP endonuclease incision and synthesis and ligation.

2. **Nucleotide excision repair (NER)** removes a short single-stranded DNA segment that contains the lesion.

(1) NER is the main repair pathway in mammals to remove bulky DNA lesions.

(2) In *E. coli*, nucleotide-excision repair relies on the ABC exonuclease composed of UvrA, UvrB and UvrC subunits.

(3) Deficiencies in NER are associated with the serious skin cancer-prone inherited disorder xeroderma pigmentosum (XP).

C. Mismatch repair (MMR)

Mismatch repair corrects mismatched bases in the newly replicated DNA strand. In *E. coli*, the transient undermethylation of the GATC sequences makes the newly synthesized strands susceptible to mismatch repair enzymes.

D. Recombination repair(RR)

Recombination repair fixes double-stranded breaks (DSBs) which require the presence of an identical or nearly identical sequence to be used as a template.

1. **Homologous recombination(HR)** occurs during the late S and G_2 phases of the cell cycle when sister chromatids are present to serve as templates. It is a high fidelity non-mutagenic process and RecA is the central core of involved proteins.

2. **Non-homologous end joining (NHEJ)** is a mutagenical nonhomologous process which directly ligates the break ends. NHEJ can rapidly joins the ends of double-stranded breaks, generating deletions or insertions.

E. SOS repair

It is a state of error-prone repair system, and is activated by bacteria that have been exposed to heavy doses of DNA-damaging agents. The SOS repair response is triggered by the interaction of RecA and the LexA repressor.

Ⅲ. Significance of DNA damage and repair

A. Double effect of DNA damage

1. One is the permanently change in the DNA, i. e. , mutation.

2. The other one is making DNA to be unused as a template for replication and transcription, leading to dysfunction or even cell death.

B. DNA repair defects

Repairde fects cause cancers, genetic diseases (xeroderma pigmentosum, ataxia telangiectasia, Cockayne syndrome, etc.), immune disease and accelerated aging.

[*Practices*]

[A1 type]

[1] The significance of mutation is that________.

A. mutation is the molecular foundation of DNA evolvement

B. all the mutations are negative

C. some mutations are fatal

D. some mutations cause diseases

E. none above is right

[2] Exposure of DNA to ultraviolet radiation can lead to the formation of____________.

A. adenine dimers B. thymine dimers C. guanine dimers

D. uracil dimmers E. cytosine dimmers

[3]____________dissociates mostly the pyridine dimers.

A. Photolyase B. Excision repairing C. Mismatch repairing

D. SOS repairing E. Realtime proofreading

[4]____________forms an apyrimidinic or apurinic site.

A. Homologous recombination B. Mismatch repair C. Base excision repair

D. Direct repair E. Nucleotide excision repair

[5]____________ is associated with the defect of DNA repairing system?

A. Xanthurenic aciduria B. Porphyria C. Xeroderma pigmentosum

D. Gout E. Icterus

[6] In base excision repair, the lesion is removed by ____________.

A. DNA glycosylase B. Excisionase C. Transposase

D. DNA polymerase E. Integrase

[7] In DNA repair, ____________ seals up broken parts of the DNA backbone.

A. transposase B. DNA polymerase C. nuclease

D. homologase E. DNA ligase

[8] Which deamination would lead to a mutation in a resulting protein if not repaired__________?

A. T to U B. C to U C. G to A

D. A to G E. U to C

[9] A 1-year-old boy is brought to the hospital due to excessive sensitivity to the sun. The pediatrician suspects a genetic disorder in the processes of____________.

A. DNA replication B. transcription C. nucleotide excision repair

D. base excision repair E. translation

[B1 type]

No. 10 ~ 11 share the following suggested answers

A. Homologous recombination B. Base excision repair C. Nucleotide excision repair

D. Mismatch repair E. Non-homologous recombination

[10] Which of the following process occurs in regions where no large -scale sequence similarity is apparent?____________

[11] Which of the following process occurs between DNA molecules of very similar sequences?____________

[X type]

[12] Mutations arise through spontaneous____________.

A. in DNA replication B. in meiotic recombination C. in RNA synthesis

D. as a consequence of the damaging effects of chemical agents on the DNA

E. as a consequence of the damaging effects of physical factors on the DNA

[13] DNA damage can be caused by____________.

A. ultraviolet radiation B. ionizing radiation C. alkylating agents

D. DNA polymerase E. aflatoxin

Answers

1	2	3	4	5	6	7	8	9	10
A	B	A	C	C	A	E	B	C	E
11	12	13							
A	ABDE	ABCDE							

Brief Explanations for Practices

4. Base excision repair remove damaged DNA bases by DNA N-glycosylases, generating abasic sites.

8. In DNA cytosine spontaneously deaminates to form uracil. This error is repaired by the uracil-DNA glycosylase system. Neither thymine nor uracil contains an amino group to deaminate. Adenine deamination forms base hypoxanthine, and guanine deamination lead to xanthine production. Only the deamination of cytosine and conversion to uridine can be expressed as a protein.

9. The boy has the disease of xeroderma pigmentosum, caused by exposure to UV light forming thymine dimmers. Nucleotide excision repair removes entire nucleotides the damaged DNA. However, base excision repair only remove a single base. Also, this disorder is not due to alterations in transcription, DNA replication, or translation.

[*Clinical Case*]

A 12-year-old boy had dark brown patches on his face, neck, limbs, and trunk from the age of 2. His skin was dry, and the neck and exposed limbs appeared like a rash, no itching. The symptoms were mild in winter and severe in summer. In the past two years, the plaques on his body had gradually increased, and the color had become darker, some of which merged into small pieces of dark brown keratinized papules and maculopapular rash with scattered telangiectasia. His eyes were lightly photophobic. There was no such person in his family, and his parents were healthy.

Initial diagnosis: Xeroderma Pigmentosum (XP)

Questions and discussion:

1. What kind of laboratory examination is needed to diagnose XP?
2. Describe the molecular mechanism of XP.

(***Fei Xiaowen, Wang Kai***)

Chapter 16

RNA Synthesis

Examination Syllabus

1. The concept of RNA biosynthesis
2. RNA synthesis process; The composition of the transcription system
3. Post-transcriptional processing

Ⅰ. Introductiont of RNA biosynthesis

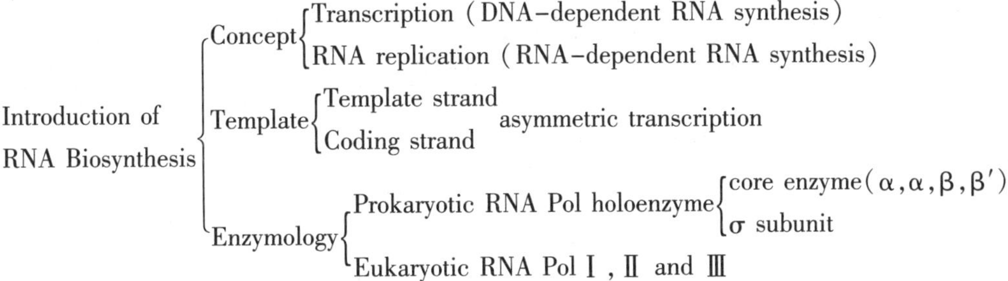

A. The concept of RNA biosynthesis

Two types: transcription (DNA-dependent RNA synthesis) and RNA replication (RNA-dependent RNA synthesis)

1. **RNA replication**: many viruses only contain RNA as genetic material, and then use an RNA-dependent RNA polymerase to replicate their genome.

2. **Transcription** is the process of producing a complementary RNA copy of a part of DNA.

B. Template of transcription

1. **Template**: During transcription, the only portion of DNA double helix is unwound and only one of the DNA strands functions as a template for RNA synthesis.

2. **Template strand** and **coding strand**: Transcription generally proceeds from the same DNA strand for a given gene, which is called the **template strand**. The left non-template DNA strand is called the **coding strand**.

3. The template strand for different genes is not identical. This selectivity of the template is called **asymmetric transcription.**

C. Enzymology of transcription

1. **RNA polymerase in Prokaryotes**: A single DNA-dependent RNA polymerase creates all types of RNA in prokaryotic organisms.

(1) **The RNA polymerase holoenzyme** has six subunits ($\alpha,\alpha,\beta,\beta',\sigma$, and ω).

(2) **The σ subunit** can dissociate from the holoenzyme after transcription has been initiated.

(3) **The RNA polymerase core enzyme** contains $\alpha,\alpha,\beta,\beta',\omega$ subunits, guide formation of phosphodiester bonds according to Watson-Crick base pair rule after transcription initiation.

(4) **Rifampicin** is a prokaryotic RNA pol inhibitor, which specificly bind to β subunit.

2. **RNA polymerases in Eukaryotes**

(1) **Three DNA-dependent RNA polymerases** (RNA pol Ⅰ, Ⅱ and Ⅲ) exist in eukaryotes, each of which contains approximately 10 subunits.

(2) **Transcription factors** are indispensable for recruitment of eukaryotic RNA polymerases to the promoter.

(3) Each eukaryotic polymerase is responsible for transcription of different classes of genes and needs a distinct set of transcription factors to guide it to the DNA template (Table 16.1).

Table 16.1 RNA polymerases in Eukaryotes

	RNA Pol Ⅰ	RNA Pol Ⅱ	RNA Pol Ⅲ
Transcript	45S rRNA	Pre-mRNA, hnRNA, lncRNA, piRNA, miRNA	tRNA, 5SrRNA, snRNA
Reaction to α-amanitin	Insensitivity	Sensitivity	Sensitivity in high
Location	Nucleolus	Nuclear	Nucleolus

(4) **α-amanitin** is a specific RNA pol Ⅱ inhibitor, which is a peptide toxin derived from the mushroom amanita phalloides. α-amanitin blocks the translocation of RNA polymerase during phosphodiester bond formation.

[*Practices*]

[A1 type]

[1] Which strand is used transcription template in DNA duplex?______

A. Template strand B. Coding strand C. Leading strand
D. Lagging strand E. Okazaki fragments

[2] Which one is correct about transcription?______

A. Requires all four dNTPs
B. Uses RNA as a template
C. Direction of synthesis is 3′→ 5′
D. Does not require a primer
E. The entire genome is transcribed

[3] The primary role of RNA polymerase Ⅱ is to synthesize ______

A. 45s rRNA B. hnRNA C. 5s rRNA
D. tRNA E. 5.8s rRNA

[4]Which one is the specific differential inhibitor of the eukaryotic nuclear RNA polymerases?____

A. Neomycin B. Puromycin C. α- Amanitin

D. Streptomycin E. Erythromycin

[5]Which is correct one about asymmetric transcription?______

A. Transcription after bidirectional replication.

B. Transcription on the same template strand appears both 5′→3′ and 3′→5′ directions.

C. Template strand isn't always on the same DNA strand.

D. After transcription and translation, the polypeptides include asymmetric "C" atom.

E. Transcription without a regular pattern.

[6]The DNA dependent RNA polymerase consists of several subunits, the core enzyme consists of______

A. $\alpha_2\beta\beta'\omega$ B. $\alpha\alpha\beta'\omega$ C. $\alpha\alpha\beta'\sigma\omega$

D. $\alpha\sigma\beta\omega$ E. $\alpha\beta\beta'\omega$

[7]Rifampicin is the specific inhibitor of RNA polymerase, the binding site is______

A. β subunit. B. α subunit. C. β′ subunit.

D. σ subunit. E. ω subunit.

[8]Which one can recognize promoter?______

A. α subunit of RNA polymerase

B. β subunit of RNA polymerase

C. β′ subunit of RNA polymerase

D. σ subunit of RNA polymerase

E. DnaB protein

Ⅱ. The process of RNA synthesis process

A. Mechanism of Transcription

1. Transcription occurs in **three distinct steps**: initiation, elongation, and termination.

2. **Transcription initiation site**: the nucleotide on the DNA template strand from which the first 5′ RNA nucleotide is transcribed, the +1 nucleotide.

3. **Promoter**: the section of DNA that controls the initiation of transcription or the specific DNA region RNA polymerase attach to.

Two promoter consensus sequences: -10 and -35 regions upstream of the initiation site. In *E. coli*, the consensus sequences: -10 region: TATAAT(**Pribnow box**)

-35 region: TTGACA.

B. Transcription in Prokaryotes

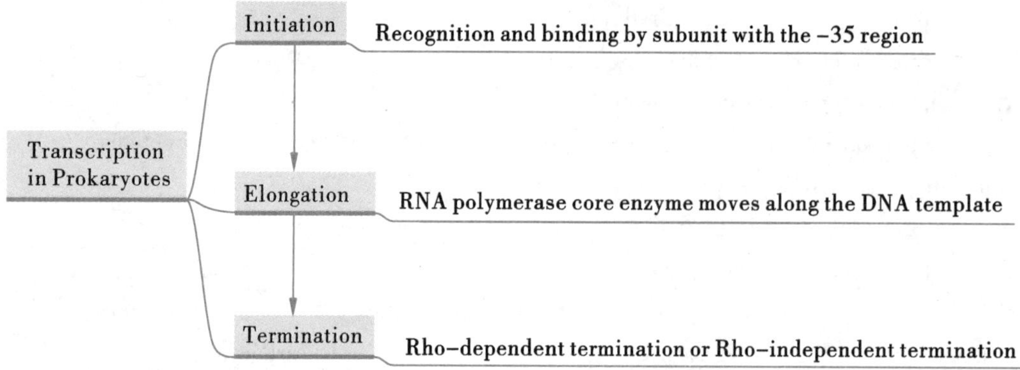

1. Transcription initiation

(1) **Closed complex formation**: RNA pol holoenzyme recognizes and binds to the promoter -35 region by σ subunit to form a closed complex and initiate transcription.

(2) **Transition from closed complex to open complex**: RNA pol holoenzym unwinds approximately 17 bp of DNA to form an open complex.

(3) **The first phosphodiester bond formation**: RNA pol then form the first phosphodiester bond between two base-paired ribonucleotides to **inititate** the new chain, in the **absence of a primer**.

(4) **Abortive initiation and promoter escape**: The transcription initiation phase ends with the production of abortive transcripts, which are short RNA transcript of approximately 10 nucleotides.

2. Transcription elongation

(1) Transcription elongation begins with the **release of the σ factor**, the RNA pol core enzyme moves along the DNA template. Ribonucleotides are added to the nascent chain according to the base-pairing rules with C pairing with G and **U pairing with A** of the DNA and A pairing with T of the DNA.

(2) As elongation proceeds, the DNA is incessantly **uncoiled** ahead of the core enzyme and **recoiled** behind it through **transcription bubble**.

(3) RNA pol **lacks proofreading function** and is more error-prone than DNA pol.

3. Transcription termination

(1) **Rho-dependent termination** is caused by collision of rho protein with RNA polymerase

● During elongation, rho protein traces after RNA pol core enzyme on the growing mRNA strand. Near the end of the gene, a stretch of **G nucleotides** on the DNA template impede the movement of RNA polymerase, resulting in a **collision of rho protein with the stalled RNA polymerase**.

● The rho protein has ATP-dependent RNA-DNA **helicase** activity that promotes the dissociation of RNA and DNA hybrids and subsequent release of the new RNA product.

(2) **Rho-independent termination** is caused by specific sequences in the DNA template strand (Figure 16.1).

● Near the end of the gene RNA pol meets with a region rich in **C-G nucleotides**, therefore mRNA strand folds back on itself, while the complementary C-G nucleotides coil together to form a stable **hairpin**.

● No sooner had RNA pol been stalled by the hairpin structure than it starts to transcribe a region rich in A-T nucleotides, forming a weak **U-A interaction** between the RNA transcript and the template DNA.

● Coupled with the impeded polymerase, this **weak interaction** induces the removal of polymerase core enzyme and release of the nascent RNA transcript.

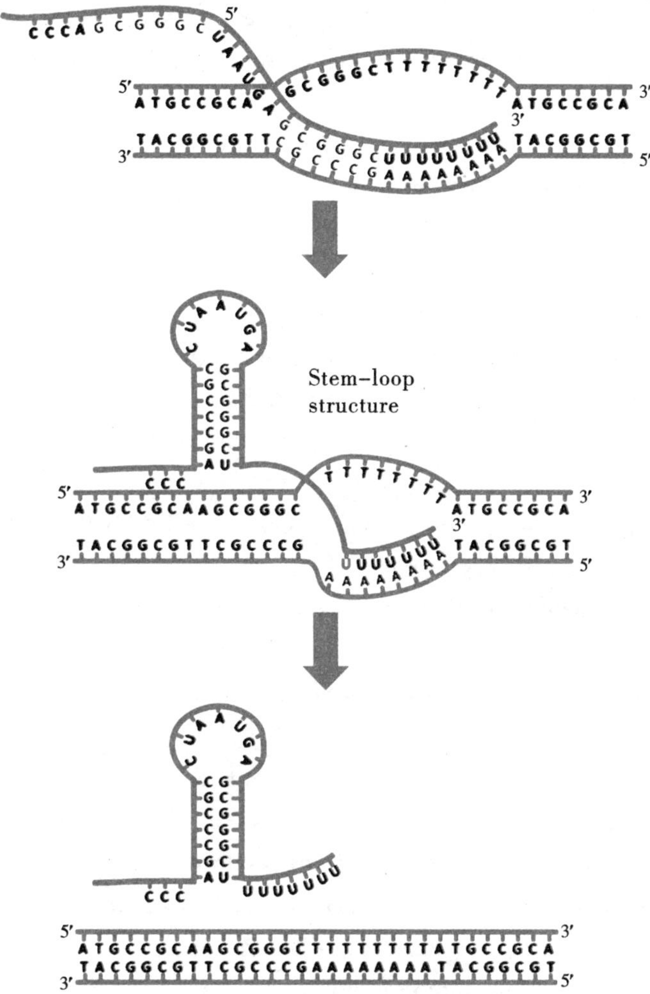

Figure 16.1 Rho–independent termination

4. Concurrent transcription and translation in prokaryotes

By the time termination takes place, the prokaryotic transcript would already have been utilized to instruct synthesis of the encoded protein because there is no membranous compartmentalization, causing these processes can occur simultaneously in prokaryotes.

C. Transcription in eukaryotes

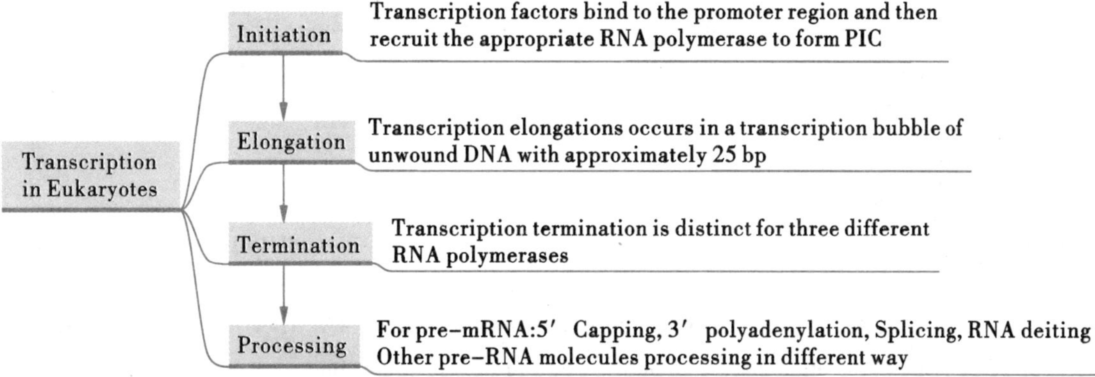

Eukaryotic transcription is carried out in the nucleus of the cell by one of three distinct RNA polymerases, depending on the RNA being produced. Unlike concurrent transcription and translation in prokaryotic organism, **transcription and translation are completely separated** by a nuclear membrane in eukaryotic cells.

1. **Transcription initiation**: Unlike the prokaryotic RNA polymerase that can interact with a DNA template on its own, **transcription factors** are required to first bind to the promoter region and then assist in the recruitment of the appropriate RNA polymerase in eukaryotic cells. The initiation process is more complicated in eukaryotic transcription.

(1) Trans-acting factors and Transcription factors

● The proteins which can directly or indirectly recognize and bind to the upstream region of the transcription start site are collectivelycalled **trans-acting factors**.

● Of these trans-acting factors, some can directly or indirectly bind to RNA polymerases, named as **transcription factors (TFs)**.

● **General transcription factor**s corresponding to eukaryotic RNA pol Ⅰ, RNA pol Ⅱ and RNA pol Ⅲ are called **TF Ⅰ, TF Ⅱ and TF Ⅲ**, respectively.

(2) **Cis-regulatory elements and Core promoter elements**

Core promoter element:

● **TATA box**, approximately 30bp upstream of the transcription start site. Only approximately 10% ~ 15% of mammalian genes

● **GC and CAAT box**es, more distally upstream region of the transcription start site.

● **enhancer** or **silencer**: a short region of DNA that can affect transcription via interaction with proteins (activators or suppressor); locate **upstream or downstream or even within introns** of the regulated gene; doesn't need to exist near the transcription start site to regulate transcription.

These consensus DNA elements are collectively called **cis-acting elements**.

(3) **Transcription pre-initiation complex (PIC)**

● The completed assembly of general transcription factors and RNA polymerase bind to specific site in the promoter, forming a **transcription pre-initiation complex (PIC)**.

● Six general transcription factors (TF Ⅱ A, B, D, E, F, and H) and RNA pol Ⅱ form a minimal PIC (Figure 16.2). Only very low or basal rate of transcription is achieved by PIC. Other proteins, such as activators and repressors, are responsible for regulating transcription rate.

2. **Transcription elongation**

(1) Transcriptional elongation is basically similar in eukaryotes and prokaryotes. However, the eukaryotic transcription machinery needs to **move histones out** of the way every time it meets with a nucleosome.

(2) Transcription elongations occur in a transcription bubble of unwound DNA with approximately 25 bp, where the RNA polymerase uses one strand of DNA as a template to catalyze the synthesis of a nascent RNA strand in the 5′ to 3′ direction.

3. **Transcription termination**

(1) **Transcription termination by RNA polymerase Ⅰ**

● RNA polymerase Ⅰ terminates transcription in response to a **specific termination sequence** in DNA being transcribed.

● A termination protein called **transcription termination factor** for RNA polymerase Ⅰ (**TTF-1**) binds the DNA at its recognition sequence and hinders further transcription.

(2) **Transcription termination by RNA polymerase Ⅱ**

● lack any specific termination sequences. RNA poly Ⅱ terminates transcription at **random locations**

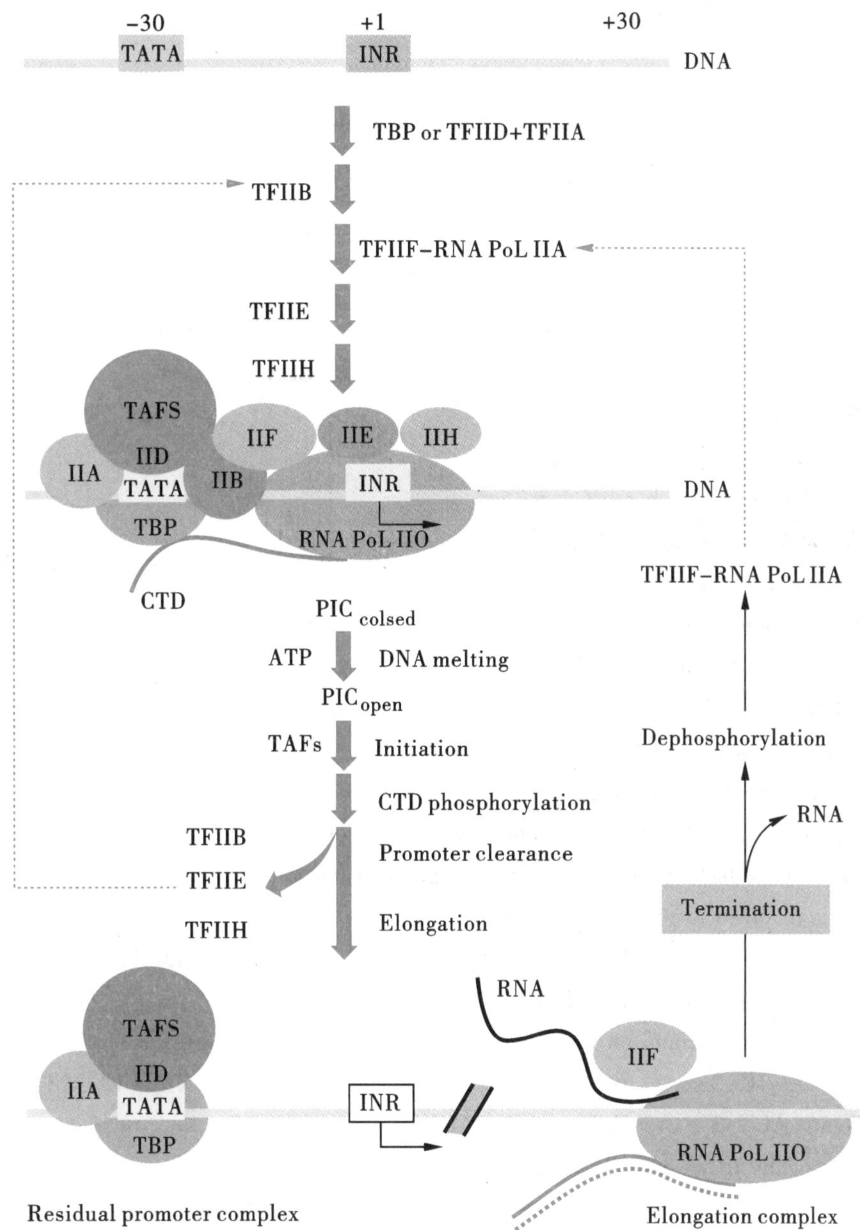

Figure 16.2 Transcription factors and PIC

past the actual end of the gene being transcribed.

●The pre-mRNA transcript is cleaved by a complex containing endonuclease between an **AAUAAA** consensus sequence and a **GU-rich element**. The sequence-specified **cleavage site** is considered as the "end" of the gene.

(3) **Transcription termination by RNA polymerase Ⅲ**

●RNA pol Ⅲ terminates transcription without entirely understood mechanism to date, possibly in response to specific termination signal in the nascent RNA.

●The RNA products transcribed by RNA pol Ⅲ have a **short stretch of 4 ~ 7 U** at their 3′ end, which somehow drives RNA pol Ⅲ releases the newly-synthesized RNA and dissociate from the template DNA strand.

[Practices]

[A1 type]

[9]What is the consensus sequence located about −10 nucleotides upstream from the start site in prokaryotes?______

A. TTAATT B. CAAT box C. GC box
D. Pribnow box E. UAG

[10]What structure can RNA polymerase Ⅱ recognize?______

A. Introns B. Exons C. Promoter
D. Operon E. Enhancer

[11]What is the promoter?______

A. The first exon of the split gene
B. The DNA sequence which repressors bind to
C. Specific DNA sequences which an RNA polymerase binds to
D. The first intron of the split gene
E. The DNA sequence which is transcribed first

[12]Which one is correct about the TATA box?______

A. Is translation start site
B. Within the first structural gene of the operon
C. Belongs to trans–acting factors
D. Located about 30 nucleotides upstream from the transcription start site in eukaryotes
E. Can exert their effects in an orientation–independent fashion

[13]Which one is involved in prokaryotic transcription termination?______

A. Signal peptide B. Rho factor (ρ) C. Release factor
D. β subunit E. σ factor

[14]The sequence of sense strand on DNA is 5′GTCAACTAG3′, the transcription products are____

A. 5′–TGATCAGTC–3′. B. 5′–GUCAACUAG–3′. C. 5′–CAGUUGAUC–3′.
D. 5′–CTGACTAGT–3′. E. 5′–GACCUAGUU–3′

[15]The function of the tRNA 3′ end is to______

A. Form local double strands.
B. Supply energy
C. —OH can bind amino acid
D. Recognize the codon on mRNA
E. Spliced components

Ⅲ. Post-transcriptional processing

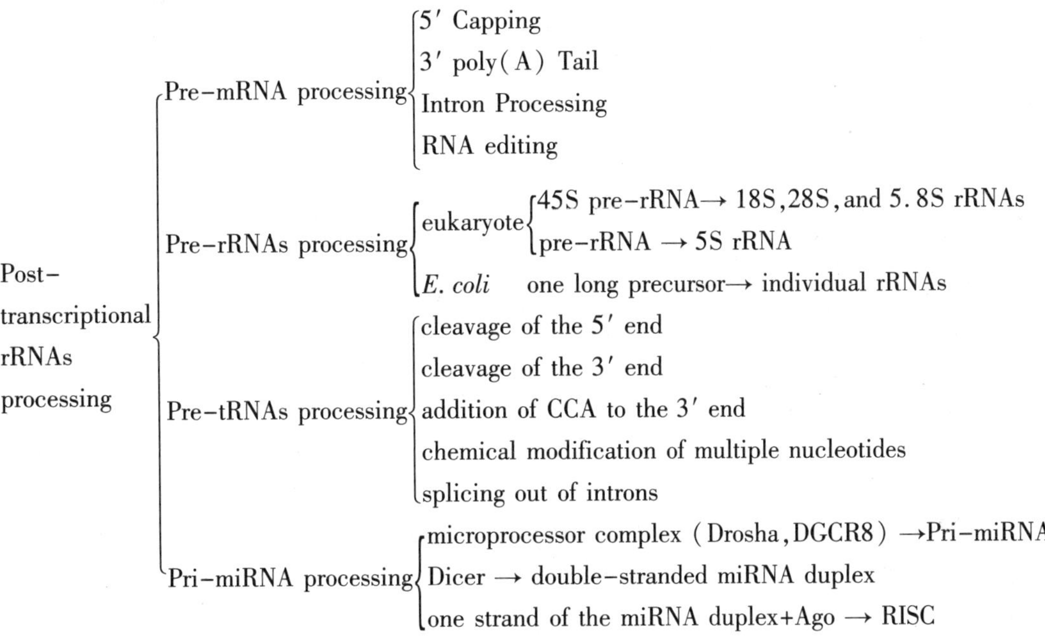

A. Pre-mRNA processing

1. 5′ Capping

(1) Capping of the pre-mRNA implicates addition of 7-methylguanosine (m^7G) to the 5′ end of the pre-mRNA by a 5′-5′ phosphate linkage while elongation is still in progress.

(2) The cap structure protects the newborn mRNA from degradation and aids in ribosome binding during translation initiation.

2. 3′ poly(A) Tail

(1) A string of approximately **200 A nucleotides**, termed as poly(A) tail, is added to 3′ end of the pre-mRNA once the pre-mRNA is cleaved between **AAUAAA and GU-rich** sequences.

(2) The poly(A) tail protects the mRNA from degradation, assists in the exportation of the mature mRNA from the nucleus, and is involved in the interaction with proteins involved in translation initiation.

3. Intron Processing

(1) Exons correspond to protein-coding sequences, while **introns** correspond to intervening sequences. Intron sequences do not encode functional proteins but may be implicated in gene regulation.

(2) **Pre-mRNA splicing**:

● Pre-mRNA splicing is conducted in a **lariat form by spliceosomes**, which are complexes containing proteins (small nuclear ribonucleoproteins, **snRNPs**) and RNA molecules (small nuclear RNAs, snRNAs).

● Pre-mRNA splicing takes place by a sequence-specific mechanism (GU at the 5′ splice site and AG at the 3′ splice site) that guarantees precise removal of introns and accurate reconnection of exons (Figure 16.3).

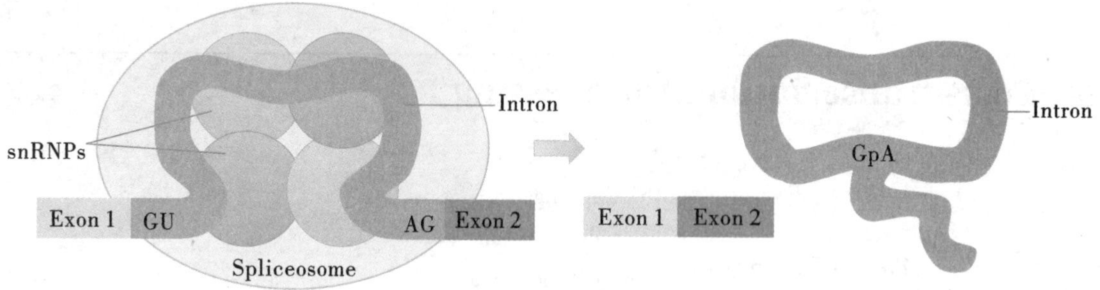

Figure 16.3 Intron Processing

4. **RNA editing**: Coding region can be changed at the mRNA level by RNA editing. In mRNA editing, a base of the mRNA is altered enzymatically.

B. Pre-rRNAs processing

1. The four rRNAs in eukaryotes are first transcribed as **two long precursor molecules**.

(1) A 45S pre-rRNA transcript is processed in the nucleolus to form the 18S, 28S, and 5.8S rRNAs in eukaryotes.

(2) The other contains just the pre-rRNA that will be processed into the 5S rRNA.

2. There are only three rRNAs in *E. coli*, which are transcribed in **one long precursor** molecule that is cleaved into the individual rRNAs.

3. Some of the bases of pre-rRNAs are methylated, which can enhance stability.

C. Pre-tRNAs processing

In all organisms, tRNAs are produced in a pre-tRNA form that needs multiple processing steps before generating the mature tRNA ready for utilization in translation.

The processing to convert the pre-tRNA to a mature tRNA involves five steps:

(1) cleavage of the 5′ end of the pre-tRNA, called the 5′ leader sequence.

(2) cleavage of the 3′ end of the pre-tRNA.

(3) addition of CCA sequence to the 3′ end of the pre-tRNA after trimming the original 3′ end.

(4) chemical modification of multiple nucleotides.

(5) splicing out of introns.

In eukaryotic organisms, the mature tRNA is created in the nucleus.

D. Pri-miRNA processing

The canonical miRNA biogenesis starts with **transcription** of miRNA genes **by RNA polymerase Ⅱ** which generates primary miRNA (**pri-miRNA**).

(1) Pri-miRNA is cleaved by **microprocessor complex**, a heterotrimeric complex containing an **RNase Ⅲ enzyme Drosha** and the double-stranded-RNA-binding protein Pasha/**DGCR8**, to generate pre-miRNA in the nucleus.

(2) In the cytoplasm, pre-miRNA is further cleaved by **Dicer**, another RNase Ⅲ enzyme, resulting in the generation of a **short (21-23nt long) double-stranded miRNA duplex with a characteristic 2nt overhang at the 3′ end**.

(3) One strand of the miRNA duplex is integrated into an Argonaute (Ago) protein, forming the minimal effector RNA-induced silencing complex (**RISC**). miRNAs can direct RISC to down-regulate gene ex-

pression by mRNA degradation and/or translational repression.

[*Practices*]

[A1 type]

[16] Which one is correct about introns?______

A. Appear in mature mRNA

B. Coding amino acid

C. Does not appear in hnRNA

D. Removed in RNA splicing

E. Is not transcribed

[17] Which process is involved in RNA editing?______

A. Post-transcriptional RNA processing

B. DNA-dependent RNA synthesis

C. Binding to the DNA template of RNA polymerase

D. Removal of introns

E. Reverse transcription

[18] The process in which hnRNA is converted to mRNA is______

A. RNA editing

B. transcription initiation

C. transcription termination

D. translation initiation

E. RNA splicing

[19] Post-transcriptional processing eukaryotic mRNAs does not include______

A. 7-methylguanosine cap is added at the 5′-terminal

B. Phosphorylation

C. Poly(A) tail is added to the 3′ end of mRNA

D. Methylation

E. The introns are removed from the primary transcript and the exons are joined

[20] Which one is wrong about RNA editing?______

A. A base is altered in the coding region of mRNA

B. RNA editing changesmRNA after transcription

C. Also termed differential RNA processing

D. Does not change the amino acid sequence

E. The coding sequence of the mRNA differs from that in the cognate DNA

[21] Which one is involved in eukaryotic rRNA precursorprocessing?______

A. Ribosome B. snRNP C. snoRNP

D. Spliceosome E. Proteasome

[22] The similarity of transcription in prokaryotes and eukaryotes is______

A. Transcriptional regulation in the operon model

B. The same RNA polymerase

C. Asymmetric transcription

D. The same promoter

E. All transcripts need post-transcriptional processing in the nucleus

[23] 5′ end of eukaryotes mRNA is______

A. Termination codon. B. Cap sequence. C. Promotor.
D. poly A tail. E. CCA sequence.

[24] Exon is______

A. DNA sequences which don't transcribe
B. Mutation of the gene
C. Split DNA fragments.
D. Coding region of the eukaryotic structure gene.
E. No coding region of the eukaryotic structure gene.

[25] 45S rRNA is______

A. rRNA of the big subunit of the ribosome.
B. rRNA of the small subunit of the ribosome
C. pre-RNA of rRNA except for 5S RNA
D. pre-RNA of rRNA except for 5.8S RNA
E. Consists of different transcription products.

Answers

1	2	3	4	5	6	7	8	9	10
A	D	B	C	C	A	A	D	D	C
11	12	13	14	15	16	17	18	19	20
C	D	B	B	C	D	A	E	B	D
21	22	23	24	25					
C	C	B	D	C					

Brief Explanations for Practices

1. Transcription is characteristic of asymmetric transcription, namely the template strand is transcribed into a RNA molecule while the coding strand is not transcribed.

2. Transcription differs from replication in that it does not require a primer. In addition, in a given cell only portions of the genome are vigorously transcribed into RNA. All four NTPs and a DNA template are required in transcription and synthesis polarity is $5'\rightarrow3'$.

3. The transcripts of RNA polymerase II include mRNA, lncRNA, miRNA, snRNA.

4. α-amanitin is a specific inhibitor of the eukaryotic RNA polymerases and the sensitivity of RNA Pol I, II, III is insensitive, high sensitivity and intermediate sensitivity, respectively.

5. Asymmetric transcription means that there is only one strand that can act as a template and the template is not always on the same strand.

6. The core RNA polymerase is $\alpha_2\beta\beta'\omega$, which associates with a sigma factor (σ) to form holoenzyme.

7. β subunit of RNA polymerase is the specific binding site of Rifampicin in prokaryotes.

8. The σ subunit enables the core enzyme to recognize and bind the promoter region.

9. The consensus sequence at the −10 region is 5′-TATAAT-3′, which was found by scientist D. Pribnow in 1975, also named Pribnow box.

10. Only after RNA polymerase specifically binds a promoter and formed the PIC, the transcription initiates.

11. RNA polymerase binds to specific sequences in the DNA called promoters, which direct the transcription of adjacent segments of DNA.

12. Many RNA Pol II promoters have a few sequence features in common, including a TATA box (eukaryotic consensus sequenceTATAAA) near about 30 nucleotides upstream from the transcription start site.

13. *E. coli* has at least two classes of termination signals: one class relies on a protein factor called ρ (rho) and the other is ρ-independent.

14. Watson-Crick base-pairing rules: A-U and G-C; And the direction of transcription is 5′→3′. The sense strand is coding strand.

15. It is the amino acid attachment site of the tRNA molecule.

16. The eukaryotic gene sequences encoding the polypeptide is split gene. The coding sequence is often interrupted by long noncoding sequences of DNA that neither appear in mature mRNA. Noncoding tracts that break up the coding region of the transcript are called introns. But introns are transcribed.

17. Coding region can be changed at the mRNA level by RNA editing. In mRNA editing, a base of the mRNA is altered enzymatically. The coding sequence of the mRNA differs from that in the cognate DNA by RNA editing.

18. RNA splicing is a process in which the introns are removed from the primary transcript and the exons are joined to form a continuous sequence that specifies a functional polypeptide.

19. Post-transcriptionally processing eukaryotic mRNAs includes the addition of 5′cap and 3′poly(A) tail, removal of introns and RNA editing.

20. The same with No. 11.

21. A series of enzymatic cleavages of the 45s RNA precursor produces the 18S, 5.8S, and 28S rRNAs, and the cleavage reactions and all of the modifications require small nucleolar RNAs (snoRNAs) found in protein complexes (snoRNPs) in the nucleolus that are reminiscent of spliceosomes.

22. The transcriptional regulation in prokaryotes is mainly executed by the operon model. The promoter of eukaryotes is more complex than that of prokaryotes. Prokaryotic mRNAs are subjected to little processing prior to carrying out their intended function in protein synthesis. Prokaryotes has no nucleus.

23. Almost all the eukaryotic mRNAs have a 7-methyguanosine triphosphate at the 5′-end. This m7GpppN structure is referred to as the 5′-cap structure.

24. Noncoding tracts that break up the coding region of the transcript are called introns, and the coding segments are called exons.

25. In eukaryotes, 45S rRNA need cleavage to produce 28S, 18S, and 5.8S rRNA.

[*Comprehensive Practices*]

[A1 type]

[1] Which one is the component of the snRNP complex?______

A. snRNA B. hnRNA C. snoRNA
D. mRNA E. tRNA

[2] The difference between transcription and DNA replication is ____________.

A. Adherence to Watson-Crick base-pairing rules
B. 3′-5′polarity
C. RNA polymerases lack a separate proofreading function
D. Nucleotides linked by 3′ → 5′-phosphodiester bonds
E. 5′-3′ polarity

[3] Which one is correct about the exon?______

A. The DNA sequence that is not transcribed

B. Noncoding region

C. The DNA sequence that is transcribed but not translated

D. The DNA sequence that is transcribed and translated

E. Splicing is the process that can remove the exons

[4] The transcript of RNA polymerase I in eukaryotes is ________.

A. snRNA B. hnRNA C. miRNA

D. 45S rRNA E. tRNA

[5] Which one is correct about ribosomes? ______

A. The amino acid carriersB. Consist of rRNA and protein

C. Consist of snRNA and protein

D. Consist of DNA and protein

E. Consist of primer, DNA and protein

[6] The primary role of RNA polymerase Ⅲ is to synthesis ______

A. 45S rRNA B. hnRNA C. lncRNA

D. tRNA and 5S rRNA E. snRNA

[7] The transcript of template 5′-ACTAGTCAT -3′ is ______

A. 5′-ACUAGUCAG-3′ B. 5′-UGAUCAGUA-3′ C. 5′- ATGACTAGT-3′

D. 5′-AUGACUAGU-3′ E. 5′-CAGCUGACU-3′

[8] Which one is correct about the TATA box? ______

A. The sequence to which RNA polymerase can bind stably

B. Transcriptional start site

C. Translational start site

D. Replication start site

E. The sequence to which ribosome can bound stably

[9] Basal Transcription factor is ________.

A. component of RNA polymerase in prokaryotes.

B. the complex of α, β, γ subunits.

C. component of RNA polymerase in eukaryotes.

D. trans-acting factor in the regulation.

E. Promotor of the eukaryotic transcription.

[10] Which one can inhibit prokaryotic RNA polymerase specifically? ______

A. Rifampicin B. Amanitin C. Pseudouridine

D. Nitronic salt E. Chloromycetin

[11] RNA holoenzyme includes σ subunit, it binds with DNA template strand at the initiation of the transcription. In the elongation stage, the σ subunit will ______

A. move with holoenzyme on DNA strand

B. serve the function as termination factor.

C. dissociate from the holoenzyme in the elongation stage.

D. in the elongation stage, the conformation will be changed.

E. bind to the template strand loosely.

[12] Feather-like structure of transcription under electron microscope indicates ______

A. double-strand template melted to single strands.

B. several initiation sites in the transcription

C. transcription product RNA and template DNA can form long hetero double strand.

D. formation of polysome appears after the transcription.

E. translation was initiated before the transcription termination.

[B1 type]

No. 13 ~ 14 shared the following suggested answers

A. TATA box B. CAAT box C. GC box

D. Pribnow box E. CCAAT box

[13] The binding site of TF Ⅱ D is ____________.

[14] Sp1 can bound to ____________.

No. 15 ~ 16 shared the following suggested answers

A. snRNA B. hnRNA C. snoRNA

D. 5S rRNA E. siRNA

[15] Which one can cut exogenous double-stranded RNA?______

[16] Which one is the component of the ribosome?______

No. 17 ~ 18 shared the following suggested answers

A. α subunit B. β subunit C. β′subunit

D. α subunit E. σ subunit

[17] Which subunit of RNA polymerase can recognize the transcription start site in prokaryotes?____________

[18] Which subunit of RNA polymerase is the specific binding site of Rifampicin in prokaryotes?____________

[X type]

[19] Post-transcriptional processing eukaryotic tRNA precursors include______.

A. enzymatic removal of nucleotides from the 5′ and 3′ ends

B. phosphorylation

C. modification of some bases such as methylation, deamination

D. the addition of 5′ cap

E. trinucleotide CCA is attached to 3′ terminal

[20] Which isinvolved in cleave mRNA?______

A. hnRNA B. miRNA C. rRNA D. tRNA E. siRNA

[21] Themechanisms by which miRNA modulates mRNA function include______.

A. translational repression

B. protein degradation

C. mRNA destabilization by mRNA deadenylation

D. mRNA degradation

E. mispairing

[22] Which protein can bind to the TATA box______?

A. RNA polymerase B. Telomerase C. Topoisomerase

D. TF Ⅱ D E. Histone

[23] Which of He following are correef abowl The double-strand structure of nucleotides______?

A. DNA/RNA can make a pairing.

B. DNA/RNA double strand is less stable than DNA/DNA double-strand.

C. There are more G–C base pairs in DNA/RNA double strand.

D. A–U base pair is the most unstable one.

E. The reason for transcription bubble formation is described in choice C.

[24] pppGpN—is______.

A. main energy supplying compounds

B. coenzyme of multiple enzymes

C. cap structure of RNA 5′–end

D. signal of transcription termination

E. conserve during the whole transcription process.

Answers

1	2	3	4	5	6	7	8	9	10
A	C	D	D	B	D	D	B	C	A
11	12	13	14	15	16	17	18	19	20
C	E	A	C	E	D	E	B	ACE	BE
21	22	23	24						
AD	AD	ABD	CE						

(*Wang Huaqin*)

Chapter 17

Protein Synthesis

Examination Syllabus

1. The concept of protein biosynthesis.
2. Protein biosynthesis system and genetic code.
3. The basic process of protein biosynthesis.

Ⅰ. The concept of protein biosynthesis

Protein biosynthesis, known as translation, is the process by which ribosomes read the genetic message in mRNA and produce a protein product according to the message's instruction.

Ⅱ. Protein biosynthesis system and genetic code

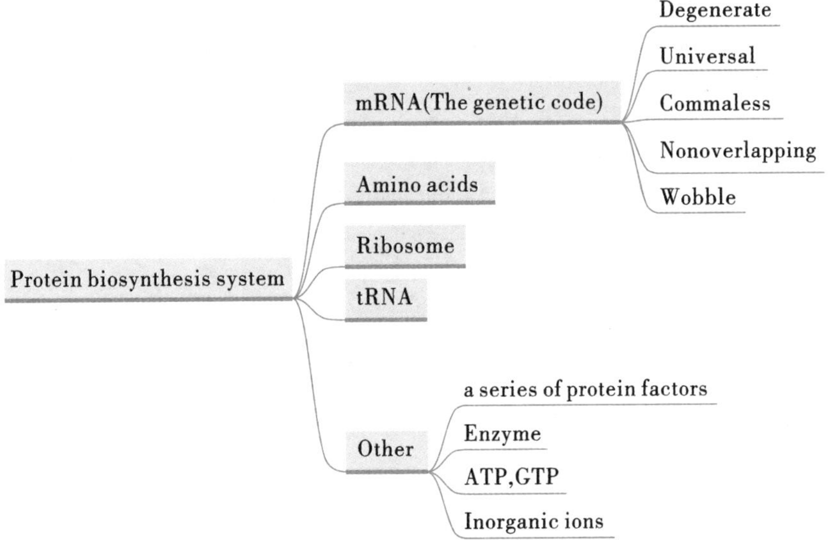

Materials of protein biosynthesis include all 20 amino acids, mRNA, tRNA, ribosome, a series of protein factors, enzyme, ATP, GTP, and inorganic ions.

A. mRNA and the genetic code

1. **mRNA is a template for protein biosynthesis.** The function of **mRNA** is to carry the genetic code from DNA to the ribosome for translation.

2. **The genetic code** is a dictionary that identifies the correspondence between a sequence of nucleotide bases and a sequence of amino acids.

(1) **Codon**: the grouping of three nucleotides (triplets of bases) in mRNA that code for one amino acid or initiation signal or termination signal.

(2) 64 codons: **initiation codon** AUG (also codes for methionine), **stop codon** UAA, UAG, UGA, 61 codons for 20 common amino acids.

3. **The nature of the genetic code** is degenerate, directional, commaless and nonoverlapping, universal and wobble.

(1) **Degenerate**: With the exception of tryptophan and methionine, every amino acid is coded by 2, 3, 4 or 6 codons. (More than one codon can code for an amino acid)

(2) **Commaless and nonoverlapping**: Codons are read 5′→3′ in mRNA without interrupted or intercross.

Open reading frame (ORF) is a run of codons that starts with AUG (start codon) in 5′-end and ends with a termination codon (UGA, UAA or UAG) in 3′-end.

(3) **Universal**: Codon assignments are virtually the same throughout all organisms. But a few differences have been found in eukaryotic mitochondria and in some prokaryotes.

(4) **Wobble**: **Anticodon** is a triplet of bases in a specific tRNA molecule. Each base in the codon base pairs with its complementary base in the anticodon. The pairing between the third base of the codon and the first base of the anticodon follows less stringent rules. **Wobble base pairing**: G—U, I—U, C, A.

B. Transfer RNA (tRNA)

1. tRNA functions as an adapter between nucleic acid and peptide sequences by picking up amino acids and matching them to the proper codons in mRNA.

2. The amino acid is covalently bound to the 3′-OH group of tRNA by aminoacyl-tRNA synthetase to form aminoacyl-tRNA. The reaction is called amino acid activation.

C. Ribosome

The ribosome is a protein synthesis machine for the cell. The ribosome is a complex of one larger and one different subunit, made up of different rRNAs and many different proteins, to translate the mRNAs of the cell into proteins.

[*Practices*]

[**A1 type**]

[1] In the characteristics of the codons, the frameshift mutation is related to ________.

A. universality

B. degeneracy

C. non-punctuation and nonoverlapping

D. wobble

E. direction

[2] In the following amino acids, which one has no corresponding genetic code? ________

A. Isoleucine B. Asparagine C. Proline
D. Hydroxylysine E. Alanine

[3]A tRNA's anticodon is 5′UGC3′, and its identifiable codon is____________.

A. 5′GCA3′ B. 5′ACG3′ C. 5′GCU3′
D. 5′GGC3′ E. 5′GCG3′

[4]The anticodons are wobble. The biological significance is____________.

A. genetic specificity
B. maintaining the stability of biological phenotypes
C. beneficial to genetic variation
D. biological generality
E. strict base pairing

[5]The specificity of amino acid activation depends on____________.

A. ribosome B. tRNA C. aminoacyl-tRNA synthetase
D. amino acid E. mRNA

[6]Which of the following codons do not represent any amino acid? ____________

A. UAA B. AUG C. AGG D. UGG E. AGA

[7]Which of the following descriptions of genetic code is correct? ____________

A. A codon can encode a variety of amino acids
B. An amino acid can have multiple codons
C. AUG is the starting code and does not encode any amino acid
D. Codons and anticodons comply with the principle of strict base pairing
E. All codons are responsible for encoding amino acids

[A2 type]

[8]Polynucleotides containing CAA repeat sequences were added to the acellular protein synthesis system. After identification, three kinds of homopolypeptides were found. They are polyglutamine, polyasparagine and polythreonine. If the codon of glutamine and asparagine is CAA and AAC, please speculate on the codon of threonine. It may be____________.

A. CCA B. CAA C. AAC D. ACA E. CAC

[B1 type]

No. 9 ~ 11 shared the following suggested answers

A. AUG B. GUA C. AGA D. UAG E. GAA

[9] The initiation codon of the genetic code is____________.

[10] The stop codon of the genetic code is____________.

[11] The codon of methionine is ____________.

Ⅲ. The basic process of protein biosynthesis

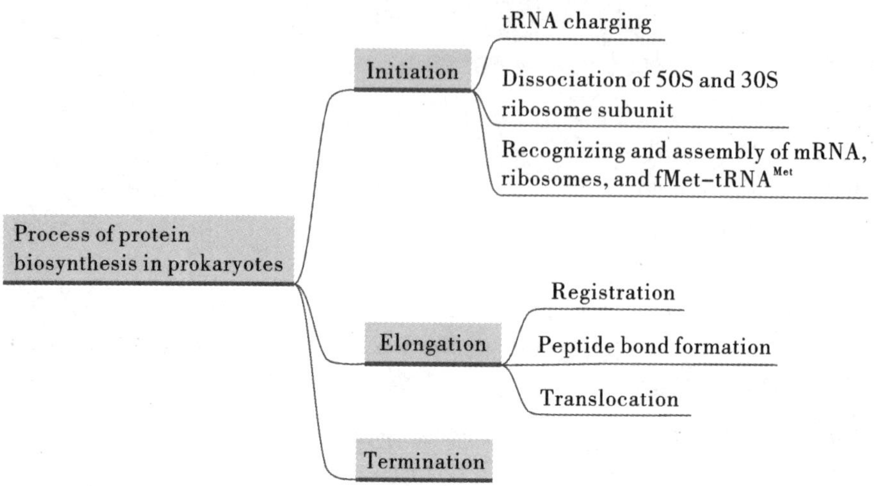

A. Protein biosynthesis in prokaryotes

1. Initiation

(1) **Before translation initiation**:

● **Amino acid activation or tRNA charging** (adding the correct amino acid to the correct tRNA): The amino acid is covalently bound to the 3′-OH group of tRNA to form aminoacyl-tRNA.

● **The dissociation of 50S and 30S ribosome subunit.**

(2) **Formation of the initiation complex**

● **Recognizing and assembly of mRNA**: **mRNA binding** to 30S subunit.

Shine-Dalgarno sequence (SD sequence, ribosomal binding site) is a G-rich mRNA sequence (consensus = AGGAGG) about 6-10 bases upstream of the initiation codon (AUG), which can bind to a complementary sequence at the 3′ end of 16S rRNA component of the 30S ribosomal subunit.

● **The binding of fMet-tRNA^{Met} and 50S ribosomal subunit.**

2. Elongation

(1) The ribosome moves from the 5′ end of the mRNA to the 3′ end, and the polypeptide chain is synthesized from the N-terminal to the C-terminal according to the order of the codon.

(2) **Ribosome cycle**: cyclic process of **entrance**, **peptide bond formation** and **translocation** repeated on the ribosome.

● **Entrance**: binding an aminoacyl-tRNA to the A site of the ribosome.

● **Peptide bond formation**: peptidyl transferase catalyzes the formation of peptide bond.

● **Translocation**: peptidyl-tRNA translocates from the A site into P site.

3. Termination

(1) When a ribosome moves onto the stop codon of mRNA, the elongation ceases, release factors (RF: RF1, RF2) can recognize the stop codon and enter the A site of ribosome.

(2) Discharge of the newly synthesized polypeptide from the complex.

B. Protein biosynthesis in eukaryotes

1. **Initiation** is much more complicated and requires a large number of protein factors.

(1) **Dissociation**: the 60S and 40S ribosome subunit are separated.

(2) **Met-tRNAi^{Met}** bind to 40S subunit with the help of specific initiation factors.

(3) **mRNA binding**: the 5′ cap of mRNA can be recognized and bound to an eIF4F cap-binding protein complex.

(4) The large ribosomal subunit joins the complex and elongation commences.

2. **Elongation**

The A site of ribosome binds the incoming aminoacyl-tRNA, the P site binds the peptidyl-tRNA. The uncharged tRNA is released from the E site. It needs eEF1 α, eEF1 β and eEF2.

3. **Termination**

Two release factors: eRF1 can recognize all three stop codons, eRF3 helps eRF1 to recognize stop codons and release the finished polypeptide.

4. **Posttranslational processing**

Newly synthesized polypeptides usually undergo structural changes called posttranslational processing, including modification and folding.

(1) **Folding**: folding can form higher-level structure of the polypeptide chain. Folding of polypeptides is usually spontaneous, however, to convert a nascent polypeptide chain into a native protein, some cofactors such as **molecular chaperones** (heat shock protein, etc.) are required.

(2) **Protein modification**: includes phosphorylation, acetylation, glycosylation and others.

(3) **Protein targeting**: after posttranslational processing, proteins are transported to their final locations where they play physiological roles. Proteins with **signal sequences** are targeted to the cell specific interval or extracellular medium.

[*Practices*]

[**A1 type**]

[12] In the process of protein biosynthesis in prokaryotes, the reaction that can occur at the E position of the ribosome is ______.

A. the entry of aminoacyl tRNA

B. transpeptidase catalytic reaction

C. releasing tRNA

D. combination with release factors

E. combination within itiation factors

[13] The protein that participates in the correct folding of the newborn polypeptide chain is ______.

A. molecular chaperone　B. G protein　C. transcription factor

D. release factor　E. initiation factor

[14] Which of the following is necessary for the protein synthesis of prokaryotes and eukaryotes? ______

A. Combination of ribosome small subunit and SD sequence

B. mRNA is transported from the nucleus to the cytoplasm

C. the peptidyl-tRNA is translocated from the A site to the P site

D. fMet-tRNA

E. Identification of 5′ cap structure by initiation factor

[15]The location of protein biosynthesis is ____________.

A. cytosol B. mitochondria C. nucleosome

D. ribosome E. nucleus

[16]The polypeptide is synthesized in the direction from ____________.

A. C-terminal to N-terminal B. N-terminal to C-terminal C. 5′→3′

D. 3′→5′ E. bi-directional

[B1 type]

No. 17 ~ 19 shared the following suggested answers

A. phosphorylation B. acetylation C. methylation

D. ubiquitination E. glucosylation

[17]The covalently chemical modification of serine, threonine and tyrosine in the polypeptide chain is ____________.

[18]What kind of post-translational modification needs the involvement of S- adenosine methionine?____________

[19] The most common post-translational modification of histone protein is____________.

Ⅳ. The relationship between protein biosynthesis and medicine

Protein biosynthesis is acting target of antibiotics and toxin. Many effective antibiotics interact specifically with the proteins and RNAs of prokaryotic ribosomes and thus inhibit bacterial protein synthesis. Most members of antibiotics do not interact with components of eukaryotic ribosomes and therefore are not toxic to eukaryotes.

A. Antibiotics (Table 17.1)

1. **Tetracycline** prevents the binding of aminoacyl-tRNAs to the A site.

2. **Chloramphenicol** and the macrolide class of antibiotics work by binding to 23S rRNA. It should be mentioned that the close similarity between prokaryotic and mitochondrial ribosomes can lead to complications in the use of some antibiotics.

3. **Puromycin** effectively inhibits protein synthesis in both prokaryotes and eukaryotes as it is a structural analog of tyrosinyl-tRNA.

Table 17.1 The principle and application of some antibiotics to inhibit protein synthesis

Antibiotics	Site of action	Principle	Application
Edeine	Small subunit of prokaryotic and eukaryotic ribosome	Impede the formation of translation initiation complexes	Antiviral drug
Tetracycline	Small subunit of prokaryotic ribosome	Inhibit the binding of aminoacyl-tRNA to A site	Antibacterial drug
Streptomycin, neomycin	Small subunit of prokaryotic ribosome	Cause errors in genetic code reading; inhibit initiation	Antibacterial drug

Continue to Table 17.1

Antibiotics	Site of action	Principle	Application
Chloramphenicol, erythromycin	Large subunit of prokaryotic ribosome	Inhibit peptidyl transferase	Antibacterial drug
Puromycin	Prokaryotic and eukaryotic ribosome	Replace tyrosinyl-tRNA into the A sites	Antineoplastic drug

B. Toxins

Diphtheria toxin: catalyze the ADP ribosylation eEF-2, **inactivate eEF-2**, and thus block mammalian protein synthesis at the elongation stage.

C. Interferon

Two mechanisms for the inhibition effects of interferon in virus translation: **deactivate eIF2; degrade viral mRNA.**

[*Practices*]

[A1 type]

[20] Which of the following drugs interferes with the translation process of eukaryotes and prokaryotes and therefore can not be used as an antibacterial drug? ____________

A. Tetracycline B. Streptomycin C. Kanamycin

D. Puromycin E. Chloramphenicol

[21] Which of the following drugs is an inhibitor of protein biosynthesis? ____________

A. 5- fluorouracil B. Kanamycin C. Methotrexate

D. Allopurinol E. Amanitin

[22] Chloramphenicol can inhibit the synthesis of bacterial protein because ____________.

A. it inhibits the function of tRNA

B. it affects the transcription of mRNA

C. it inhibits the binding between small subunit and large subunit of the ribosome

D. it inhibits the activity of initiation factors

E. it inhibits the activity of large subunit of the ribosome

[A2 type]

[23] A pharmaceutical company found a new antibiotic that inhibits the synthesis of bacterial protein. The researchers found that when the antibiotic was added to the protein synthesis system in vitro, the mRNA sequence AUGUUUUUUUAG in the system was only translated into a two peptide, fMet-Phe. The protein synthesis step that is most likely to be inhibited by the antibiotic is____________.

A. aminoacyl tRNA binding to the A bit of ribosome

B. peptidyl transferase activity

C. ribosome translocation

D. initiation

E. termination

[B1 type]

No. 24 ~ 27 shared the following suggested answers

A. tetracycline B. streptomycin C. interferon

D. puromycin E. chloramphenicol

[24] A drug that can phosphorylate initiation factors and inhibit protein synthesis is ___________.

[25] A drug that can bind to 30S subunit of the ribosome and inhibit protein synthesis is___________.

[26] A drug that can bind to 50S subunit of the ribosome and inhibit protein synthesis is _________.

[27] A drug that can break out the peptidyl from the ribosome and disrupt the synthesis of protein is ___________.

Answers

1	2	3	4	5	6	7	8	9	10
C	D	A	B	C	A	B	D	A	D
11	12	13	14	15	16	17	18	19	20
A	C	A	C	D	B	A	C	B	D
21	22	23	24	25	26	27			
B	E	C	C	A	E	D			

Brief Explanations for Practices

1. Codons which represent triplets of bases in mRNA are read without interrupted or intercross. Frameshift Mutations result from deletion or insertion of nucleotides in DNA. Thus the mRNA sequence change results in a down stream amino acid change.

2. There are 64 codons, including initiation codon, stop codon and codons for 20 common amino acids involved in the composition of protein. Isoleucine, asparagines, proline and alanine belong to 20 common amino acids, but hydroxylysine is not involved in the composition of protein.

3. Each base in the codon base pairs with its complementary base in the anticodon. Base pairing is between the third base of the codon and the first base of the anticodon.

4. The correct addition of amino acids in the translation requires the mutual recognition of the base pairs depending on the codon on the mRNA and the anticodons on the tRNA. The wobble base pairs ensure the relative stability of the amino acids determined by the genetic code, which is beneficial for maintaining the relative stability of biological phenotypes.

5. The amino acid is covalently bound to the 3′-OH group of tRNA by aminoacyl synthetase to form aminoacyl-tRNA. The specificity of amino acid activation depends on aminoacyl synthetase.

6. The stop codon UAA, UAG and UGA do not represent any amino acid.

7. Degeneracy means that single amino acid is coded by several different codons so that a codon cannot encode a variety of amino acids. AUG is the starting code and can encode methionine. The pairing between the third base of the codon and the first base of the anticodon follows less stringent rules. The stop codons are not responsible for encoding amino acids.

8. For polynucleotides containing CAA repeat sequences, continuous reading can be carried out from the first C, second A, or third A, respectively. The first codon is CAA, encoding glutamine. The second codon is AAC, encoding asparagines. The third codon is ACA, encoding threonine.

9. The initiation codon is AUG.

10. The stop codon is UAA, UAG, UGA.

11. The initiation codon is AUG, and it is also the codon of methionine.

12. The A site of ribosome binds the incoming aminoacyl-tRNA, the P site binds the peptidyl-tRNA.

The uncharged tRNA is released from the E site in prokaryotes.

13. The major function of molecular chaperones is to assist the correct folding of nascent polypeptide chains by blocking their hopeless entangling or insignificant intermolecular interactions, which are mentioned in chapter 1 (Structures and functions of protein). G Protein is part of signal transduction pathway. Initiation factor and release factor are involved in protein synthesis. The function of transcription factor is regulating gene expression.

14. fMet-tRNA, combination of ribosome small subunit and SD sequence are necessary for the protein synthesis of prokaryotes. Identification of 5′ cap structure by initiation factor and mRNA transportation from the nucleus to the cytoplasm occurred in eukaryotes. While in the elongation stage of translation in prokaryotes and eukaryotes, the peptidyl-tRNA is translocated from the A site to the P site.

15. The ribosome is a protein synthesis machine for the cell.

16. The polypeptide is synthesized in the direction from N-terminal to C-terminal.

17. Serine, threonine and tyrosine have hydroxyl group which can be phosphorylated by protein kinase.

18. Under the catalysis of methyltransferase, S-adenosine methionine can provide methylate group for protein methylation.

19. The most common post-translational modification of histone protein is acetylation.

20. Puromycin effectively inhibits protein synthesis in both prokaryotes and eukaryotes as it is a structural analog of tyrosinyl-tRNA.

21. Methotrexate and 5- fluorouracil are antitumor drugs leading to the inhibition of the biosynthesis of DNA. Allopurinol leads to the inhibition of uric acid synthesis. Only kanamycin is an antibiotic which can inhibit bacterial protein biosynthesis.

22. Chloramphenicol works by binding to 23S rRNA, which is located in 50S subunit of the ribosome.

23. According to the sequence of mRNA, it can be used as a template to synthesize a three peptide fMet-Phe-Phe in the cell. Now, the only product is a two peptide, meaning that the initiation of protein synthesis was not affected, and the Phe-tRNA can bind to the A site of the ribosome, and form the first peptide bond by transpeptidase resulting a two- peptide product, fMet-Phe. No synthesis of three peptides indicates that the translocation of the ribosome is suppressed.

24. Interferon activates eIF-2 kinase in the presence of virus double-stranded RNA, causing eIF-2 phosphorylation and deactivation, thus inhibiting the synthesis of virus protein.

25. Tetracycline prevents the binding of aminoacyl-tRNAs to the 30S subunit of the ribosome.

26. Chloramphenicol works by binding to 23S rRNA, which is located in 50S subunit of the ribosome.

27. Puromycin is a structural analog of tyrosinyl-tRNA. In eukaryotic and prokaryotic ribosomes, it can transfer the peptidyl to its amino group, leading to premature polypeptide release.

[*Comprehensive Practices*]

[**A1 type**]

[1] The process of transforming the nucleotide sequence of DNA into the sequence of amino acids in the protein includes ____________.

A. Replication and transcription

B. Replication and reverse transcription

C. Transcription

D. Translation

E. Transcription and translation

[2] The number of codons encoding 20 common amino acids is ____________.

A. 64　B. 61　C. 60　D. 20　E. 16

[3] The I on the first base of tRNA anticodons can be paired with the A, C, and U on the third base of mRNA codon. That means ____________.

A. Degeneracy　B. Universality　C. Specificity
D. Wobble　E. Direction

[4] Which of the following is the first reaction of amino acids incorporated into the polypeptide chain in the prokaryotic cell? ____________

A. Binding of fMet-tRNA to the ribosome
B. Binding of mRNA and ribosome 30S subunit
C. Binding of 30S subunit and 50S subunit of the ribosome
D. Aminoacyl-tRNA synthetase catalyzes the activation of amino acids
E. Binding of fMet-tRNA and mRNA

[5] A mRNA's codon is 5′UCG3′, and its identifiable anticodon is ____________.

A. 5′GCA3′　B. 5′CGA3′　C. 5′CGU3′
D. 5′CGG3′　E. 5′AGC3′

[6] Both UGC and UGU can encode cysteine. That means the ____________ of codons.

A. degeneracy　B. universality　C. specificity
D. wobble　E. direction

[7] Which of the following characteristics of aminoacyl-tRNA synthetase is correct? ____________

A. The catalytic reaction requires GTP
B. Exists in the nucleus
C. There is absolute specificity only for tRNA
D. There is absolute specificity only for amino acids
E. There is specificity for both amino acids and tRNA

[8] The determinant of the sequence of amino acids in a protein molecule is ____________.

A. tRNA
B. the species of amino acids
C. the sequence of nucleotides in mRNA molecules
D. Ribosome
E. rRNA

[9] The composition of initiation complex of translation is ____________.

A. DNA+RNA+RNA polymerase
B. ribosome+Met-tRNA
C. DNA+protein+RNA
D. Initiation factor+ribosome
E. ribosome+Met-tRNA+mRNA

[10] The enzyme that can catalyze peptide chain extension in protein biosynthesis is ____________.

A. carboxyl peptidase　B. transpeptidase　C. aminoacyl-tRNA synthetase
D. aminopeptidase　E. transaminase

[11] In the process of protein biosynthesis, which of the following substances can release polypeptide chains from the ribosome? ____________

A. Release factors　B. Transpeptidase　C. Ribosome depolymerization
D. Stop codon　E. tRNA

[12] Which of the following drugs can inhibit the binding of aminoacyl-tRNA to small subunits of the ribosome? ____________

A. Lincomycin B. Streptomycin C. Tetracycline

D. Puromycin E. Chloramphenicol

[13] The target site that interferon can inhibit the translation process is____________.

A. IF-1 B. IF-3 C. eIF-2 D. eEF-1 E. eEF-2

[14] The substance that is not needed for protein biosynthesis is ____________.

A. multiple protein factors B. mRNA C. ribosome

D. amino acids E. DNA

[15] The correct description of translation is that____________.

A. all 64 codonscan represent amino acids

B. each addition of an amino acid requires registration, peptide bond formation, and translocation

C. the synthesisof the polypeptide chain requires CTP

D. ribosomal large subunit and small subunit have been combined together

E. polypeptide synthesis is bi-directional

[16] After protein biosynthesis, the protein that has undergone post-translational modification is ____________.

A. alanine B. tyrosine C. methionine

D. hydroxyproline E. cysteine

[17] Which of the following molecules is involved in protein folding? ____________

A. DNA binding protein B. Proinsulin C. Signal peptide

D. Heat shock protein E. Histone

[18] In the process of protein biosynthesis, the regulation site is mainly in the ____________.

A. termination stage

B. initiation stage

C. elongation stage

D. the whole process of synthesis

E. amino acid activation

[**A2 type**]

[19] A tRNA, originally supposed to transport cysteine ($tRNA^{cys}$), was wrongly carried on alanine at the time of activation, producing alanine-$tRNA^{cys}$. If the error is not corrected in time, what is the fate of the alanine residue? ____________

A. It is added into the protein randomly corresponding to random codon

B. It is added into the protein corresponding to the alanine codon

C. It is added into the protein corresponding to the cysteine codon

D. It will be catalyzed by the enzyme in the cell and transformed into cysteine

E. Because it cannot be used for protein synthesis, it will always bind to tRNA

[20] A 17-year-old patient suffered from microcytic anemia. The detection showed that β chain of hemoglobin contains 172 amino acid residues, rather than normal 141 amino acid residues. It may be caused by which of the following genetic mutations? ____________

A. UAA→CAA B. UAA→UAG C. GAU→GAC

D. GCA→GAA E. CGA→UGA

[B1 type]

No. 21 ~ 23 shared the following suggested answers

A. ATP B. CTP C. GTP D. TTP E. UTP

[21] The activation of amino acids requires ____________.

[22] The extension of polypeptide chain requires ____________.

[23] Entry of amino acid-tRNA to A site requires ____________.

No. 24 ~ 26 shared the following suggested answers

A. enzyme activation

B. glycosylation of amino acid residues

C. allosteric regulation of enzyme

D. add cap structure in the 5′ end of mRNA

E. formation of aminoacyl-tRNA

[24] The post-translational processing is ____________.

[25] The post-transcriptional processing is ____________.

[26] The activation of amino acids is____________.

Answers

1	2	3	4	5	6	7	8	9	10
E	B	D	D	B	A	E	C	E	B
11	12	13	14	15	16	17	18	19	20
A	C	C	E	B	D	D	B	C	A
21	22	23	24	25	26				
A	C	C	B	D	E				

(*Zhang Baifang*, *Yu Hong*)

Chapter 18

Regulation of Gene Expression

[*Examination Syllabus*]

1. Gene expression: concept, types (constitutive gene expression, induction and repression), characteristics (temporal specificity, spatial specificity).

2. **Regulation of gene expression**: concept, significance, different stages, basic elements.

3. **Mechanism of the regulation of gene expression**: in prokaryotes (lactose operon), in eukaryotes (*cis*–acting element, *trans*–acting factor).

I. Gene expression

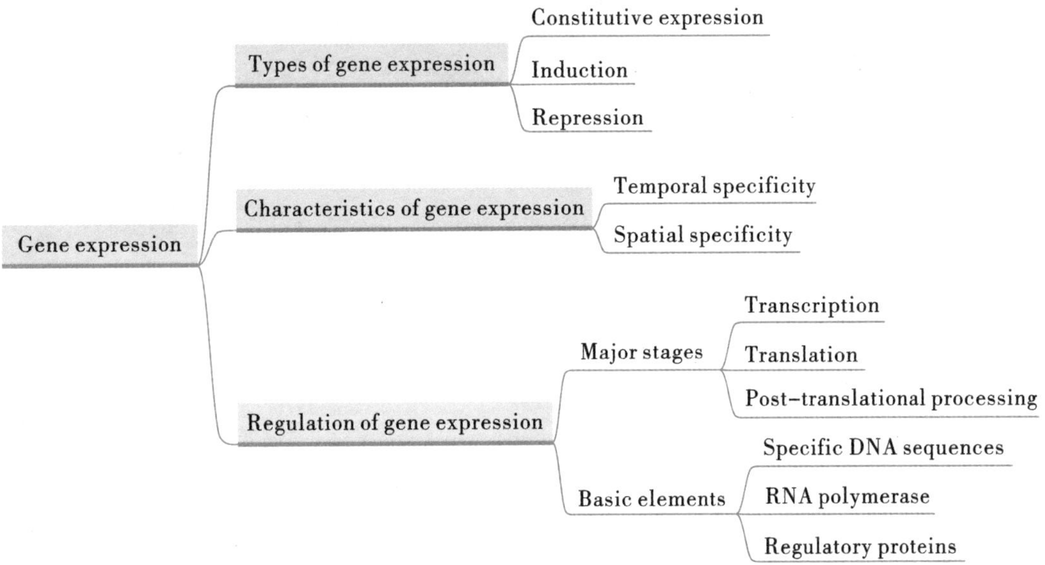

A. Types of gene expression

1. **Gene expression** refers to the process in which the genetic information stored in DNA sequence was presented into protein molecules with certain biological function. The process includes transcription and translation.

2. **Constitutive gene expression** means that some gene is usually expressed in almost all cells of an organism and less susceptible to the environment. This kind of gene is called house-keeping gene.

3. **Induction** means that some gene can be induced to express with the stimulation of certain environmental signals. This kind of gene is called inducible gene.

4. **Repression** means that some gene can be repressed to express with certain environmental stimuli. This kind of gene is called repressible gene.

B. Characteristics of gene expression

1. **Temporal specificity of gene expression** refers to the expression of some gene that happens in an order of timing, also known as stage specificity.

2. **Spatial specificity of gene expression** means that the expression of certain gene happens at specific spatial organization, also known as cell specificity or tissue specificity.

C. Regulation of gene expression

1. **Regulation of gene expression** (also known as **gene regulation**) includes a wide range of mechanisms that are used by cells to increase or decrease the production of specific gene products (protein or RNA).

2. **The significance of gene regulation** is making a living organism quickly adapt to the changing environment and maintain their growth, proliferation, cell differentiation and development.

3. **The multi-stages of gene regulation** include the change of genetic information (such as DNA rearrangement and methylation), transcription (especially transcription initiation), translation and post-translational processing.

4. **The basic elements** involved in gene regulation include specific DNA sequences (such as promoter), transcriptional regulatory proteins, DNA-protein interactions, protein-protein interactions, RNA polymerase, etc.

Ⅱ. Mechanism of gene regulation

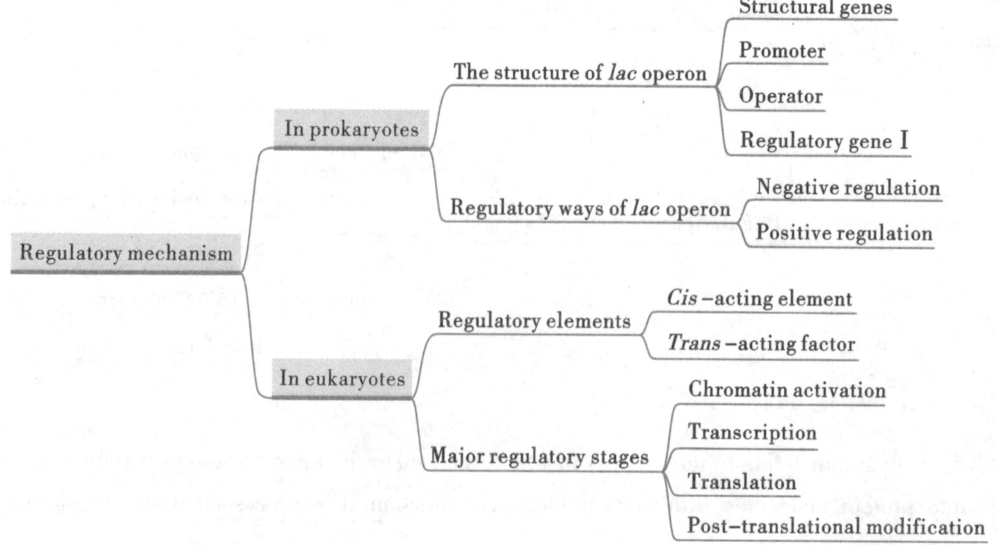

A. Mechanism of prokaryotic gene regulation

1. **The operon** is the basic unit of transcriptional regulation in prokaryotes, which is made up of structural genes and regulatory sequences.

2. **Structural genes** refer to several genes that are functionally related and share one promoter and one common sequence for transcription termination. Therefore, only a single mRNA is transcribed to produce all proteins of the operon, which is also known as polycistronic mRNA.

3. **Lactose operon** (***lac*** **operon**) is required for the transport and metabolism of lactose in *E. coli* and some other bacteria.

4. **The structure of** ***lac*** **operon**

(1) **Three adjacent structural genes** (*lac Z*, *lac Y*, and *lac A*): encode β-galactosidase, β-galactosidepermease and transacetylase respectively, which are required for the metabolism of lactose.

(2) **Promoter (P)**: can be bound by RNA polymerase to initiate transcription process.

(3) **Operator (O)**: can be recognized and bound by specific repressor protein.

(4) **Regulatory gene I** (***lac I***): encode repressor protein and mediate negative regulation.

5. **Regulatory ways of the** ***lac*** **operon**

(1) **The negative regulation of** ***lac*** **operon** is mediated by repressor. In the absence of lactose, the repressor binds to the operator and blocks the binding of RNA polymerase to the promoter, thus inhibiting the transcription of structural genes. When lactose is available, galactose serves as an inducer and binds to the repressor. As a result, the repressor leaves the operator so that RNA polymerase can bind to the promoter and start transcription of structural genes.

(2) **The positive regulation of** ***lac*** **operon** is mediated by the cAMP-CAP complex. In the absence of glucose, **cAMP** (**cyclic adenosine monophosphate**) levels rise and cAMP binds to **CAP** (**catabolite activator protein**). Then the complex binds to the regulatory region on the operon, stimulating the binding of RNA polymerase to the promoter and enhancing transcription.

B. Mechanism of eukaryotic gene regulation

1. **Regulatory elements in eukaryotic gene expression**

(1) ***Cis*****-acting elements** refer to the regions of non-coding DNA which regulate the transcription of neighboring genes, including promoter, enhancer and silencer.

(2) **Promoter** is a region of DNA sequence that is recognized by RNA polymerase and determines the start site and frequency of transcription.

(3) **Enhancer** is a short region of DNA that can be bound by proteins to greatly increase the efficiency of transcription.

(4) **Silencer** is a region of DNA sequence capable of binding transcriptional regulatory factors such as repressor, which consequently results in the inhibition of transcription.

(5) ***Trans*****-acting factor** is a group of nuclear protein factors which interact with *cis*-acting elements and RNA polymerase to regulate gene expression, also known as transcription factors (TF), transcriptional regulators or *trans*-acting protein.

2. **Regulated stages in eukaryotic gene expression** include chromatin activation, transcription initiation, post-transcriptional modification, translation initiation, post-translational modification, etc. Among all these steps, transcription initiation is the key point in gene regulation.

[Practices]

[A1 type]

[1] Which one of the following is not correct about house-keeping genes? ____________

A. They are continuously expressed in almost all cells of an organism.

B. They are continuously expressed at almost all growth stages of an organism.

C. They are continuously expressed in almost all individuals of a species.

D. They are continuously expressed at a certain growth stage of an organism.

E. They are expressed in almost all cells throughout the life process of an organism.

[2] Alpha fetal protein coding genes are expressed in fetal liver cells, but not expressed in adult liver cells. This kind of gene expression belongs to____________.

A. Spatial-specific expression

B. Temporal-specific expression

C. Organ-specific expression

D. Tissue-specific expression

E. Cell-specific expression

[3] Which one of the following doesn't show the stage specificity of gene expression? ____________

A. A gene is expressed in differentiated skeletal muscle cells and not expressed in undifferentiated cardiomyocytes.

B. A gene is expressed in undifferentiated skeletal muscle cells, and not expressed in differentiated skeletal muscle cells.

C. A gene is expressed in differentiated skeletal muscle cells and not expressed in undifferentiated skeletal muscle cells.

D. A gene is expressed during embryonic development and not expressed after birth.

E. A gene is not expressed during embryonic development and expressed after birth.

[4] In order to repair the damage caused by the ultraviolet radiation in bacteria, the gene encoding DNA repair enzymes are activated. Which one of the following terms can best describe this phenomenon? ____________

A. DNA damage B. DNA repair C. DNA expression

D. Induction E. Repression

[5] Regulation of gene expression mainly refers to ____________.

A. regulation of DNA replication

B. post-transcriptional modification

C. regulation of reverse transcription

D. formation of protein folding

E. regulation of transcription

[6] ____________ is bound by RNA polymerase and initiates transcription.

A. Promoter B. Enhancer C. Silencer

D. Operon E. Attenuator

[7] In *lac* operon, *lac I* gene encodes____________.

A. β-galactosidase

B. β-galactoside permease

C. transacetylase

D. one kind of activator

E. one kind of repressor

[8] In *lac* operon, the repressor binds to ____________.

A. P fragment B. O fragment C. CAP binding site

D. I gene E. Z gene

[9] A prokaryotic operon usually consists of ____________.

A. one promoter and one structural gene

B. one promoter and several structural genes

C. several promoters and one structural gene

D. several promoters and several structural genes

E. two promoters and several structural genes

[10] The regulation of gene expression by operon occurs at ____________ level.

A. replicational B. transcriptional C. translational

D. post translational E. reverse transcriptional

[11] In *lac* operon, the positive regulation of CAP occurs ____________.

A. when glucose level is high and cAMP level is low

B. when glucose level is low and cAMP level is low

C. when in the absence of glucose, cAMP level is low

D. when in the absence of glucose, cAMP level is high

E. when both glucose and cAMP levels are high

[12] ____________ induces the transcription of *lac* operon.

A. Allo lactose B. Glucose C. Lactose

D. Arabinose E. Fructose

[13] The following statements are the characteristics of the structure of the eukaryotic gene. Which one is not true? ____________

A. The protein-coding genes in eukaryotic genome are not continuous.

B. Eukaryotic genome is huge.

C. Eukaryotic genome contains a lot of repeated sequences.

D. Most of the transcripts are polycistronic.

E. Eukaryotic DNA is complexed with histones to constitute chromatin in the nucleus.

[14] ____________ doesn't belong to gene expression.

A. The transcription of tRNA

B. The transcription of mRNA

C. The translation of mRNA into protein

D. The transcription of rRNA

E. The replicatios of RNA

[15] In *E. coli*, under high-lactose, high-glucose conditions, ____________ could lead to maximal transcription activation of the *lac* operon.

A. a mutation in the CAP-binding site leading to enhanced binding

B. a mutation in the *lac I* gene

C. a mutation in the operator sequence

D. a mutation leading to lower binding of repressor

E. a mutation leading to lower cAMP levels

[16] ____________ determines tissue-specific expression.

A. Promoter B. Enhancer C. Silencer

D. Operator E. Transcription factor

[17] *Cis*-acting elements include____________.

A. transcriptional repressor B. transcriptional activator

C. σ factor D. ρ factor E. enhancer

[18] *Trans*-acting factors include____________.

A. elongation factor B. enhancer C. operator

D. promoter E. transcription factor

[19] Which one of the following best defines *trans*-acting factor? ____________

A. It is a kind of protein encoded by a certain gene that regulates the transcription of any gene.

B. It is a kind of protein encoded by a certain gene that regulates the transcription of another gene.

C. It refers to all kinds of protein molecules involved in the regulation of transcription.

D. It refers to all kinds of protein molecules involved in the regulation of translation.

E. It refers to all kinds of nuclear factors involved in the regulation of gene expression.

[B1 type]

No. 20 ~ 21 share the following suggested answers.

A. constitutive expression B. induction C. repression

D. coordinate expression E. SOS response

[20] After ultraviolet radiation, the expression of a gene that encodes DNA repair enzymes in bacteria is enhanced. This phenomenon belongs to____________.

[21] The expression of house-keeping genes belongs to____________.

No. 22 ~ 23 share the following suggested answers.

A. CAP-binding site B. promoter C. operator

D. sequences of structural genes E. sequences coding for repressor

[22] CAP binds to____________.

[23] Repressor protein binds to____________.

No. 24 ~ 25 share the following suggested answers.

A. Terminator B. Exon C. TATA box

D. Operator E. Intron

[24] ____________is involved in the regulation of eukaryotic gene transcription.

[25] ____________is involved in the regulation of prokaryotic gene transcription.

[X type]

[26] The characteristics of eukaryotic gene structure include____________

A. the genome is huge.

B. the genes are not continuous.

C. the transcripts are monocistronic.

D. it contains a lot of repeated sequences.

E. the transcripts are polycistronic.

[27] ____________can affect the level of gene expression.

A. Transcription initiation

B. Post-transcriptional processing

C. mRNA degradation

D. Translation and modification of protein

E. Post-translational processing

[28] The characteristics of the expression of house-keeping gene include____________

A. the expression level is usually high.

B. the expression is less susceptible to environmental changes.

C. the expression is susceptible to environmental changes.

D. the stage-specificity is not obvious.

E. the expression level is usually low.

[29] The structure of *lac* operon consists of____________.

A. three structural genes B. one operator C. one promoter

D. a regulatory gene E. one enhancer

[30] ____________is the basic element that affects transcription initiation.

A. DNA sequence B. RNA sequence C. Regulatory protein

D. Protein-DNA interaction E. The activity of RNA polymerase

[31] ____________can describe the specificity of gene expression in mammals.

A. Temporal specificity B. Spatial specificity C. Stage specificity

D. Tissue specificity E. Cell specificity

[32] ____________belongs to *cis*-acting element.

A. Promoter B. Enhancer C. Silencer

D. Operator E. Transcription factor

[33] *Trans*-acting factor refers to____________.

A. transcriptional regulatory proteins

B. a class of protein factors that play a role in the nucleus

C. RNA polymerase D. DNA polymerase E. DNase I

[34] ____________is related to the negative regulation of gene expression in *lac* operon.

A. *Lac* I gene B. *Lac* Z gene C. *Lac* repressor protein

D. O fragment E. Promoter

[35] When there is only glucose in the medium, the activity of galactosidase in *E. coli* is low. This phenomenon is due to which one of the following reasons? ____________

A. Repressor protein binds to the operator.

B. Glucose makes the repressor protein fall off from the operator.

C. Repressor protein blocks RNA polymerase from binding to the promoter.

D. Structural genes can't be fully expressed.

E. The *lac* operon is in a repressed state.

[36] Which one of the following proteins is involved in the regulation of the *lac* operon? ____________

A. RNA polymerase B. Repressor protein C. CAP

D. cAMP E. IPTG

[37] The statement ____________is correct about the structure and function of enhancer.

A. enhancer can replace promoter in biological process.

B. enhancer can exert its function without promoter.

C. enhancer can exert its function at a location far away from the transcription start site.

D. enhancer can bind with the tissue-specific transcription factor.

E. enhancer belongs to house-keeping gene.

[38] Which one of the following belongs to the characteristics of eukaryotic gene transcription? ____________

A. The chromatin changes obviously.

B. The negative regulation by repressor mainly occurs in the process of transcription.

C. Positive regulation mainly occurs in the process of transcription.

D. Transcription and translation are coupled (gene transcription and translation happen at the same location simultaneously).

E. Transcription and translation are not coupled (gene transcription and translation happen at separate locations).

[39] ____________ could affect the activity of RNA polymerase.

A. Promoter sequence B. Enhancer sequence C. Specific transcription factor

D. Repressor E. None of the above

[40] Which one of the following is the characteristic of the structure and function of promoter or the sequence of promoter? ____________

A. It includes the transcription start site.

B. It includes RNA polymerase-binding site.

C. It determines the frequency of transcription.

D. It determines the accuracy of transcription initiation.

E. It determines the termination of a gene.

[41] ____________ plays a role in the positive regulation of *lac* operon.

A. The increase of cAMP level B. The increase of glucose level C. The decrease of cAMP level

D. The decrease of glucose level E. The high level of lactose

Answers

1	2	3	4	5	6	7	8	9	10
D	B	A	D	E	A	E	B	B	B
11	12	13	14	15	16	17	18	19	20
D	A	D	E	A	E	E	E	B	B
21	22	23	24	25	26	27	28	29	30
A	A	C	C	D	ABCD	ABCDE	ABD	ABCD	ACDE
31	32	33	34	35	36	37	38	39	40
ABCDE	ABC	A	ACD	ACE	ABC	CD	ACE	ABCD	ABCD
41									
AD									

Brief Explanations for Practices

15. In order to transcribe the *lac* operon, the repressor must bind to galactose and leave the operator region, and the cAMP-CAP complex must bind to the operator in order for RNA polymerase to bind. Raising cAMP, even though glucose is present, will allow the cAMP-CAP complex to bind and recruit RNA polymerase.

36. RNA polymerase, repressor protein, CAP and cAMP are involved in the regulation of *lac* operon, but cAMP is not protein. IPTG is an analogue of galactoside and can serve as inducer to stimulate the transcription of *lac* operon.

37. Enhancer can't replace promoter. Enhancers are the binding sites for tissue-specific transcription factors, so they can determine the tissue-specific expression of gene. Enhancers not only can act upstream or downstream of the genes, but can exert their regulatory function remotely. Only with the presence of promoter, can enhancer exert its function. Enhancer doesn't belong to house-keeping gene.

38. The chromatin structure changes obviously in the process of eukaryotic gene transcription. Positive regulation mainly occurs in the process of eukaryotic transcription. Since eukaryotic cell has cell nucleus, transcription and translation processes are not coupled.

39. Promoter is a region of DNA sequence where RNA polymerase binds. In *lac* operon, when the repressor binds to operator, it blocks the binding of RNA polymerase to the promoter. Enhancer can increase the efficiency of transcription which requires the involvement of RNA polymerase, so enhancers and transcription factors are also related to the activity of RNA polymerase.

41. In the absence of glucose, cAMP levels increase. cAMP binds to CAP to enhance transcription of *lac* operon.

(***Zang Mingxi***)

Chapter 19

Signal Transduction

[*Examination Syllabus*]

1. **Signaling molecules**: concept, classification, second messengers

2. **Receptors**: classification and function characteristics

3. **Membrane receptor-mediated signal transduction mechanism**: G protein-coupled receptor (GPCR)-mediated signal transduction pathway; single transmembrane receptor (enzyme-coupled receptor)-mediated signal transduction pathway

4. **Intracellular receptor-mediated signal transduction mechanism**: concept and classification; mechanism

[*Major points*]

Overview of signal transduction

1. **Concept**: the process inwhich chemical signals released by one cell evoke a receptor-mediated response in another.

2. **Two phases**

(1) **Intercellular signal transduction**: characterizes the passage of a signal from one cell to target cells via the extracellular environment.

(2) **Intracellular signal transduction**: comprises the biochemical decoding of that signal on receipt by the target cells, a process that involves stepwise regulation of intracellular signaling molecules that ultimately results in cellular response, altering some metabolic processes or biological activities of the cell such as gene regulation, membrane transport, cell motility, etc. (Figure 19.1)

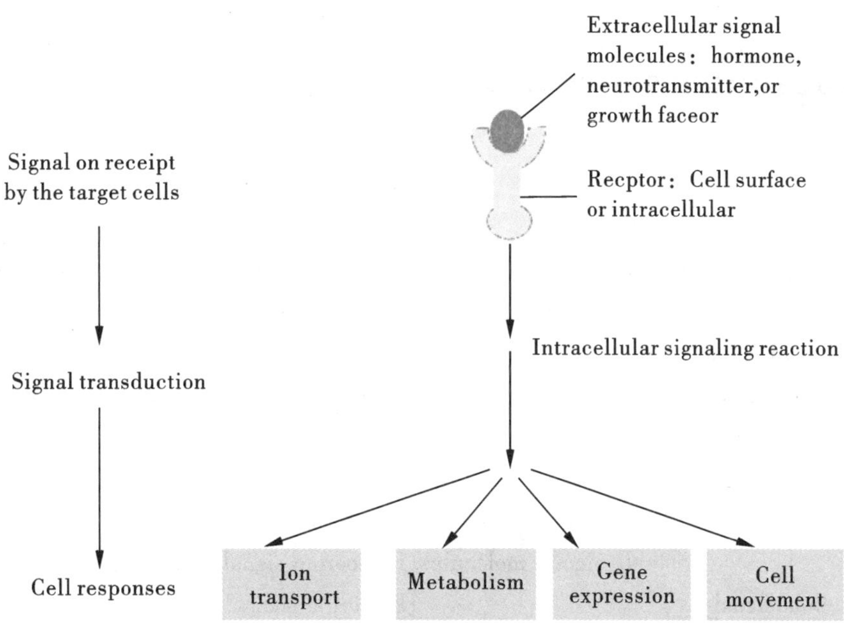

Figure 19.1 Schematic diagram of the basic mode of intracellular signal transduction

3. **Intracellular signal transduction pathway** or **signaling pathway**: the orderly arrangement of a set of signal transduction molecules which converts and transmits signals in turn through mutual recognition and interaction in cellular signal transduction.

4. **Signal transduction network**

(1) In the same cell, the signal converted by **one receptor** molecule can be transmitted through one or **more** intracellular signal transduction **pathways.**

(2) The signal converted by **different receptor** molecules can also be transmitted through the **same** signal transduction **pathway**.

(3) Because the **same** signal transduction **molecules** in cells can be recruited by **different** signal transduction **pathways**, there may be **cross-talking** among different signal transduction pathways, thus forming a complex signal transduction network.

5. **Molecular mechanisms of signal transduction**

(1) Changes of the **active state** of signal transduction molecules.

(2) Changes of the **concentration** or content of signal transduction molecules.

(3) Changes of the **conformation** of signal transduction molecules.

(4) Changes of the **localization** or distribution of signal transduction molecules.

6. **Basic rules of signal transduction pathway**

(1) Transmission and termination of signals.

(2) Cascade amplification effect of signals.

(3) Complexity and diversity of signal transduction pathways.

(4) Universality and specificity of signal transduction pathway.

Ⅰ. Extracellular signal molecules

A. Concept

For multicellular organisms, the extracellular signals received by the individual cells in the body can be a contact-dependent signal and also be various chemical and physical signals (such as electricity, magnetism, light, sound, or radiation signal, etc.) in a microenvironment. However, **the extracellular signals are mainly chemical signals** established by continual variation and evolution for adaption to the environment.

B. Classification

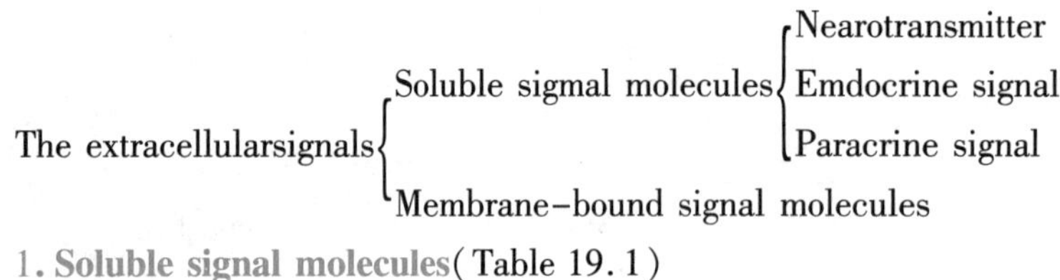

1. **Soluble signal molecules** (Table 19.1)

Table 19.1 Classificationof soluble signal molecules

	Neurocrine	Endocrine	Paracrine and autocrine
Name ofthe chemical signal	neurotransmitter	hormone	cytokines
Distance	nm	m	mm
Receptor	membrane receptors	membrane or intracellular receptors	membrane receptors
Examples	acetylcholine, glutamic acid	insulin, thyroid hormone, growth hormone	epidermal growth factor, interleukin, nerve growth factor

2. **Membrane-bound chemical signals**

(1) In multicellular organisms, adjacent cells can transmit signals through the specific recognition and interaction of cell surface molecules. One of them sends the signal through cell surface molecule, which bindswith the specific molecule (receptor) of another celland transmits message into the "target" cell.

(2) These cell surface molecules of the "sender" cell are called as **membrane-bound signaling molecules** or **contact-dependent signaling molecules**.

(3) The communication mode of transmitting signal through the specific recognition and interaction of adjacent cell surface molecules is called **contact-dependent signaling**. The interaction of ICAM and surface molecular interaction between T and B lymphocyte all belong to this mode.

Ⅱ. Receptors and signal transduction molecules

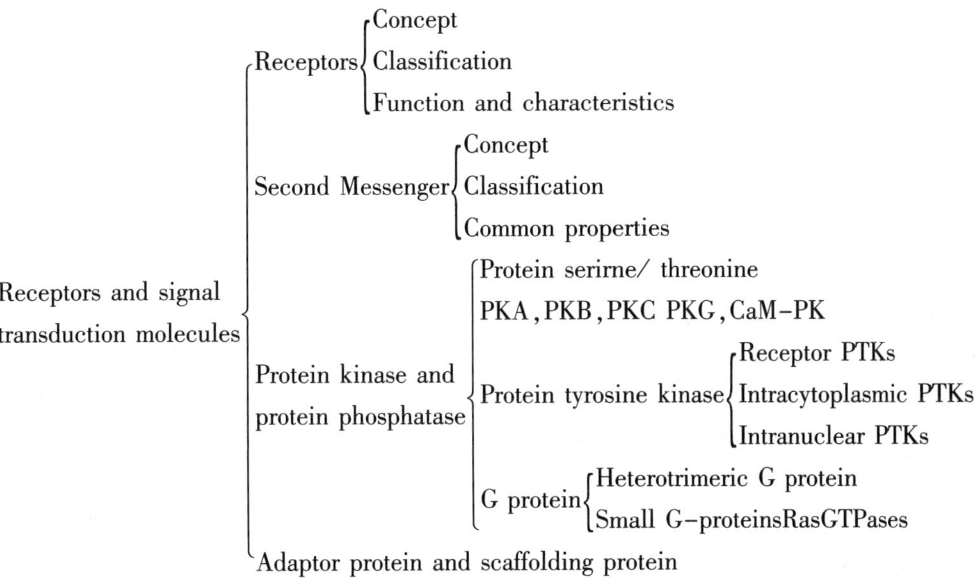

A. Receptors

1. Concept: A receptor is a protein molecule (or individual glycolipid) on the cell membrane or inside the cell which recognizes and bind to chemical signals from outside a cell and transmits the extracellular chemical signals into the cell inside to cause biological effects.

2. Classification (Table 19.2)

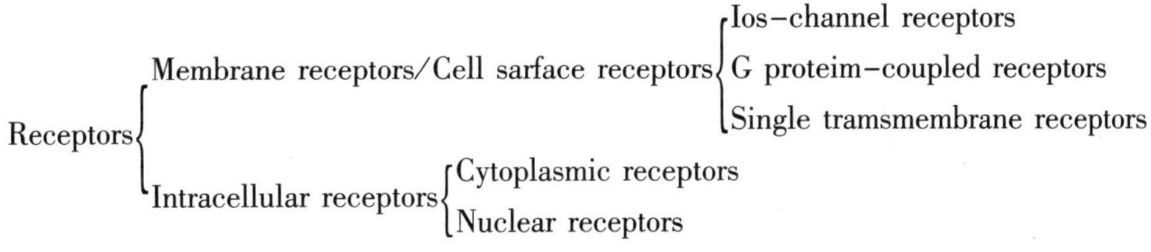

Table 19.2 The characteristics of three membrane receptors

Characteristics	Ion-channel receptors	G protein-coupled receptors	Single transmembrane receptors
Endogenous ligand	neurotransmitter	neurotransmitter, hormone, chemotactic factor, external stimulus (light, smell)	growth factor cytokines
Structure	channel formed by oligomers	monomer	monomer with or without catalytic activity
Number of transmembrane segments	4	7	1

Continue to Table 19.2

Characteristics	Ion-channel receptors	G protein-coupled receptors	Single transmembrane receptors
Function	ion channel	activating G protein	activating protein tyrosine kinase; activating protein filaments/threonine kinases
Cell response	depolarization and hyperpolarization	depolarization and hyperpolarization, regulate function and expression level of protein	regulate protein function and expression level, regulate cell differentiation and proliferation

3. **Function and characteristics of receptors**

(1) **Ligand**: chemical signals that bind to receptors. Soluble and membrane-bound signal molecules are common ligands.

(2) **Ligandbinding**

● **Water-soluble** signaling molecules and **membrane-bound** signal molecules usually can't get into the target cells. So, they transmit signals by binding to the **membrane receptors** on the target cell surface.

● **Liposoluble** chemical signals transmit signals by binding to **intracellular receptors**.

(3) **Function of receptors**: Both membrane receptors and intracellular receptors can recognize and convert ligand signals to be the recognizable intracellular signals and transmit them to other molecules, ultimately result in a series of biological effects such as intercellular adhesion and endocytosis, etc.

(4) **Characteristics of the binding of the receptor to the ligand**: high specificity, high affinity, saturability and reversibility.

B. Second Messenger

1. **Concept**: intracellular signaling molecules released by the cell in response to exposure to extracellular signaling molecules—the first messengers.

2. **Classification**

(1) Hydrophobic molecules (water-insoluble molecules): DAG and phosphatidylinositols; can diffuse from the plasma membrane into the intermembrane space where they can reach and regulate membrane-associated effector proteins.

(2) Hydrophilic molecules (water-soluble molecules): cAMP, cGMP, IP_3 and Ca^{2+}, located within the cytosol.

(3) Gases: NO, CO and H_2S; can diffuse both through the cytosol and across cellular membranes (Table 19.3).

Table 19.3 The main second messenger

Name	Regulated concentration	Representative target molecule	Related cell function
cAMP	AC catalyzescAMP synthesis. cAMP-dependent PDE degrades cAMP by hydrolyzing cAMP into 5′-AMP.	PKA, ion channel	metabolism, transcription, sense of taste, sense of smell
cGMP	GC catalyzes cGMP synthesis. cGMP-dependent PDE degrades cGMP by hydrolyzing cGMP into 5′-GMP.	PKG, ion channel	myocardial and smooth muscle contractions, sense ofvision

Continue to Table 19.3

Name	Regulated concentration	Representative target molecule	Related cell function
IP_3	PLC hydrolyzes PIP_2 to form IP_3.	IP_3 receptor (Ca^{2+} channel)	transcription, cytoskeletal recombination, proliferation
DAG	PLC hydrolysis to PIP2	PKC	transcription, cytoskeletal recombination, proliferation
Ca^{2+}	Extracellular Ca^{2+} influx and intracellular Ca^{2+} pool release	PKC, calmodulin	transcription, cytoskeletal recombination, proliferation
PIP_3	PI_3K catalytic PIP_2 phosphorylation to form	PKB	metabolism, cell adhesion
NO	NOS	GC, cytochrome	myocardial and smooth muscle contractions, oxidative stress

cAMP: cyclic AMP; cGMP: cyclic GMP; AC: adenylatecyclase; GC: guanylatecyclase; PDE: phosphodiesterase; PK: protein kinase; IP3: inositol-1,4,5-triphosphate; DAG: diacylglycerol; PLC: phospholipase; PIP3: phosphatidylinositol-3,4-triphosphate; NOS: NO synthase

3. The common properties of second messengers

(1) They can be synthesized/released and broken down again in specific reactions by enzymes or ion channels;

(2) In intact cells, the concentration (such as cAMP, cGMP, DAG, IP3, etc.) or distribution (such as Ca^{2+}) of some second messengers can change rapidly under the action of extracellular signals;

(3) Their production/release and destruction can be localized, enabling the cell to limit space and time of signal activity;

(4) The molecular analogues can mimic the action of extracellular signals, while blocking the molecular changes can block the cell's response to extracellular signals;

(5) There are specific target molecules in the cell, they can act as allosteric effector on target molecules;

(6) They are not at the center of the energy metabolic pathway.

C. Protein kinase and protein phosphatase

Phosphorylation and de-phosphorylation of protein, catalyzed by protein kinase (PK) and protein phosphatase, are the most important way of rapidly regulating the activity of intracellular signal transduction molecules.

1. Protein kinase (Table 19.4)

Table 19.4 Some important protein kinases in cells

Type	Name	Regulator	Substrate	Relatedpathways
Protein serirne/ threonine kinase	PKA	cAMP	glycogen synthase, CREB	Metabolism, transcription
	PKB	PIP_3	glycogen synthase kinase, caspase 9	Metabolism, apoptosis, cell proliferation
	PKC	DAG, Ca^{2+}	Membrane calcium ion channel, c-Fos	Transcription, cytoskeletal reorganization, cell proliferation

Continue to Table 19.4

Type	Name	Regulator	Substrate	Relatedpathways
	PKG	cGMP	Myosin, NOS	Myocardial and smooth muscle contractions
	CaM-PK	Ca^{2+}-CaM		Muscle contraction, stress
	MAPK	Ras		Cell proliferation and differentiation, stress, inflammation, etc
	CDK	Cyclins		Cell cycle
	TGF-β receptor	TGF-β		cellular proliferation and differentiation
Protein tyrosine kinase (PTKs)	Receptor PTKs: EGFR、InsR	EGF, insulin	Autophosphorylation, IRS-1	Cell proliferation, differentiation, metabolism
	Intracytoplasmic PTKs: Src, Syk, JAK, Tecfamily	Receptor activation	T-cell receptor, B-cell receptor	Cell activation
	Intranuclear PTKs: Abl			Cell cycle

Notes: MAPK, mitogen activated protein kinase; CDK, cyclin dependent kinase; NOS, nitric oxide synthase; CaM, calmodulin; CREB, cAMF response element binding factor; TGF, transforming growth factor; EGF, epidermal growth factor; IRS-1, insulin receptor substrate-1

2. Protein phosphatase

(1) Protein phosphatase **dephosphorylates** the phosphorylated proteins, whose role is exactly opposite to protein kinase.

(2) **Types**: serine/threonine phosphatases, protein tyrosine phosphatases, and dual effect phosphatases (removing the phosphate groups from tyrosine and serine/threonine residues simultaneously).

D. G protein (guanine nucleotide-binding proteins or GTP binding proteins)

1. **Function**: act as a switch in signal transduction. the conformational change and the activity:

(1) Active state: GTP binding, 'on'

(2) Inactive state: GDP binding, 'off'

2. Theclasses of G proteins

(1) Heterotrimeric G protein

● **Subunits**: α, β and γ (β, γ, dimeric complex)

Gα: **GPCR** (**G protein-coupled receptors**) binding;

GTP/GDP binding; GTPase;

downstream signal transduction molecules binding

Gβγ binding

● Signaling molecules (**ligand**) bind to a domain of the **GPCR** located outside the cell which activates GPCR, and an intracellular GPCR domain then in turn activates a particular **G protein**.

GPCR function as a guanine nucleotide exchange factor (GEF) thatexchanges GDP for GTP:

G protein(inactive)→**G protein**(active)→**Effectors**

(GDP binding) (GTP binding) (downstream molecules)

"Off" "On"

Different α Subunit— Different Effector

(2) **SmallG-proteins**: 20kD to 25kD, also known as Small GTP_{ases}; The best-known members are the Ras GTP_{ases} (Ras subfamily GTP_{ases}).

E. Adaptor protein and scaffolding protein

1. Protein interaction domain

(1) **The structural basis of signal transduction pathways and networks**: In some processes of signal transduction, signal transduction molecules interact with each other, gather and form **signal transduction complexes(signaling complexes)**.

(2) The basis of signal transduction complex formation is **protein interactions**. The structural basis of protein interaction is **protein interaction domain**(Table 19.5).

(3) **Characteristics of protein interaction domain**:

● There is high homology in different signal transduction molecules.

● There are more than two protein interaction domains in one signal transduction molecule which can combine two or more signal molecules simultaneously.

● The same protein-like interaction domain can exist in different molecules, which selectivity binds to the downstream signal molecule.

● They have no catalytic activity.

Table 19.5 Protein interaction domains and their recognizedmotifs

Protein interaction domain	Abb.	Molecule categories	Identifying motif
Src homology 2	SH2	protein kinase, phosphatase, adapter proteinetc.	Phosphotyrosine motif
Src homology 3	SH3	adapter protein, phospholipase, protein kinase, etc.	Rich Proline motif
Pleckstrin homology	PH	protein kinase, Cytoskeleton regulates molecules, etc.	phospholipids derivatives
Protein tyrosine binding	PTB	protein kinase, phosphatase	Phosphotyrosine motif

2. Adaptor protein

(1) **Adaptor proteins** are proteins that are accessory to main proteins in a signal transduction pathway, and contain a variety of protein-binding modules that link the upstream and downstream signal transduction molecules together and facilitate the creation of larger signaling complexes.

(2) **Function**: mediate specific protein-protein interactions that drive the formation of protein complexes; lack any intrinsic enzymatic activity

(3) **Example**: MYD88 and Grb2.

3. Scaffold proteins

(1) **Function**: interact and/or bind with multiple members (proteins) of a signaling pathway, tethering them into complexes.

(2) In such pathways, they regulate signal transduction and **help localize** pathway componentsto spe-

cific areas of the cell to avoid cross response and maintain the **specificity** of signal transduction pathway. It also increases the **complexity and diversity** of regulation.

Ⅲ. Membrane receptor-mediated intracellular signal transduction

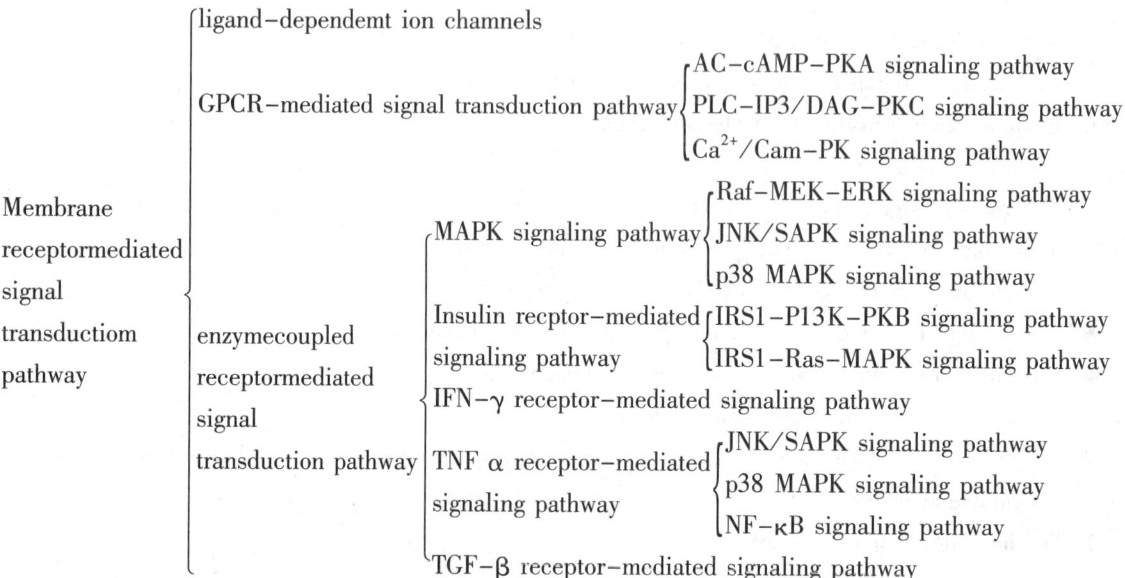

A. GPCR-mediated signal transduction

GPCR (G protein-coupled receptor) commonly mediates the transduction of neurotransmitters, peptide hormones, chemokines and exogenous physical and chemical factors in taste, vision, smell and ol faction.

The basic pattern/basic process of GPCR-mediated signal transduction pathway:

(1) Extracellular signal binds to GPCR;

(2) GPCR activates G protein;

(3) Activated G protein activates or suppresses downstream effector molecules;

(4) Effector molecules cause a rapid change in the content or distribution of intracellular second messengers;

(5) The second messenger activated the corresponding target molecule (mainly protein kinase) through allosteric regulation and then activated protein kinase changed some metabolic enzymes and transcription factors related to gene expression through phosphorylation, and ultimately results in various cell response responses (Table 19.6).

Table 19.6 GPCR-mediated signaling pathway

Signaling pathway	Target Protein	Ligand	Mechanism	Biological effects
AC-cAMP-PKA signaling pathway	PKA	Glucagon, adrenaline (β1, β2), adrenocorticotropic hormone, parathyroid hormone, prostaglandin E1 and E2, growth hormone inhibin, dopamine, histamine H2 receptor and serotonin	Ligand + GPCR → G protein → AC activation → cAMP ↑ → PKAactivation → phosphorylation of target proteins	Regulate metabolism (glucose, cholesterol and lipid metabolism); Regulate gene expression (activate transcription factors and regulate gene expression by CREB.); Regulate cell polarity (activate ion channels and regulate cell membrane potential)
PLC-IP3/ DAG-PKC signaling pathway	PKC	Angiotensin II, thyrotropin - releasing hormone (TRH), norepinephrine, antidiuretic hormone, adrenaline (α1, α2), gonadotropin - releasing hormone, gastric secreting hormone-releasing peptide, acetylcholine M1	Ligand + GPCR → G protein→PI-PLC activation → DAG + IP3 →PKC activation → phosphorylation of target proteins	Regulate multiple physiological functions; Regulate gene expression (phosphorylates the transcription factor such as *c-FOS* of immediate-early gene and accelerates its expression, promoting cell proliferation.)
Ca^{2+}/CaM - PK signaling pathway	CaM-PK		Ligand + GPCR → G protein → promote Ca^{2+} influx to the cytoplasm → activate Ca^{2+}/CaM comple x → activate downstream molecules (CaM-PK) →phosphorylation of target proteins	Play roles in muscle contraction and movement, glucose metabolism, synthesis and release of neurotransmitters, cell secretion and division, etc. Cal-PK II can modify and activates synaptophysin I, skeletal muscle glycogen synthase, tyrosine hydroxylase, tryptophan hydroxylase and so on.

CaM: calmodulin; CaM-PK: CaM-dependent protein kinase

B. Enzyme-coupled receptor-mediated signal transduction

Enzyme-coupled receptors:

- single transmembrane receptors or **protein tyrosine kinase (PTK)** -associated receptors.
- **Characteristics:** having an intrinsic enzymatic activity on their cytosolic domain or by being coupled with an enzyme.
- **Two main domains:** ligand-binding site on the outside and an enzyme binding (catalytic) domain on the inside of the cell.
- **Function:** Enzyme-coupled receptor mainly receives extracellular signals such as **growth factors and cytokines**, transduces signals through the interaction of protein molecules, and regulates the function

and expression of proteins and the **proliferation and differentiation** of cells.

1. MAPKsignaling pathways

(1) MAPK signaling pathways are the signal transduction pathways represented by mitogen-activated protein kinase (**MAPK**), which are characterized by MAPK cascade activation.

(2) MAPK signaling pathways mainly include **ERK** (Ras-Raf-MEK-ERK signaling pathway), **p38 MAPK** signaling pathway and **JNK/SAPK** signaling pathway (Figure 19.2).

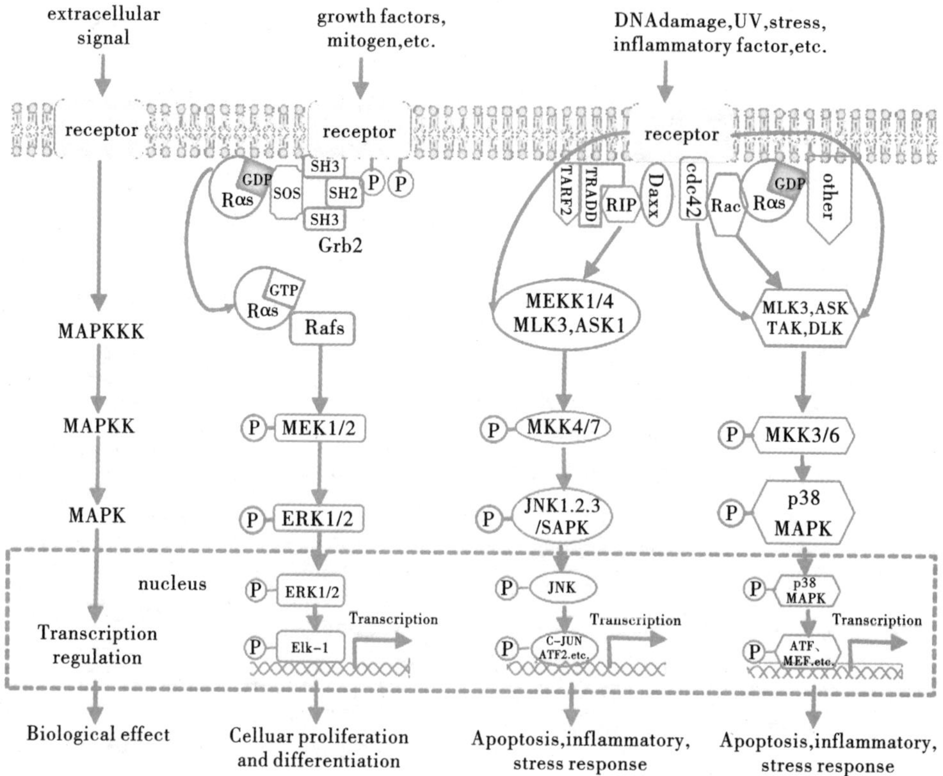

Figure 19.2 MAPK Pathway mediated by protein kinase coupled receptor

2. Insulin receptor-mediated signaling transduction pathway

(1) Insulin receptor (IR): **receptor PTK**

● **Subunits**: two α subunits: have ligand-binding sites

two β subunits: have intrinsic protein tyrosine kinase (**PTK**) activity.

● **Function**: Insulin transmits the signal to the cell through the insulin receptor on the cell surface and regulates the expression of **insulin-sensitive metabolic enzymes and specific genes**, which plays a very important physiological role in cell metabolism and the regulation of gene expression.

(2) Insulin receptor-mediated signal transduction pathways mainly are **IRS1-PI3K-PKB and IRS-Ras-MAPK signaling pathways** (Figure 19.3). The biological effect of insulin in cells is related to the substrates of **PKB**.

PKB→inactivate the glycogen synthase kinase 3 (**GSK3**) →intracellular glycogen synthesis ↑

PKB→promote the translocation of glucose transporter 4 (**GLUT4**) →GLUT4 on the cell membrane ↑, glucose intake ↑

3. Interferon γ receptor-mediated signaling pathway

(1) Interferon (IFN-γ) is produced by activated T cells, which can promote antigen presentation and specific immune recognition and promote the secretion of antibodies by B cells.

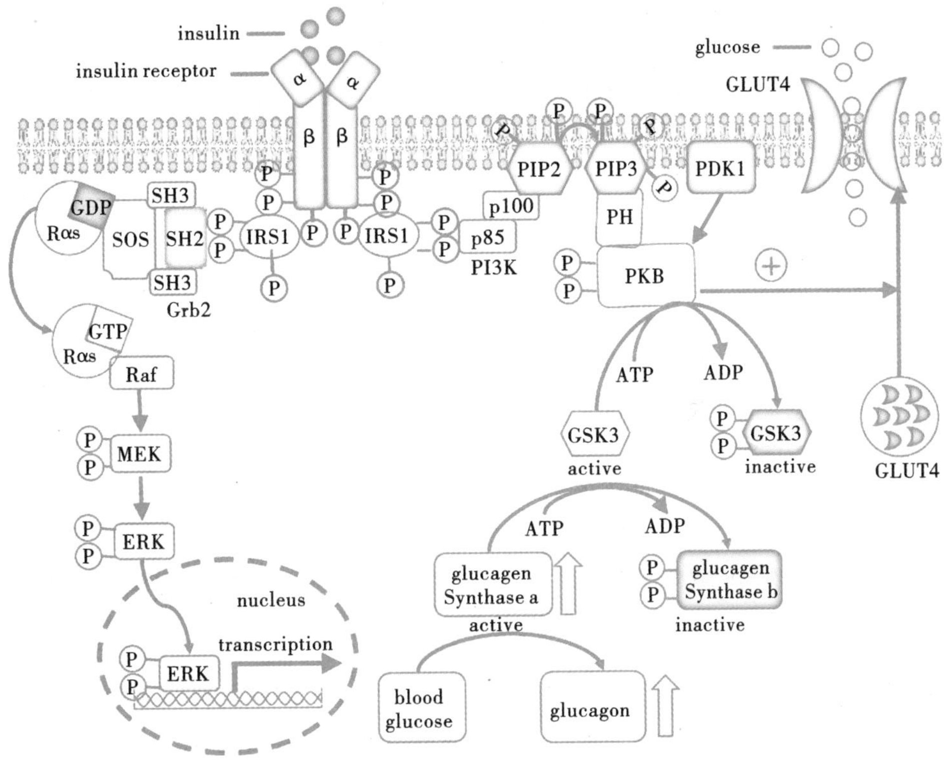

Figure 19.3 Insulin receptor-mediated signal transduction pathway

(2) IFN-γ binds to its receptor and causes the dimerization of the receptor, which activates the JAK-STAT system and transmits interferon stimulation signals into the nucleus by STAT.

●JAK (Janus kinase) is a class of non-receptor protein tyrosine kinase in the cytoplasm, which binds to the proximal membrane of cytokine receptor.

●STAT is a signal transduction and transcriptional activation factor. Each STAT molecule has a SH2 structure. In the JAK-STAT pathway, the activated receptors can be combined with different JAK and STAT, so the transduction signal of this pathway is diverse and flexible.

4. TNFα receptor-mediated Signaling pathway

(1) The tumor necrosis factor α (TNFα) can mediate the transduction of signals *through its receptor TNF receptor (TNF-R)*.

(2) TNFR-associated factor (TRAF) is involved in the signal transduction of this pathway. There are at least six subtypes of TRAF in eukaryotic cells, each of which can bind to different receptors and transmit different signals to the cells. After binding TNFα to TNF-R1, TNF-R is mainly transmitted through NF-κB, JNK and p38 MAPK signal pathway.

5. TGF-β receptor-mediated signaling pathway

(1) The receptor of transforming growth factor β (TGF-β) is also a single transmembrane receptor. However, the receptor itself has a catalytic domain of protein serine/threonine kinase. The receptors belonging to TGF-β family include receptors of BMP and activin. The cytokines of this family are involved in the regulation of *proliferation, differentiation, migration and apoptosis*.

(2) *TGF-β binds to* typeI andtype Ⅱ receptors respectively to form heteropolymer. Activated type II receptor phosphorylates and activates type I receptor. Type I receptors phosphorylate and transfer Smad2, Smad3 and Smad4 to form trimer into cells, bind to Smad binding elements, regulate the transcription rate of the corresponding genes, and ultimately affect cell differentiation.

Ⅳ. Intracellular receptor-mediated signaling pathway

1. Ligand: **liposoluble chemical signals** such as steroid hormones, thyroxines, prostaglandins, vitamin A and its derivatives, vitamin D and their derivatives.

2. The **basic processes**

(1) In the absence of hormone, the intracellular receptor forms a complex with an inhibitory protein molecule, **heat shock protein (HSP)**, to prevent the receptor to translocate into nucleus and bind to DNA.

(2) **Signal transduction mechanism**

When the hormones enter the cell

● some bind to their receptor located in the nucleus to form a hormone-receptor complex

● while others first bind to their receptor in the cytoplasm and then enter the nucleus as a hormone-receptor complex. After the hormone binding to the intracellular receptor, the conformation of the receptor changes, leading to the depolymerization of HSP, exposing the nucleus transfer site and DNA binding site of the receptor, transferring the hormone-receptor complex into the nucleus and binding to the **hormone response element (HRE)** in the promoter region of its target gene (Figure 19.4).

● Hormone receptor complexes that bind to hormone response elements interact with basic **transcription factors** and other **transcriptional regulators** located in the promoter region, which open or close down their downstream genes and regulate the expression of their target genes at the transcriptional level, and then to change the gene expression profile of the cell.

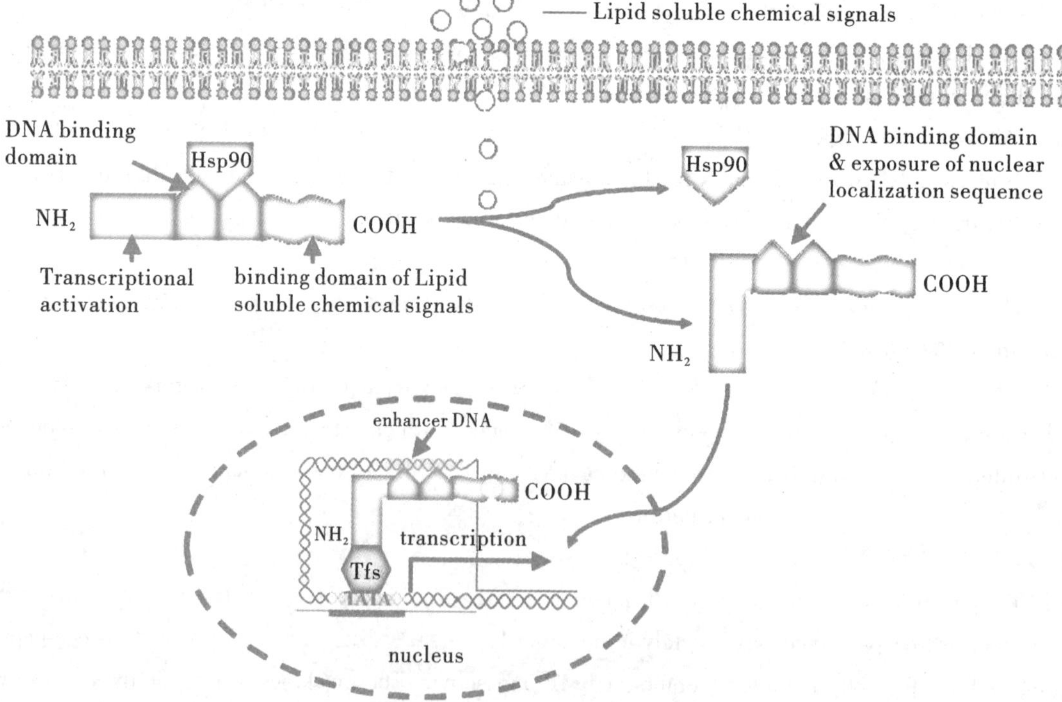

Figure 19.4 Intracellular receptor-mediated signal transduction pathway

V. Abnormal intracellular signal Transduction and Medicine

A. Abnormal intracellularsignal transduction

1. Abnormal activation and disability of receptors

2. Abnormal activation and disability of intracellular signal transduction molecules

Changes in the structure and chemical modification of intracellular signal transduction molecules lead to their abnormal activation and maintenance of the active state. The structure changes or decreased expression of intracellular signal transduction molecules may lead to abnormal deactivation.

B. Abnormal signal transduction and disease

1. Abnormal function or phenotype of cells caused by abnormal signal transduction

(1) Gain of abnormal proliferative ability of cells, eg. RAS gene mutation

(2) Abnormal secretory function of cells, eg. growth hormone secretion

(3) The changes of membrane permeability, eg. cholera toxin.

2. Loss of normal cell functions caused by abnormal signal transduction

(1) Loss of normal secretory function

(2) Loss of the normal responsiveness or the physiological regulation

C. Signal transduction molecules are important targets for drug action

At present, some drugs have been used in the clinic, especially in the treatment of cancer and autoimmune disorders.

[*Practices*]

[A1 type]

[1] Which one of the following is a liposoluble signal molecule?__________

A. Insulin B. Thyroid hormone C. EGF

D. Ag II E. ICAM

[2] Which one of the following is a membrane-bound chemical signal?__________

A. Insulin B. Thyroid hormone C. EGF

D. Ag II E. ICAM

[3] Which one of the following is not a second messenger?__________

A. NO B. CO C. H_2S

D. Ca^{2+} E. Na^+

[4] Which of the following is a nuclear receptor?__________

A. Acetylcholine receptor B. Glucocorticoid receptor C. Interferon-γ receptor

D. Insulin receptor E. Isoprenaline receptor

[5] Nuclear receptors are essentially ligand activated__________.

A. transcription factor B. effector C. ion channel

D. protein tyrosine kinase E. protein serine/threonine kinase

[6] The main way of signal transduction system regulating target proteins is through __________.

A. methylation and demethylation

B. acetylation and deacetylation

C. phosphorylation and dephosphorylation

D. glycosylation and desaccharification

E. ubiquitination and deubiquitination

[7] Which protein kinase of the following is the target molecule of cAMP?__________

A. PKA B. PKB C. PKC D. PKD E. PKG

[8] Which protein kinase of the following is the target molecule of cGMP?__________

A. PKA B. PKB C. PKC D. PKD E. PKG

[9] The most common mutation of small G protein Ras in the tumor can cause__________.

A. Ras reduce

B. Ras inactivation

C. dissociation disorder of Ras and GDP

D. reduced activity of Ras GTPase

E. the reduced ability of Ras to activate the ERK pathway

[10] Which protein kinase of the following doesn't belong to protein serine/threonine kinase?

____

A. PKA B. PKB C. PKC D. PTK E. CaM-PK

[11] Which of the following cannot activate NF-κB? __________

A. TNFα B. Virus C. Reactive oxygen species

D. Endotoxin E. Thyroxine

[12] Which signaling pathway is closely related to cell proliferation and hypertrophy and tumor development? __________

A. Guanosine acyclic enzyme pathway

B. Nuclear receptor pathway

C. Adenylatecyclase pathway

D. Receptor PTK pathway

E. Non-receptor PTK pathway

[13] Which one as following about ERK family is true? __________

A. Participate in the regulation of cellular inflammatory response

B. Participate in the regulation of cell proliferation and differentiation

C. Participate in cell stress regulation

D. The MAPK subfamily found earlier

E. The MAPK subfamily found later

[14] Which signaling pathway of the following is involved in cellular inflammation and stress response? __________

A. TGF-β receptor-mediated signaling pathway

B. IFN-γ receptor-mediated signaling pathway

C. IRS1-Ras-MAPK signaling pathway

D. IRS1-PI3K-PKB signaling pathway

E. JNK/SAPK signaling pathway

[15] Which cellular process does TGF-β receptor signaling pathway participate mainly? __________

A. Participate in the regulation of cellular inflammatory response

B. Participate in the regulation of cell proliferation and differentiation

C. Participate in cell stress regulation

D. Participate in the regulation of sugar metabolism

E. Participate in the regulation of lipid metabolism

[16] Which dimer of the following is the main form of NF-κB playing a physiological role in the body? ________

A. p50-p65 dimer
B. p52-p65 dimer
C. p50-RelB dimer
D. p52-RelB dimer
E. p65- RelB dimer

[17] Which one of the following isn't the intracellular receptor-mediated signaling pathway? ________

A. Glucocorticoid receptor-mediated signaling pathway

B. NF-κB receptor-mediated signaling pathway

C. Prostaglandins receptor-mediated signaling pathway

D. TGF-β receptor-mediated signaling pathway

E. Thyroxin receptor-mediated signaling pathway

[18] Which one of the following isn't the liposoluble chemical signals? ________

A. Steroid hormones

B. Prostaglandins

C. Platelet-derived growth factor

D. Vitamin A and its derivatives

E. Vitamin D and their derivatives

[19] Which of the following diseases isn't related to the abnormal G protein? ________

A. Cholera

B. Pseudohypoparathyroidism hypothyroidism

C. Gigantism

D. Acromegaly

E. Myasthenia gravis

[20] The molecular mechanism of the abnormal G protein caused by cholera toxin is ________.

A. the A subunit of cholera toxin makes G protein α subunit mutant

B. the B subunit of cholera toxin makes G protein α subunit mutant

C. the A subunit of cholera toxin causes ADP-ribosylation of G protein α subunit

D. the B subunit of cholera toxin causes ADP-ribosylation of G protein α subunit

E. the A and B subunit of cholera toxin makes inactivation of G protein α subunit

[21] In hereditary hypoparathyroidism, α subunit mutation of G protein in signaling pathway of parathyroid hormone could result in ________.

A. conformational change

B. invariability of protein primary structure

C. abnormal activation of the function

D. abnormal inactivation of the function

E. None of the above

[B1 type]

No. 22 ~ 25 share the following suggested answers.

A. protein kinase A (PKA)

B. protein kinase C(PKC)

C. Ca^{2+}/ calmodulin dependent protein kinase

D. Caspases

E. JAK-STAT pathway

[22] Diacylglycerol can activate ____________.

[23] cAMP can activate ____________.

[24] IL-1 receptor can activate____________.

[25] Interferon receptor can activate ____________.

No. 26 ~ 29 share the following suggested answers.

A. serine/threonine protein kinase type receptor

B. tyrosine kinase receptor

C. nuclear receptor

D. ion channel receptor

E. death receptors

[26] TGFβ receptor belongs to____________.

[27] Insulin receptor belongs to____________.

[28] TNFα receptor belongs to____________.

[29] Androgen receptor belongs to ____________.

[X type]

[30] The increase in the number of signal transduction proteins regulating cell proliferation in tumors is due to its____________.

A. increase of gene copy number

B. abnormal overexpression of gene

C. degradation decreases or slows down

D. constitutively activated mutation

E. inhibitory mutation

Answers

1	2	3	4	5	6	7	8	9	10
B	E	E	B	A	C	A	E	D	D
11	12	13	14	15	16	17	18	19	20
E	D	B	E	B	A	D	C	E	C
21	22	23	24	25	26	27	28	29	30
D	B	A	D	E	A	B	E	C	ABC

Brief Explanations for Practices:

19. When cholera toxin composed of A and B two subunits enters into human small intestinal epithelial cells, A subunit directly acts on α subunit of G protein and results in its ADP-ribosylation modification. The intrinsic GTPase activity of α subunit is lost after modification and be unable to recover to the GDP binding form, so that α subunit of G protein is continuously activated, causing severe diarrhoea and dehydration. Pseudohypoparathyroidism is a genetic disease caused by a decrease in the responsiveness of target organs to

parathyroid hormone (PTH), in which the pathogenesis of type PHP1A is due to a single gene mutation that encodes the GSα allele. The mRNA level of GS α in patients was 50% lower than that in normal controls, leading to signal transduction decoupling between PTH receptor and AC. Growth hormone (GH) is a polypeptide hormone secreted by the adenohypophysis, which function is to promote body growth. When GS α was continuously activated, the AC activity is increased, and the content of cAMP is increased, the growth and secretory function of pituitary cells are active. The hypersecretion of GH could stimulate bone overgrowth, causing acromegalysis in adults and giant in children.

(***Li Dongmin***)

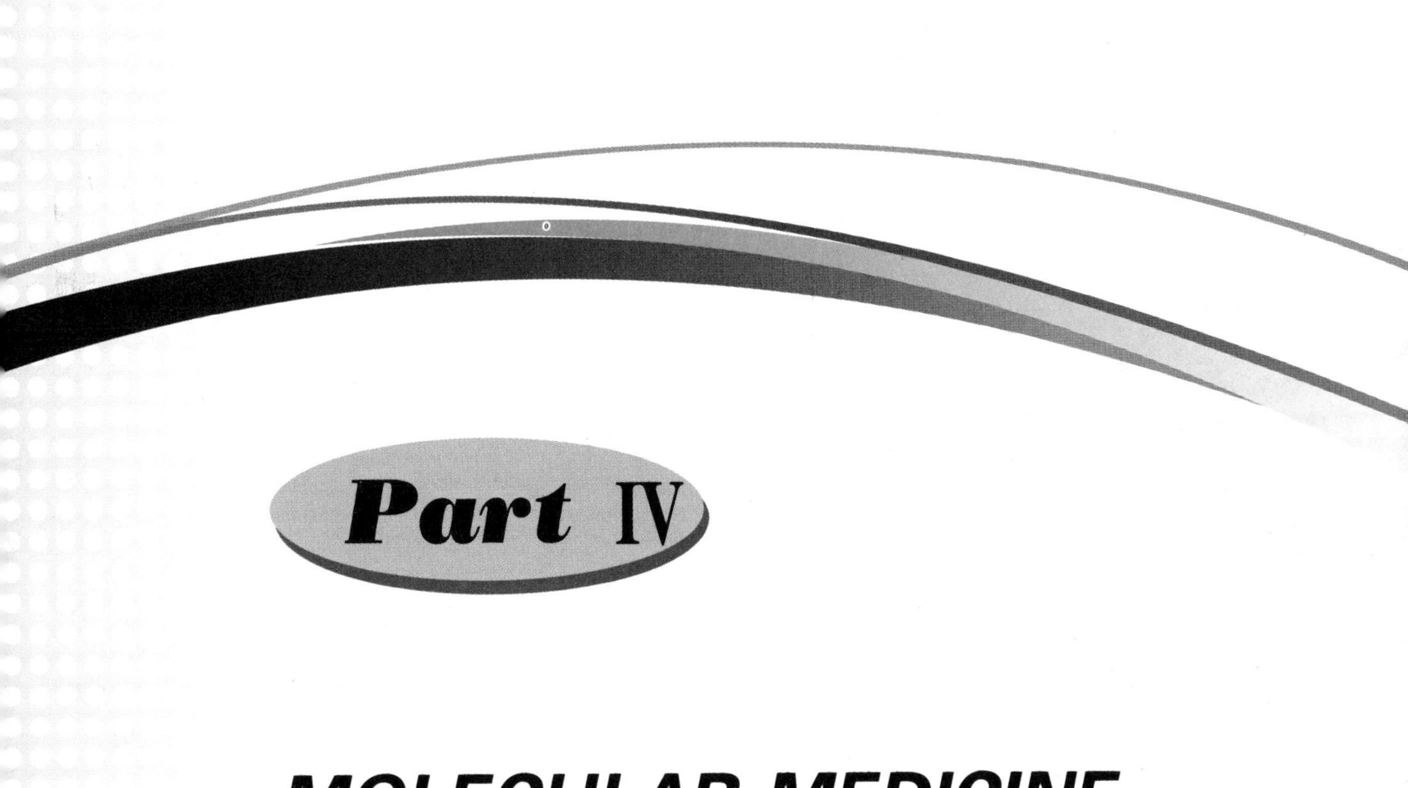

Part IV

MOLECULAR MEDICINE

Chapter 20

The Basics of Omics

Examination Syllabus

Genome and genomics, proteomics: Definition

[Major points]

-ome in molecular biology is a new suffix referring to a *complete set of molecules in an organism cell. For example genome, proteome etc.*

Omics are the studies for the fields with -ome, which include structural, functional, molecular interaction and bioinformatics studies of a whole set of molecules, such as genomics and proteomics.

Omics accoding to the central dogma

Genomics→Transcriptomics→Proteomics→Metabolomics

Ⅰ. Genomics

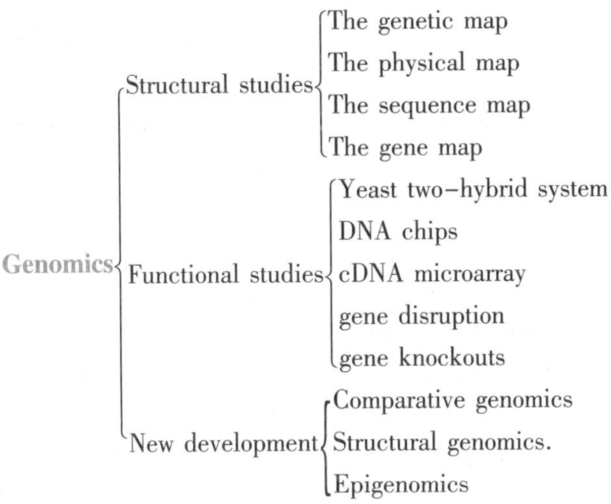

Genome: all genetic materials of an organism, including both genes and the sequences between genes, may be subdivided into the nucleic genome, mitochondria genome and chloroplast genome.

C-value: the amount of DNA in haploid of a somatic cell in eukaryotes and in prokaryotes C-value is the same size as genome.

C-value paradox: the genome size does not correlate with organism complexity.

Human Genome Project (HGP):

The HGP goals include mapping, sequencing, and identifying genes, storing and analyzing data, and addressing the ethical, legal, and social issues.

A. The genome structure

1. **The genetic map (linkage map)**: a map based on the recombination frequencies between genetic markers.

cM (centiMorgan) is the unit in the genetic map. One cM of two genetic markers equivalents to 1% of there combination frequency.

Higher recombination rate means the two markers further apart.

2. **The physical map**: a series of continuous overlapping cloned genomic DNA segments extending from one telomere of the chromosome to the other.

3. **The sequence map**: the most detailed, highest resolution map in the whole genome.

4. **The gene map (transcriptional map)**: The transcriptional map is a dynamic set of mRNA sequences. In different tissue, under certain conditions the transcriptional map is different.

B. Functional genomics (post-genomics)

Functional genomics studies focus on gene transcriptions, translations, and protein-protein interactions on the genome wide scale.

C. Comparative genomics

Comparative genomics works by aligning sequences of different organisms to identify patterns that operate over both large and small distances in genome.

D. Structural genomics

1. The **traditional structural genomics** refers to that studies genome structure, including all the genomic maps structures.

2. The **late meaning of the structural genomics** is to describe the 3-dimensional structures of every protein encoded among the whole genome.

3. **Main methods**: X-ray crystallography, nuclear magnetic resonance (NMR), sequence-based modeling, homology modeling, etc.

E. Epigenomics

1. Epigenetics focuses on the analysis of epigenetic changes across the entire genome, which are relevant toof chemical changes that influence gene expression but do not affect the sequence of DNA.

2. The epigenomics study includes DNA methylation, histone modification, chromatin accessibility, gene silencing, etc.

3. The epigenomics also study the effects of environment, life style, age, diet, hormones, toxins and disease state on the gene expression and development.

Ⅱ. Proteomics

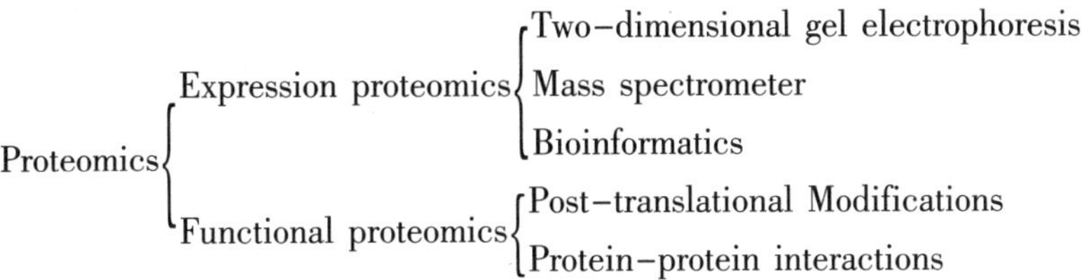

Proteome is a whole set of proteins expressed by a genome, cell, tissue or organism.

Proteomics is the study onthe dynamic proteome.

Proteomics usually study the set of expressed proteins in a given cell type or organism, at a given time, under defined conditions.

Proteomics study the overall level of intracellular protein composition, structure, and its own unique activity patterns, protein-protein interaction, protein function, protein modifications, and protein localization, etc.

A. Expression proteomics

1. Expression proteomics is the quantitative study of protein expression between samples under different conditions (for example: healthy and cancer)

2. **Two-dimensional gel electrophoresis** (2-DE). The main method used in expression proteomics is the separation of proteins by two-dimensional gel electrophoresis.

In the first dimension, the proteins are separated by isoelectric focusing. In the second dimension, proteins are separated by molecular weight using SDS-PAGE.

B. Post-translational modifications (PTM)

Post-translational modifications are key mechanisms to increase proteomic diversity. The post-translational modifications include phosphorylation, glycosylation, ubiquitination, S-nitrosylation, methylation, acetylation, lipidation etc.

C. Interaction proteomics

1. The characterization of protein-protein interactions is used to determine protein functions and to demonstrate how proteins assemble in larger complexes.

2. **The biochemical methods**: Co-immunoprecipitation, affinity electrophoresis, phage display, Chemical cross-linking, etc.

3. **New methods**: protein microarrays, analytical ultracentrifugation, light scattering, fluorescence spectroscopy, protein-fragment complementation assay, etc.

Ⅲ. Other kinds of Omics

1. **Transcriptome** is the set of all RNA molecules in one cell or a population of cells.

2. **Metabolome** refers to the complete set of small-molecule metabolites to be found within a biological

sample.

3. **Metabolomics** refers to the systematic identification and quantification of these small-molecule metabolic productsof a biological system (cell, tissue, organ, biological fluid, or organism) at a specific time point.

4. **Glycomics** is the study of the complete set of glycan structures expressed by specific cells, tissues or organisms.

5. **Lipidomics** is the study of the structure and function of the complete set of lipids produced in a given cell or organism as well as their interactions with other lipids, proteins and metabolites.

Ⅳ. Applications of omics in medicine and health

1. **The application of genomics**: genomic medicine, nutritional genomics, metagenomics, conservation genomics and synthetic biology.

2. **The application of proteomics**: human plasma proteome, cancer, drug development.

[*Practices*]

[A1 type]

[1] Which of the following word does not belong to omics?__________

A. Transcriptomics B. RNomics C. Economics
D. Lipidomics E. Proteomics

[2] The Human Genome Project officially started in__________.

A. 1984 B. 1985 C. 1986
D. 1990 E. 1992

[3] The major goal of HGP is to draw four maps except__________.

A. genetic map B. physical map C. chemical map
D. sequence map E. genes map

[4] The working draft of DNA sequence of the human genome was announced in the year__________.

A. 1990 B. 1995 C. 2000
D. 2001 E. 2003

[5] The 3 billion base pair human genome is the DNA size of __________ in a somatic cell.

A. Haploid B. Diploid C. Triploid
D. Tetraploid E. Can't be decided.

[6] The relationship of the distance between the two genetic markers and their recombination rate is as follow__________

A. The higher recombination rate means they are further apart.
B. The lower recombination rate means they are further apart.
C. The higher recombination rate means they are closer.
D. The higher recombination rate means they are more linked to each other.
E. The higher recombination rate means their genes are bigger.

[7] The genetic map is also called__________.

A. gene map B. transcript map C. physical map
D. linkage map E. recombination map

[8] Unit one cM in the genetic map means __________.

A. One centiMeter. B. One centiMolar. C. One centiMolecule.
D. One centiMorgan. E. One centiMendel.

[9] Which map in the genome structure is the most detailed? ____________
A. Cytogenetic map B. Linkage map C. Physical map
D. Chemical map E. Sequence map

[10] The gene map is also called ____________.
A. the genetic map B. the linkage map C. the transcriptional map
D. the physical map E. the sequence map

[11] Now the genes in the human genome are estimated about____________.
A. 10,000 B. 20,000 C. 50,000
D. 100,000 E. 1,000,000

[12] A cDNA library contain clones representing which of the following? ____________
A. Genomic DNA B. Repeated DNA sequences C. Introns
D. mRNA E. Micro RNA (miRNA)

[13] Which type of genomics studies the transcripts and proteins expressed by a genome? ____________
A. Structural genomics B. Sequencing genomics C. Comparative genomics
D. Functional genomics E. Epigenomics

[14] Which technique doesn't belong to the functional genomicsapproaches? ____________
A. Genomic DNA sequencing B. DNA chips C. cDNA microarray
D. Gene disruption E. Gene knockouts

[15] Which of the following can be used to screen gene expression. ____________?
A. Southern blot B. Western blot C. cloning library
D. cDNA library E. DNA microarrays

[16] Which type of genomics study the similarities and differences among the genomes of multiple organisms?____________
A. Epigenomics B. Metagenomics C. Functional genomics
D. Structural genomics E. Comparative genomics

[17] A plan to determine massive protein 3-D structures belongs to____________.
A. Sequencing genomics B. Physical genomics C. Functional genomics
D. Structural genomics E. Comparative genomics

[18] Which method is the most efficient way to study the protein 3-D structures? ____________
A. Fluorescencespectrometry B. UVspectrometry C. Circular Dichroism(C. D.)
D. NMR E. X-ray diffraction

[19] Proteomics is the large-scale study of ____________, particularly their structures and functions.
A. DNA B. RNA C. mRNA
D. proteins E. metabolism

[20] The first dimension of separation for two-dimensional electrophoresis is based on____________.
A. molecularweight B. amino acids sequence C. nucleic acids sequence
D. folding E. isoelectric point

[21] The separation of the two-dimensional electrophoresis is based on____________.
A. molecularweight and pH
B. isoelectric point and pH
C. molecularweight and isoelectric point

D. isoelectric point and conformation

E. molecular weight and conformation

[22] Which of the following technique doesn't belong to expression proteomics? ____________

A. Two-dimensional gel electrophoresis

B. Mass spectrometry

C. Bioinformatics

D. Yeast two-hybrid system

E. Quantitative proteomics

[23] The post-translational modifications don't include____________.

A. Phosphorylation　B. Glycosylation　C. Acetylation

D. Amination　E. Methylation

[X type]

[24] The HGP has not finished the following maps____________.

A. genetic map　B. physical map　C. sequence map

D. genes map　E. transcriptional maps

[25] The Human Genome Project goals include____________.

A. mapping human genome

B. sequencing human genome

C. identifying human genes

D. recombination human genes

E. developing faster, more efficient sequencing technologies

[26] Does the human genome include which of the following? ____________

A. Nucleic genome　B. Chloroplast genome　C. Mitochondria genome

D. Golgi genome　E. Ribosomal genome

[27] The goal of functional genomics includes____________.

A. "When and where are genes expressed?"

B. "How do genes and gene products interact?"

C. "How do gene expression levels differ in various cell types and states?"

D. "What are the functional roles of different genes and in what cellular processes do they participate"

E. "How are genes regulated? Where are the active gene promoters in a particular cell type?"

[28] The epigenomics study includes ____________.

A. DNA sequencing　B. DNA methylation　C. histone modification

D. chromatinaccessibility　E. gene silencing

[29] The epigenomics also study the effects of gene expression by____________.

A. environment　B. life style　C. age

D. diet　E. disease

[30] The techniques used in the expression proteomics include____________.

A. two-dimensional gel electrophoresis

B. mass spectrometry

C. bioinformatics

D. fluorescence spectrometry

E. NMR

[31] Post-translational modifications include____________.

A. phosphorylation B. carboxylation C. glycosylation
D. amination E. acetylation

[32] The methods to investigate protein–protein interactions include____________.
A. co–immuno precipitation B. affinity electrophoresis C. phage display
D. chemical cross–linking E. amino acids sequencing

[33] Some new methods to investigate protein–protein interactions include____________.
A. protein microarrays B. analytical ultracentrifugation C. light scattering
D. fluorescence spectroscopy E. protein–fragment complementation assay

Answers

1	2	3	4	5	6	7	8	9	10
C	D	C	C	A	A	D	D	E	C
11	12	13	14	15	16	17	18	19	20
B	D	D	A	E	E	D	E	D	E
21	22	23	24	25	26	27	28	29	30
C	D	D	DE	ABCE	AC	ABCDE	BCDE	ABCDE	ABC
31	32	33							
ACE	ABCD	ABCDE							

(*Wang Yanfei*)

Chapter 21

Gene Manipulation and Omics-based Technologies

[*Examination Syllabus*]

1. **Isolation and purification of nucleic acid** (genomic DNA, plasmid and RNA): basic principle.
2. **PCR**: principle.
3. **Nucleic acid molecular hybridization**: principle, main types and applications.
4. **Sequencing technologies**: the principles and basic methods.
5. **Gene function analysis techniques and omics technologies**: basic methods.

Ⅰ. Extraction and purification of nucleic acids

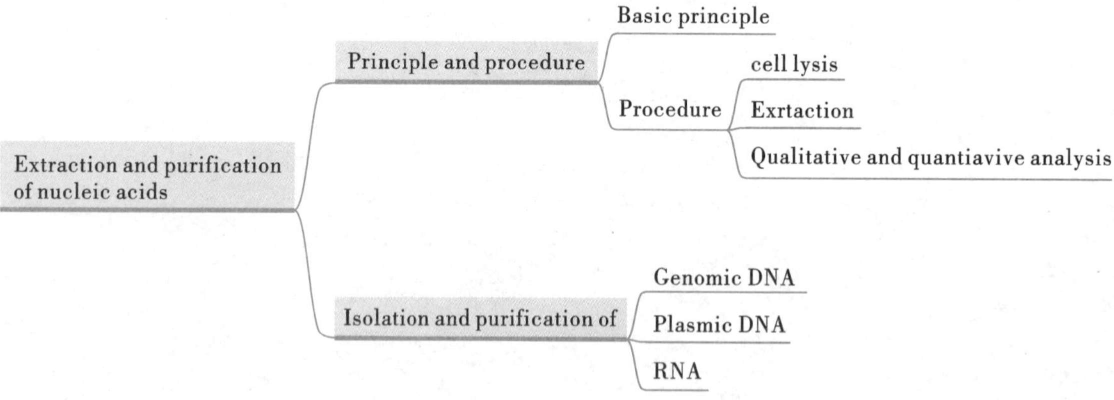

A. Principle and procedures

1. **Basic principle**

(1) Keep nucleic acids intact and functional and avoid the denaturation and inactivation of target nucleic acids.

(2) Eliminate contamination by other molecules such asprotein.

2. **Procedures**

(1) Cell lysis: physical method (ultrasonic method, homogenization) and chemical method (chemical

reagents: detergents such as SDS and enzymes)

(2) Extraction:

● Nucleic acids can usually dissolve in water, diluted acid and base, but not dissolve in organic solvent such as ethonal, acetone, butanol and so on.

● Organic solvent such as phenol and chloroform and detergents such as SDS or proteinase are usually used to denature and eliminate protein by further centrifuge and keep nucleic acids in the water phase (supernatant). Nucleic acids in the water phase are further precipitated by absolute ethyl alcohol or isopropanol.

(3) Quantitative analysis: ultraviolet spectrophotometer;

Qualitative analysis: electrophoresis.

B. Isolation and purification of genomic DNA

1. Isolation and purification

(1) SDS treatment: lyse cell, denature protein, dissociate chromosome and release DNA.

(2) Proteinase K: increase the production of DNA by hydrolyzing protein.

(3) Purification: add phenol/chloroform to eliminated denatured protein and organic chemicals.

(4) Precipitation: add ethanol or isopropanol

2. Quantitative and qualitative analysis

(1) A good result of ultraviolet spectrophotometer can show an evident absorption peak in 260nm and the ration of A_{260}/A_{280} is between 1.7-1.9.

(2) Agarose gel electrophoresis of genomic DNA can show the integration of DNA molecule, in which a single high molecular weight DNA band should be observed without degraded fragments. Degradation of genomic DNA can cause smear of bands on gel electrophoresis.

C. Isolation and purification of plasmid DNA

1. **The key point of plasmid DNA extraction** is to separate itself from genomic DNA and protein within the cell.

2. **SDS-alkaline lysis method**: a typical and commonly used way.

(1) SDS can lyse bacteria by its detergent property, and alkaline denature and precipitate protein and genomic DNA.

(2) In the subsequent neutralization by adding acid, plasmid can renature to covalently closed circular double-strand conformation due to its different physical-chemical properties with genomic DNA, while the genomic DNA form precipitation with protein due to its irreversible denaturation.

(3) The isolated plasmid can be further participated and purified by ethanol or isopropanol.

3. **Spin-column type of plasmid extraction kit**: makes use of adsorption chromatography and adopt silicon matrix membrane to bind with plasmid DNA specifically under high salinity and reduced pH condition.

4. Quantitative and qualitative analysis

(1) Quantitative analysis: the ration of A_{260}/A_{280} is between 1.7-1.8. Protein contamination can decrease the ration of A_{260}/A_{280}.

(2) Qualitative analysis: agarose gel electrophoresis, the cccDNA usually moves fastest and runs in the far front. While open-loop DNA runs slowest, and linear DNA is between them.

D. Isolation and purification of RNA

1. The extraction of RNA is to separate it from DNA and protein.

2. **Guanidiniumisocyanate-phenol-chloroform method**

(1) Guanidiniumisocyanate: lyse cells and inhibit RNase activity.

(2) Under the acidic condition, DNA can dissolve in phenol while RNA can not and phenol is also protein denaturant.

(3) Further extraction by adding chloroform makes DNA dissolvable in the organic phase and RNA dissolvable in the water phase and protein are denatured and participated, which makes RNA isolated from DNA and protein. (Figure 21.1)

(4) RNA can be precipitated by isopropanol or ethanol and dissolved in distilled water.

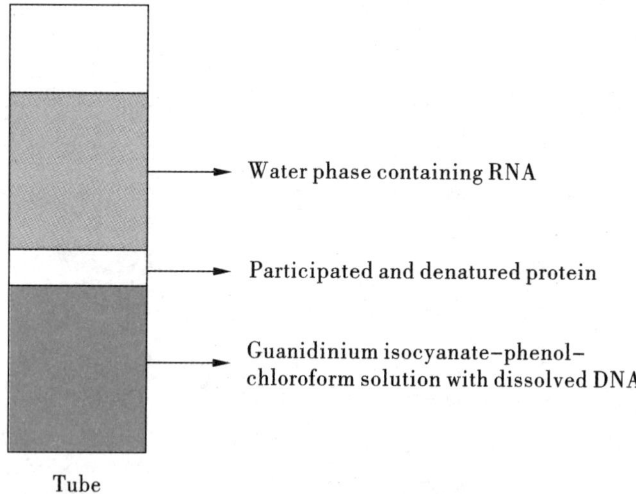

Figure 21.1 RNA isolation by guanidiniumisocyanate reagent.

3. **The most important notes in RNA isolation** is to avoid RNase contamination during the whole operation. Because RNase is widely present, so all the supplies should be treated by DEPC-treated H_2O and all the reagent should be prepared by DEPC-treated H_2O.

4. **Quantitative and qualitative analysis**

(1) The ration of A_{260}/A_{280} is between 1.8-2.1. An ideal ratio of RNA extraction is 2.0.

(2) There are two distinct bands of 28S rRNA and 18S RNA in electrophoresis, and the 28S rRNA band is approximately twice as intense as the 18S RNA band. (Figure 21.2)

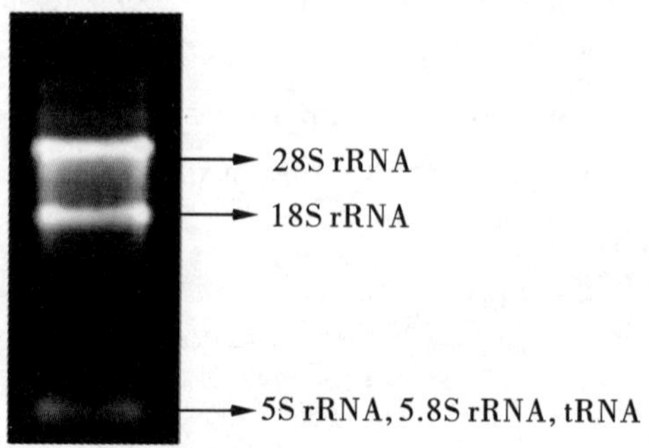

Figure 21.2 Agarose electrophoresis of isolated RNA

[*Practices*]

[A1 Type]

[1] What's the function of proteinase K in genomic DNA extraction? ____________

A. Hydrolyze protein to facilitate the separation of DNA.

B. Hydrolyze RNA to reduce the contamination of RNA.

C. Binding with genomic DNA.

D. Binding with RNA.

E. Increase the concentration of extracted DNA.

[2] Which of the following is an ideal extraction product in plasmid isolation? ____________

A. Most products are open-loop plasmid DNA.

B. Most products are covalently closed circular double-strand DNA (cccDNA).

C. Most products are linear plasmid DNA.

D. Mixture of cccDNA and linear plasmid DNA.

E. None of the above.

[3] What does it possibly mean if the ratio of 28S rRNA band to 18S rRNA band is less than 2? ____________

A. DNA is degraded during isolation

B. Protein is degraded during isolation.

C. There is protein contamination in RNA.

D. RNA is partly degraded during isolation.

E. There is DNA contamination in RNA.

[B1 Type]

No. 4 ~ 5 share thefollowing suggested answers.

A. 1.7 ~ 1.8   B. 2.0   C. 1.5 ~ 1.8

D. >2.0   E. <1.6

[4] An ideal ratio of A_{260}/A_{280} in RNA extraction is: ____________

[5] What's the normal range of A_{260}/A_{280} in the product of plasmid DNA extraction? ____________

[X Type]

[6] Which of the following factors can influence the quality of isolated genomic DNA? ____________

A. Protein contamination.

B. Samples are under repeated freezing and thawing.

C. Violent operation.

D. DNase is not inhibited.

E. RNase is not inhibited.

Ⅱ. PCR technology

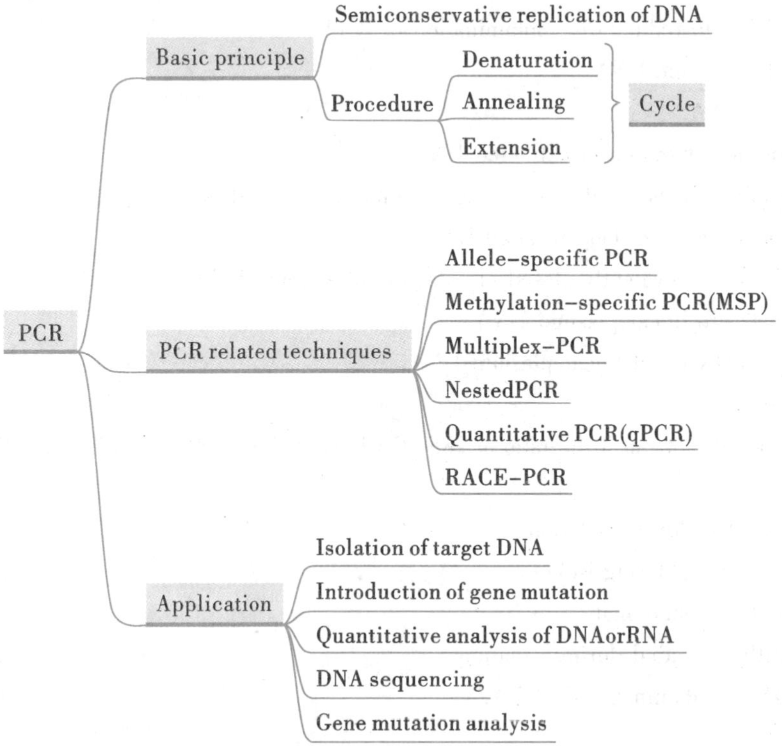

A. Basic principle of PCR technology

1. **Polymerase chain reaction** (PCR) is a cheap and reliable way to replicate a specific DNA segment.

PCR uses each strand of the double-helix DNA molecule as a template and oligo deoxynucleotides as primers to synthesize DNA under the catalysis of DNA polymerase according to the principle of semi-conservative replication of DNA.

2. **PCR procedures** include cycles of three steps: denaturation, annealing and extension.

(1) **Denaturation**: heat reaction system to 94 ℃-98 ℃ for 20-30 s. Heating breaks hydrogen bonds between double strands of DNA and makes each strand as template.

(2) **Annealing**: the reaction temperature is decreased to 50-65 ℃ for 20-40 s, allowing annealing of the primers to each of the single-stranded DNA templates.

(3) **Extension**: In the third step, DNA polymerase binds to the primer-template hybrid to start catalyzing the polymerization of the new strand at 72 ℃.

The above three steps form a cycle of PCR, and the newly generated DNA strands become the template for next round of replication, leading to produce a large amount of DNA target fragments after 25-30 cycles. The exponential amplification of PCR can generate 2^n (n = cycle numbers) copies of DNA target fragments from one copy in theory.

3. **A basic and typical PCR reaction system** requires components and reagents including DNA template, specific primers, thermostable DNA polymerase (such as Taq polymerase), 4 dNTPs and Mg^{2+} containing buffer.

4. **Quantitative and qualitative analysis**

(1) Agarose gel electrophoresis can show the size, density and specificity of PCR product. Most PCR can generate 0.1 ~ 10kb DNA fragment.

(2) By comparing with DNA marker, the size of the PCR product can be determined. The number of bands under electrophoresis can help to ensure the specificity of PCR.

B. PCR related techniques

1. **Allele-specific PCR**: It is based on single-nucleotide variations (SNVs) including differences between alleles.

2. **Methylation-specific PCR (MSP)**: a sensitive method to make a methylation analysis on genomic DNA of interest.

3. **Multiplex-PCR**: The result of multiplex-PCR can be indicated by amplicons of varying sizes that are specifically produced by different DNA primers. Multiplex-PCR can be applied in many detections such as detecting gene deletions, pathogen identification and so on.

4. **Nested PCR**: an improved PCR to reduce the non-specific amplification by designing two sets of primers and performing PCR twice. Nested PCR is very useful for such conditions as rare DNA templates or high PCR background.

5. **Quantitative PCR (qPCR)**: It is a big modification technique based on conventional PCR, which initiates monitoring the amplification process of target DNA rather than detecting the end products. The common methods for real-time detection of PCR products are by using.

(1) non-specific fluorescent dyes such as widely used SYBR green.

(2) sequence-specific DNA probes labeled with a fluorescent reporter.

6. **RACE-PCR**: 5′-RACE or 3′-RACE can be used to identify unknown 5′ end or 3′ end part of an RNA transcript especially a mRNA molecule. RACE is combined with high throughput sequencing to develop Deep-RACE which can help map the location of target RNA on the genome.

C. Application of PCR

1. **Isolation of target DNA**. PCR can produce large amounts of target DNA copies by specific amplification from genomic DNA.

2. **Introduction of gene mutation**. By designing 5′ sequence of primers, insertion, deletion and point mutation can be introduced into target DNA sequence during PCR.

3. **Quantitative analysis of DNA or RNA**. As described in qPCR method, the specific nucleic acid number can be absolutely measured and gene expression can be compared by relative quantification.

4. **DNA sequencing**. Amplification function of PCR produces large amounts of pure DNA sequences from very few starting materials, which facilitates further DNA sequencing.

5. **Gene mutation analysis**. PCR can be combined with other techniques to detect gene mutation with high sensitivity, such as RFLP (Restriction Fragment Length Polymorphism), SSCP (single-strand conformation polymorphism), ASO (allele-specific oligonucleotide) analysis, gene chip technology, and so on.

[*Practices*]

[A1 Type]

[7] What're the classical steps of PCR? ____________

A. Hot start, annealing and extension.

B. Denaturation, annealing and extension.

C. Annealing and extension.

D. Only extension.

E. Increase the concentration of extracted DNA.

[8] What's the basic principle of quantitative PCR (qPCR)? ____________

A. Determine the end product of fluorescent-labeled in PCR.

B. Monitoring the amplification process of target DNA rather than detecting the end products.

C. Can only quantify mRNA expression.

D. Has the same sensitivity with conventional PCR.

E. None of the above.

[X Type]

[9] Which of the following are applications of PCR in molecular biology? ____________

A. Quantitative analysis of DNA or RNA.

B. DNA sequencing.

C. Protein analysis.

D. Gene mutation analysis.

E. Isolation of target RNA

Ⅲ. Nucleic acid hybridization

A. Principles of nucleic acid hybridization

1. Nucleic acid hybridization is based on the **denaturation and renaturation** of nucleic acid and the property of **complementary binding** between A-T, G-C and A-U in nucleic acids.

2. A hybridization **probe** is a fragment of single DNA or RNA with length of 100-1000 bases long that have been labeled. The probe can be used to detect the presence of some specific nucleic acid sequence in DNA or RNA sample by binding to its complementary sequence.

B. Nucleic acid hybridization techniques and applications

1. Many molecular biology techniques are based on nucleic hybridization, including Southern blot, Northern blot, PCR and DNA sequencing.

2. Blotting is a technique of transferring macromolecule such as DNA, RNA or protein from gel to support blotting membrane. After blotting, the DNA, RNA and protein on the blotting membrane can be further detected or analyzed by labeled hybridization probes (antibodies for protein detection).

3. The major property and functions of blotting are summarized in table 21.1.

Table 21.1 The major property and functions of blotting

Blotting	Basic process	Function
Southern blot (DNA blotting)	Enzyme digestion, electrophoresis, transfer to membrane, probe hybridization, washing and detecting.	Classical method of DNA analysis. It is mainly applied in quantitative and qualitative analysis of genomic DNA, recombinant plasmid and phage.

Continue to Table 21.1

Blotting	Basic process	Function
Northern blot (RNA blotting)	The process and principle are similar to Southern blot. It is RNA to be transferred in Northern blot. Due to the small size of RNA molecule, the enzyme digestion step can be omitted.	The most reliable method of analyzing mRNA expression level because of its high specificity and low false – positive rate. It is mainly applied in quantitative and qualitative analysis of some specific mRNA expression in a tissue or cell. Differential expression of mRNA in tissues and cells can also be compared by Northern blot.
Western blot (Protein blotting)	Different from Southern blot and Northern blot, it is protein to be transferred under electricity motivation. In addition, it is labeled antibody to be bound with protein in detection.	It is a reliable method of protein analysis. It can be applied in detecting the presence of some specific protein, making a semi – quantitative analysis of some protein expression and detecting the interaction between proteins.

[*Practices*]

[A1 Type]

[10] Nucleic acid hybridization is the complementary binding between single strands of__________

A. DNA and DNA. B. RNA and RNA. C. DNA, RNA, DNA and RNA.
D. DNA and RNA. E. DNA and protein.

[11] What does it mean if a labeled DNA probe is detected positive in a sample? __________

A. There is complementary DNA in the sample.
B. It indicates the presence of some specific nucleic acid sequence of DNA or RNA in the sample.
C. There is complementary RNA in the sample.
D. Some specific protein is present in the sample.
E. None of the above.

[B1 Type]

No. 12 ~ 13 share the following suggested answers.

A. Northern blot B. Southern blot C. Western blot
D. nucleic acid hybridization E. dot blot

[12] A classical method of DNA analysis is: __________

[13] A reliable qualitative and semi-quantitative analysis method of protein expression is: __________

[X Type]

[14] Which of the following techniques combine the nucleic acid hybridization with blotting?

____

A. Northern blot. B. Southern blot. C. Western blot.
D. PCR. E. DNA sequencing

Ⅳ. DNA sequencing

DNA sequencing
- First generation sequencing
  - Sanger method
  - Maxam-Gilbert sequencing
- Second-generation sequencing: Next-generation or highthroughput sequencing
- Third-generation sequencing
  - Nanopore sequencing
  - Single molecule real time sequencing

DNA sequencing is one of important technology in molecular biology and will be widely used in personalized medicine. Sequencing method includes:

1. First-generation sequencing

Maxam-Gilbert sequencing and the Sanger method, represents the first generation of DNA sequencing methods.

(1) **Maxam-Gilbert sequencing** is based on base-specific partial chemical modification of DNA and subsequent cleavage of the DNA backbone at sites adjacent to the modified nucleotides.

(2) **Dideoxy chain-termination method/Sanger method** is based on the selective incorporation of chain-terminating dideoxynucleotides by DNA polymerase during in vitro DNA replication. ddNTPs terminate DNA strand elongation. The ddNTPs may be radioactively or fluorescently labeled for detection in automated sequencing machines.

2. Second-generation sequencing

It often referred to as Next-generation sequencing (NGS), or high-throughput sequencing, has applied to genome sequencing, genome resequencing, transcriptome profiling (RNA-Seq), DNA-protein interactions (ChIP-sequencing), and epigenome characterization, including Illumina (sequencing by synthesis), 454 pyrosequencing and SOLiD (sequencing by oligo ligation and detection).

3. Third-generation sequencing

It is a class of DNA sequencing methods under active development. It works by reading the nucleotide sequences at the single-molecule level instead of small DNA segments, containing Single-molecule real-time sequencing (SMRT) and Nanopore sequencing.

[*Practices*]

[A1 Type]

[15] Which ofthe following description about Sanger sequencing method is wrong? ____________

A. ddNTP lacks 3′-OH group

B. ddNTP competes with dNTP during DNA strand extension

C. Sequencing reaction contains a DNA template, primer, DNA polymerase, dNTPs, and one of four ddNTPs

D. dNTPs are labeled with a different fluorescent dye.

E. Currently, the Sanger method remains widely used.

[16] Which statement is wrong about NGS? ____________

A. NGS is of high throughput.

B. NGS is widely used for RNA-seq and microRNA-seq.

C. NGS is fast

D. The cost per base is low for NGS

E. The read size of NGS is short.

[B1 Type]

No. 17 ~ 18 share the following suggested choices.

A. sequencing by synthesis

B. sequencing by ligation

C. dideoxy chain-termination

D. chemical modification based cleavage

E. exonuclease based synthesis

[17] The principle of the Illumina platform is: ____________

[18] The principle of SOLiD platform is: ____________

[X Type]

[19] Which of the following techniques belongs to NGS? ____________

A. Pyrosequencing.

B. Sequencing by synthesis

C. Sequencing by ligation

D. Dideoxynucleotide termination method

E. Maxam-Gilbert sequencing

V. Gene function analysis techniques

The gene functional analysis methods based on bioinformatics as well as experimentation are involved here.

A. Bioinformatic analysis of nucleic acid and protein

1. The analysis of nucleic acid sequence: Similarity search: Basic local alignment search tool (BLAST), multi-sequence alignment (MSA), ORF analysis, chromosome location, gene structure analysis (http://genome.ucsc.edu/cgi-bin/hgBlat) are recommended, regulatory region analysis, such as FirstEF (http://rulai.cshl.org/tools/FirstEF/), TFsearch (http://diyhpl.us/~bryan/irc/protocol-online/protocol-cache/TFSEARCH.html) and MATCH (http://gene-regulation.com/pub/programs.html#match.

2. The analysis of protein sequence

Besides Blast, highly integrated protein analysis website (https://www.expasy.org/) can display a complete analysis for a protein property, including transmembrane prediction, prediction of signal peptide, subcellular localization, protein-protein interaction network and etc.

3. Signaling pathway analysis

Cell signaling pathway is part of any communication process that governs the basic activities of cells and coordinates all cell actions. Currently, there are three most popular tools for signaling pathway analysis, IPA, GSEA and DAVID (https://david.ncifcrf.gov/).

B. Experimental techniques of gene functional analysis

There are two strategies to conduct the gene functional analysis, namely, the loss of function and the gain of function.

1. Loss of function study: Site-directed mutagenesis, gene knock-out, gene interference (RNAi, miRNA, nuclease, antisense RNA and etc).

2. **Gain of function study**: transgene and gene knock-in.

3. **Genome editing**: CRISPR (Clustered regularly interspaced short palindromic repeats)/Cas and transcription activator-like effector nuclease (TALEN), Zinc finger nuclease (ZFN)

4. **Protein-protein interaction study**: immunoprecipitation, GST-pulldown and yeast two-hybrid.

5. **Protein-DNA interaction study**: CHIP.

Ⅵ. Functional genomics

It refers to as a field of molecular biology that study gene (and protein) functions and interactions by making use of the vast amount of data given by genomic and transcriptomic projects. It focuses on the dynamic aspects such as gene transcription, translation, regulation of gene expression and protein-protein interactions. A key characteristic of functional genomics studies is high-throughput method based genome-wide approach to these questions.

A. Goals of functional genomics are to generate and synthesize genomic and proteomic knowledge into an understanding of the dynamic properties of an organism.

B. Techniques and general applications

1. **DNA level**: the ENCODE (Encyclopedia of DNA elements) project should be mentioned.

2. **RNA level**: transcriptome profiling is conducted by RNA-Seq and MicroRNA sequencing, NGS is the gold standard tool for them.

3. **Protein level**: protein-protein interactions can be done by the yeast two-hybrid system (Y2H). Affinity purification and mass spectrometry (AP/MS) is able to identify proteins that interact with one another in complexes.

4. **Animal level**: loss of gene function can be investigated by systematically "knocking out" genes one by one.

[*Practices*]

[A1 type]

[20] Whichof the following description about gene targeting is wrong? ______

A. The most commonly used technique for KO is the Cre-lox recombination system.

B. CRISPR/Cas system is very effective for genome editing without off-target effect.

C. TALEN and ZFN are prominent tools in the field of genome editing.

D. Knock-in involves a gene inserted into a specific locus and is thus a "targeted" insertion.

E. Conditional KO is that the elimination of specific gene occurred in certain tissues.

[X type]

[21] Whichof the following techniques can be used for loss of function study? ______

A. Site-directed mutagenesis. B. Gene targeting C. TELEN

D. CRISPR/Cas E. RNAi

Ⅶ. Techniques to investigate macromolecular interaction

1. Many different methods have been developed to investigate the DNA-Protein, RNA-Protein and Protein-Protein Interactions, each method has its limitation, and researchers always use several methods to supplement each other.

2. **Detect DNA-Protein interaction**: Electrophoretic Mobility Shift Assay (EMSA), Yeast one-hybrid assay (Y1H), Chromatin immunoprecipitation (CHIP), Systematic evolution of ligands by exponential enrichment (SELEX), Proximity ligation assay (PLA) and et al.

3. **Detect RNA-Protein interaction**: RNA EMAS, RNA pull-down assay, RNA immunoprecipitation (RIP), Ultraviolet-induced crosslinking immunoprecipitation (CLIP) and its variants.

4. **Detect Protein-Protein interaction**: yeast two-hybrid method, Proximity ligation assay (PLA), Co-immunoprecipitation (Co-IP), Pull-down assay and et al.

5. Real-time biomolecular interaction (BIA) is a very useful investigating the macromolecular interaction.

[*Practices*]

[A1 type]

[22] Which of the following could not be used in DNA-Protein interaction analysis? ____________

A. Ultraviolet-induced cross-linking immunoprecipitation

B. Electrophoretic Mobility Shift Assay

C. Chromatin immunoprecipitation.

D. Systematic evolution of ligands by exponential enrichment

E. Proximity ligation assay

[23] Which of the following techniques could be used in protein-protein interaction analysis? ____________

A. Yeast two-hybrid method B. Proximity ligation assay C. Co-immunoprecipitation

D. Pull-down assay E. All of the above

[X Type]

[24] Which of the following are applications of real-time biomolecular interaction analysis? ____________

A. Kinase analysis B. Biomolecular interaction analysis

C. Drug screening D. Antibody development E. Macromolecular Interaction

Ⅷ. Omics Techniques

1. Omics techniques give an opportunity for accurate and comprehensive interpretation of biological processes and functions that involves various fields of studies: genomics, transcriptomics, proteomics epigenomics and metabolomics.

2. **Genomics**: Sanger sequencing, DNA-microarrays and Next-Generation sequencing.

3. **Transcriptomics**: RNA-microarray or fragmenting the cDNA and building a library to sequence by synthesis (RNA-sequencing); running the microarray or sequence through the platform of choice; performing QC

4. **Proteomics**: Two-dimensional gel electrophoresis (2DE), 2D Fluorescence Difference Gel Electrophoresis (2-D DIGE), two-dimensional gel electrophoresis-liquid chromatography with mass spectrometric (2D LC-MS), Mass spectrometry (MS).

[*Practices*]

[A1 type]

[25] Which of the following is the definition of Genomics? ____________

A. The large scale study of proteins, including their structure and function, within a cell or organism

B. The study of global metabolite profiles in a system (cell, tissue or organism) under a given set of conditions

C. The study of the structure, function and expression of all the genes in an organism.

D. The study of the mRNA within a cell or organism

E. None of the above

[X Type]

[26] Which of the following method can be used in Genomics? ____________

A. Sanger sequencing B. DNA-microarrays C. Next-Generation sequencing

D. Mass spectrometry E. None of the above

Answers

1	2	3	4	5	6	7	8	9	10
A	B	D	B	A	ABCD	B	B	ABD	C
11	12	13	14	15	16	17	18	19	20
B	B	C	AB	D	E	A	B	ABC	B
21	22	23	24	25	26				
ABCDE	A	E	ABCDE	C	ABC				

Brief Explanations for Practices

1. The function of proteinase K in genomic DNA isolation is to degrade protein and help to separate DNA from protein.

2. Covalently closed circular double-strand DNA (cccDNA) is the natural and functional form of plasmid DNA. High-quality isolation of plasmid DNA should obtain a large amount of cccDNA with less open-loop plasmid DNA and no linear plasmid DNA.

3. The ratio of 28S rRNA band to 18S rRNA band in good isolation of RNA is 2.0. Decreased ration means possible degradation of RNA.

4. The normal ratio of A_{260}/A_{280} in RNA extraction should between 1.9-2.03. An ideal ratio of A_{260}/A_{280} in RNA extraction is 2.

5. The normal ratio of A_{260}/A_{280} in plasmid DNA extraction should between 1.7-2.0. The higher of the ratio, the better.

6. Many factors can influence the quality of isolated genomic DNA, including protein residual, repeated freezing and thawing of samples that results in degradation of DNA, violent operation that results in a break of DNA, and active DNase degrading genomic DNA.

7. Classical PCR consists of three steps of denaturation, annealing and extension.

8. Different from conventional PCR, qPCR quantify RNA by monitoring the amplification process of tar-

get DNA. It is a real-time PCR.

9. PCR is widely applied in molecular biology, including quantitive analysis of DNA or RNA, DNA sequencing, protein analysis, gene mutation analysis and so on.

10. Nucleic acid hybridization can occur between DNA, RNA, DNA and RNA. As long as the sequences are complementary to each other.

11. A labeled DNA probe is detected positive in a sample, which means there is a complementary sequence in the sample. Because the DNA probe is designed specifically, so the binding indicates the presence of some specific nucleic acid sequence of DNA or RNA.

12. A classical method of DNA analysis is Southern blot.

13. Western blot is a commonly used method to detect protein expression. By using specific antibody, protein expression can be qualitatively and quantitatively analyzed.

14. Northern blot and Southern blot are techniques combining the nucleic acid hybridization with blotting. Western blot is used to analyze protein. PCR is based on nucleic acid hybridization but not with blot.

15. In the Sanger sequencing method, ddNTPs instead of dNTP are labeled with different fluorescent dye

16. NGS is high throughput and widely used for RNA-seq and microRNA-seq. It is fast, low cost. The read size depends on different sequencing chemistry.

17. Illumina platform is based on Sequencing by synthesis

18. SOLiD platform is based on sequencing by oligo ligation and detection.

19. NGS includes Illumina, 454 pyrosequencing and SOLid and so on.

20. The major weakness of CRISPR/cas9 technique is its off-target effect.

21. These techniques, including Site-directed mutagenesis, Gene targeting, TALEN, CRISPR/Cas and RNAi can be used for loss of function study.

22. Ultraviolet-induced crosslinking immunoprecipitation is always used in RNA-Protein interaction analysis.

23. All of the techniques could be used in Protein-Protein interaction analysis.

24. All of these are applications of real-time biomolecular interaction analysis.

25. Genomics is the study of the structure, function and expression of all the genes in an organism.

26. Mass spectrometry is an essential tool in proteome analysis; others could be used in Genomics.

(***Li Jiao*, *Lü Lixia*, *Liu Baoqin***)

Chapter 22

Recombinant DNA Technology

Examination Syllabus

1. **Recombinant DNA technology**: concept, basic principle, process.

2. **Gene engineering and medicine**: discovery of disease-related genes, biopharmaceutical, gene diagnosis, gene therapy.

[*Major points*]

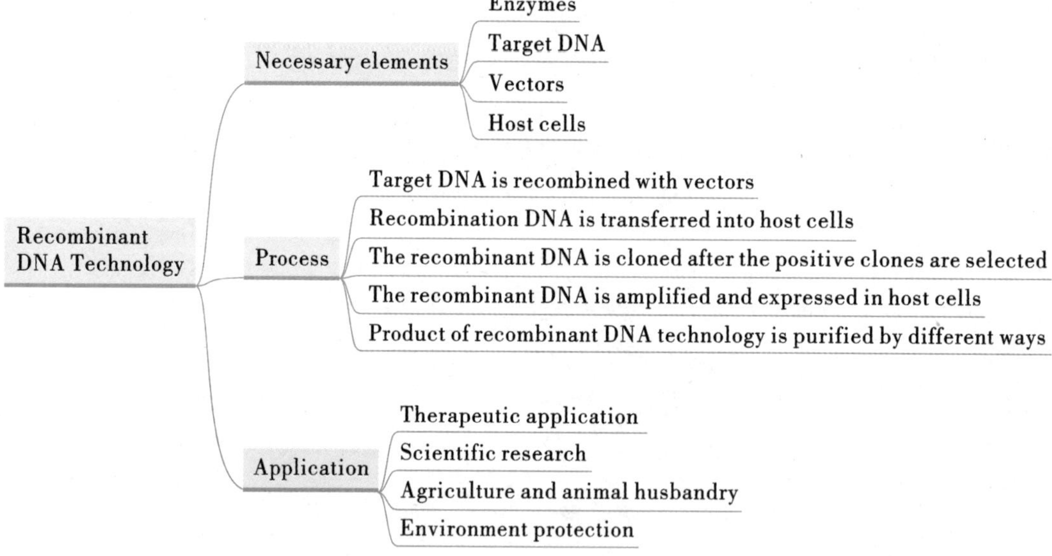

Recombinant DNA Technology
Necessary elements
Enzymes
Target DNA
Vectors
Host cells
Process
Target DNA is recombined with vectors
Recombination DNA is transferred into host cells
The recombinant DNA is cloned after the positive clones are selected
The recombinant DNA is amplified and expressed in host cells
Product of recombinant DNA technology is purified by different ways
Application
Therapeutic application
Scientific research
Agriculture and animal husbandry
Environment protection

Ⅰ. The necessary elements for recombinant DNA technology

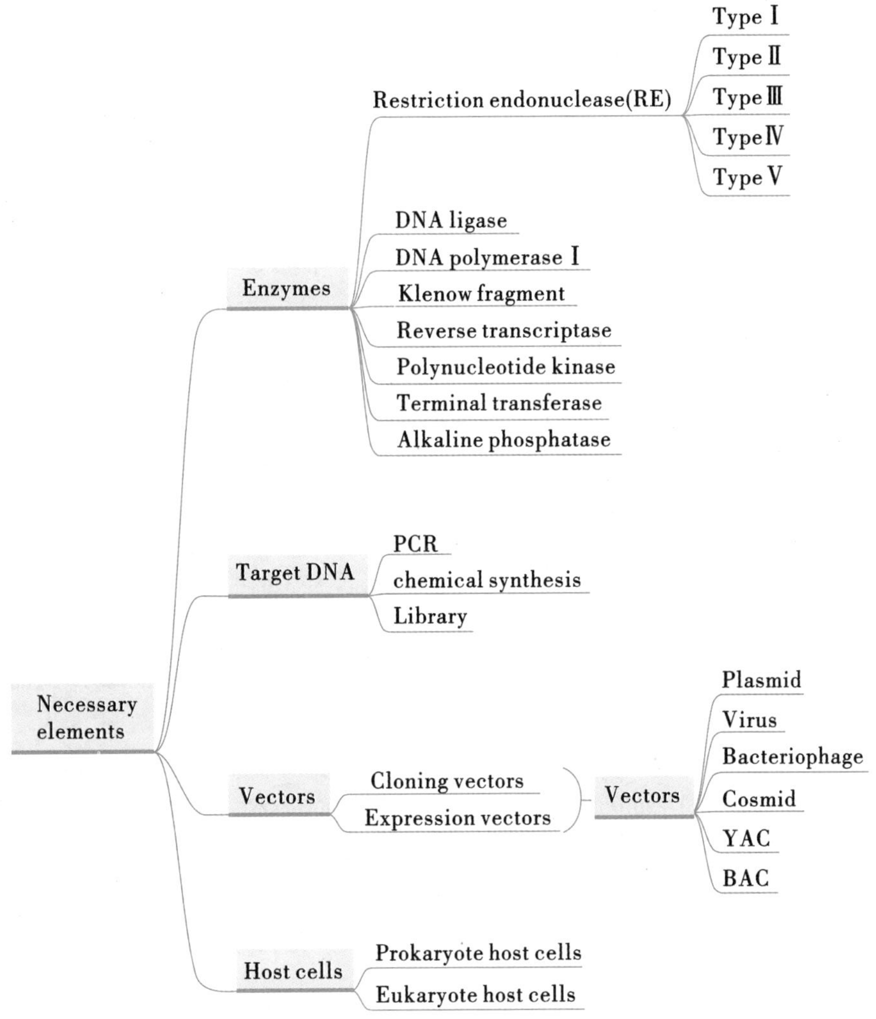

A. Enzymes

1. Enzymes used in DNA recombination(Table 22.1)

Table 22.1 Enzymes used in DNA recombination

Enzymes	Functions
Restriction enzyme	Restriction enzyme cuts DNA at the recognition site
DNA ligase	DNA ligase links target DNA sequence with vector by catalyzing the formation of a phosphodiester bond between the 5′-phosphate of a nucleotide on one fragment and the 3′-hydroxyl of another
DNA polymerase Ⅰ	DNA pol Ⅰ has 5′-3′polymerization activity, 3′-5′ and 5′-3′exonuclease activity. It can be used in synthesis of double cDNA or linkage of DNA fragments, making a probe with high specific activity by using the nick translation method, analysis of DNA sequence, and filling the 3′ terminal.

Continue to Table 23.1

Enzymes	Functions
Klenow fragment	It is large fragment of DNA pol I, which has 5′-3′polymerization activity and 3′-5′exonuclease activity, but lack 5′-3′exonuclease activity. It is commonly used in the synthesis of 2nd strand of cDNA, and marking at the 3′ of double-strand.
Reverse transcriptase	Reverse transcriptase is DNA polymerase using RNA as template, is used in the synthesis of cDNA, and in filling the nicks, marking or analyzing DNA sequence instead of DNA pol I.
Polynucleotide kinase	Polynucleotide kinase catalyzes the 5′phosphorylation of polynucleotides or probe marking.
Terminal transferase	Terminal transferase adds homopolymer tail at 3′ of polynucleotides.
Alkaline phosphatase	Alkaline phosphatase cuts the phosphate group at terminal.

2. **Main restriction enzymes**(Table 22.2)

Table 22.2 Mainrestriction enzymes

type	Type Ⅰ	Type Ⅱ	Type Ⅲ
Abundance	Less common than type II	Most common	Rare
Composition	Three - subunit complex: individual recognition (HsdR), endonuclease (HsdS), and methylase (HsdM) activities	Endonuclease and methylase are separate, homodimers or homotetramers	Hetero-oligomeric, multifunctional proteins composed of two subunits, Res (DNA recognition and modification) and Mod (DNA cleavage)
Recognition site	Cut both strands at a nonspecific location up to 1000 bp away from recognition site	Cut both strands at a specific site, usually undivided, palindromic, 4 to 8 bp	Two separate non - palindromic sequences that are inversely oriented
Cleave site	Remote from a recognition site	Within or at short specific distances from a recognition site	20to 30 bp after the recognition site
Cofactor requirements	ATP, SAM, Mg^{2+}	Mg^{2+}	ATP, Mg^{2+}, SAM (not necessary)
Function	Both restriction and modification activities	Single function (restriction) enzymes	part of a complex with a modification methylase
Use in recombinantDNA research	Not useful	Very useful	Not useful

3. **Sticky (cohesive or staggered) and blunt ends**

Some REs show a sticky double-stranded cut, which leaves single-stranded overhangs, and maybe a 3′ or a 5′ overhang. Whereas a blunt double-stranded cut does not leave an overhang.

B. Target DNA

Target DNA inserted into the vector can be obtained in many ways.

1. **Target DNA can be synthesized by PCR.**

2. Chemical syntheses is another way to get target DNA

3. Target DNA can be screened from the libraries, such as genomic library and cDNA library.

C. Vectors

1. Classification by function

(1) **Cloning vectors** carry the target DNA and amplify it in host cells.

Important features of cloning vectors:

● must have one origin of replication at least.

● contain a number of unique RE cleavage sites that are present only once in the vector, called multiple cloning sites (MCS).

● carry a selectable marker, usually in the form of antibiotic resistance genes or genes for enzymes missing in the host cell.

(2) **Expression vectors** having expression elements are used for target gene expression.

Expression vectors should have expression elements for foreign gene expression in host cells besides basic elements for vectors.

2. Classification by origins

(1) **Plasmids** are the most commonly used vectors.

Plasmids are autonomously replicating circular extra-chromosomal DNA. pBR322 is one of first used vectors. pUC18 is a representative pUC plasmid vectors derived from pBR322. pGEM-3Z is a cloning vector which can transcript in vitro. T-A cloning vectors are linear plasmids with an unpaired T at the 3′, which is convenient for pairing with the unpaired A at the 5′ of PCR products.

(2) **Bacteriophage** is a virus that infects and replicates within bacteria whereby foreign DNA with a size of 5 -11 kb may be inserted, which is suitable for cDNA cloning. Cosmids mean plasmids with cos site of phage λ, which are designed for large DNA sequence cloning (40-45 kb).

(3) **Cosmids** mean plasmids with *cos* site of phage λ, so they behave both as plasmids and as phages. As a type of cloning vectors, cosmids are usually used in genetic engineering, for example, to build genomic library.

(4) **Yeast artificial chromosomes (YACs) and bacterial artificial chromosomes (BACs)** are designed for cloning huge pieces of DNA (up to hundreds of thousands of base pairs). YACs are a type of yeast linear cloning vectors. BACs are cloning vectors constructed on the base of F-plasmid.

3. Expression vectors

Expression vectors are classified as two categories, prokaryotic expression vectors and eukaryotic expression vectors.

(1) **Prokaryotic expression vectors** are mostly plasmids. They have the basic transcriptional regulatory elements of operon besides the essential elements of cloning vectors, such as promoter, terminator, and Shine-Dalgarno sequence.

(2) **Eukaryotic expression vectors** have the transcriptional regulatory elements such as promoter, enhancer, polyA signal, the ribosome binding site near the initiation codon, the marker for screening and so on, besides the basic element of cloning vectors.

(3) **Shuttle vectors** can replicate and amplify its carrying gene in prokaryotes and expressed it in eukaryotes.

D. Host cells

1. Prokaryotic host cells

Prokaryotic host cells are commonly used. *E. coli* is the most common prokaryotic host cells because of the following merits.

(1) Clear genetic background.

(2) Short growth cycle.

(3) Perfect system of vector and host.

(4) Stable recombinant DNA in *E. coli*.

The shortcoming is *E. coli* can produce endotoxin.

2. Eukaryotic host cells

Eukaryotic host cells have some advantages in mammalian protein expression.

The host cells used in eukaryotic expression system mainly include mammalian cells, saccharomyces (yeast), and insect cells. The recombinant protein produced by mammalian cells is more similar to natural proteins, and their activity is better than those proteins produced by saccharomyces (yeast) and insect cells.

Ⅱ. The process of DNA recombination

A. Target DNA is recombined with vectors

1. The first step is preparing exogenous DNA and vector for recombination.

(1) Obtain the target DNA sequence by PCR or other ways.

(2) Select an optimum vector.

2. The second step is the recombination of exogenous DNA and vector.

There are two ways to obtain the recombination molecule, one is enzyme-cut (single RE, double REs cutting) and link up way, and the other is homologous recombination way.

(1) Link exogenous DNA and vector with ligase.

(2) Link exogenous DNA and vector by recombination.

B. Recombination DNA is transferred into host cells

1. Transformation is typically used to describe non-viral DNA transfer in bacteria and non-animal eukaryotic cells, including plant cells.

(1) Chemical transformation.

(2) Electroporation.

2. Transfection is the process of deliberately introducing naked or purified nucleic acids into eukaryotic cells.

(1) Physical treatment, such as electroporation, cell squeezing, nanoparticles, magnetofection, sonoporation, optical transfection, etc.

(2) Chemical materials, such as liposome and calcium phosphate.

(3) Biological particles, such as viruses.

C. The recombinant DNA is cloned after the positive clones are selected

Ways to screen the host cells introduced the recombinant DNA.

1. **Blue-white screening**
2. **Utilization of resistance genes**
3. **Insertion of the target gene into a certain gene**
4. **DNA sequencing**

D. The recombinant DNA is amplified and expressed in host cells

1. **Prokaryote expression system.**
2. **Eukaryote expression system**(Table 22.3).

Table 22.3 Comparison of prokaryote and eukaryote expression system

type	Prokaryote expression system	Eukaryote expression system
Vectors	Elective markers, a strong promoter, suitable sequences can control translation, a well-designed MCS	Selective markers for eukaryote cells, promoter, signals for transcription and translation termination, signal for adding poly(A) tail to mRNA, and chromosomal integration site etc.
Host cells	Prokaryote cells such as *E. coli*.	Mammalian cells, insect cells, yeasts
Advantages	①The protein products can be harvested within a short time. ②The cost is relatively low.	①This system can not only express cDNA, but also express genes amplified from eukaryote genome. ②The proteins will be modified properly and distribute and accumulate in proper location in cells. ③The expressed protein will not form inclusion bodies, and will not degrade easily.
Disadvantages	①prokaryotic expression system only can be used to express the eukaryote proteins by their cDNA products, but not the gene directly amplified from their genome. ②The eukaryote proteins can't be folded or glycosylated correctly in prokaryotic expression system due to lack of proper post-translational modification. ③ The expressed proteins always form insoluble inclusion bodies. ④The sustained expression or overexpression of some genes may cause toxic effects on host cells.	①The technology is more difficult. ②The expression time is longerc. The cost is higher.

E. Product of recombinant DNA technology is purified by different ways

1. The raw material should be pretreated and roughly separated.

(1) Pretreatment of biological samples by ultrasonication, repeated freezing and thawing, surfactant cracking and enzymolysis.

(2) Roughly separation of pretreated samples by centrifugation, dialysis and precipitation.

2. Chromatography and electrophoresis are the main methods in purification.

3. An integrated strategy is adopted in protein purification.

The purification strategy is based on the following items.

(1) The source, physical and chemical characteristics of recombinant protein.

(2) The purification purpose of the target protein.

(3) The protein purity and cost of the strategy.

Ⅲ. The application of recombinant DNA technology

1. Widely used in therapeutic application.
2. Necessary for scientific research.
3. Required for agriculture and animal husbandry.
4. Used in environment protection.

[*Practices*]

[A1 type]

[1] ____________ contributes mostly to recombinant DNA technology.

A. Restriction endonuclease B. DNA ligase C. DNA polymerase I

D. reverse transcriptase E. terminal transferase

[2] ____________ cut both strands at a specific site, usually undivided, palindromic, 4 to 8 bp.

A. Type Ⅰ RE B. Type Ⅱ RE C. Type Ⅲ RE

D. Type Ⅳ RE E. Type Ⅴ RE

[3] ____________ has no MCS.

A. pBR322 B. pUC18 C. Bacteriophage

D. Bacterial artificial chromosomes E. pGEM-T vector

[4] Which of the follaring is not merits of eukaryote expression system ____________.

A. This system can express genes amplified from eukaryote genome.

B. The proteins will be modified properly.

C. The proteins can accumulate in the proper location of cells.

D. The expression time is short.

E. The protein will not degrade easily.

[5] ____________ is not methods to select positive clones.

A. Blue-white screening.

B. Utilization of resistance genes.

C. Transduction.

D. Insertion of the target gene into a certain gene.

E. DNA sequencing.

[B1 type]

No. 6 ~ 9 share the following suggested answers

A. Plasmid B. Bacteriophage C. Cosmid

D. Yeast artificial chromosome

E. Bacterial artificial chromosome

[6] Which one is the most commonly used vector? ____________

[7] Which one could be inserted the large fragment of DNA, from 100 to1000 kb? ____________

[8] Which one could be have both as plasmids and as phages? ____________

[9] T–A cloning vectors is belonging to ____________.

No. 10 ~ 12 share the following suggested answers

A. Transcription B. Translation C. Infection

D. Transformation E. Transfection

[10] ____________ is the process by which foreign DNA is introduced into a cell by a virus or viral vector.

[11] ____________ is the process of deliberately introducing naked or purified nucleic acids into eukaryotic cells.

[12] ____________ is used to describe non–viral DNA transfer in bacteria and non–animal eukaryotic cells, including plant cells.

[X type]

[13] ____________ is not merits of *E. coli* as host cells.

A. The growth cycle is short.

B. They have a clear genetic background.

C. They can produce many kinds of endotoxin.

D. The recombinant DNA is stable in *E. coli*.

E. The expressed proteins always form insoluble inclusion bodies.

[14] Which of the following are reasonable in recombinant DNA technology ____________?

A. sticky ends are better than blunt ends

B. the shuttle vectors can be propagated in more than one type of host cells

C. eukaryote expression system is better than prokaryote expression system

D. blue–white screening is usually used to distinguish recombinant transformants from nonrecombinant transformants

E. chromatography and electrophoresis are the main methods in purification

[15] The recombinant DNA technology can be applied in ____________.

A. biopharmaceutical

B. gene diagnosis and gene therapy

C. scientific research

D. agriculture and animal husbandry

E. environment protection

Answers

1	2	3	4	5	6	7	8	9	10
A	B	A	D	C	A	D	C	A	C
11	12	13	14	15					
E	D	CE	ABDE	ABCDE					

(***Meng Liesu***)

Chapter 23

Oncogenes and tumor suppressor genes

Examination Syllabus

1. Oncogene and tumor suppressor gene: concept, mechanisms of oncogene activation
2. Growth factor: concept, mechanisms of function

[Major points]

Genes involved in tumorigenesis include those whose products:

(1) directly regulate cell proliferation (either promoting or inhibiting)

(2) are involved in the control of programmed cell death or apoptosis

(3) participate in the repair of damaged DNA.

Depending on how they affect each process, these genes can be grouped into two general categories: proto-oncogenes (growth-promoting, differentiation-inhibiting) and tumor suppressor genes (growth-inhibiting, differentiation-promoting).

Two categories of genes play major roles in triggering cancer. Dominant gain-of-function mutations in proto-oncogenes and recessive loss-of-function mutations in tumor-suppressor genes are oncogenic.

Ⅰ. Oncogenes

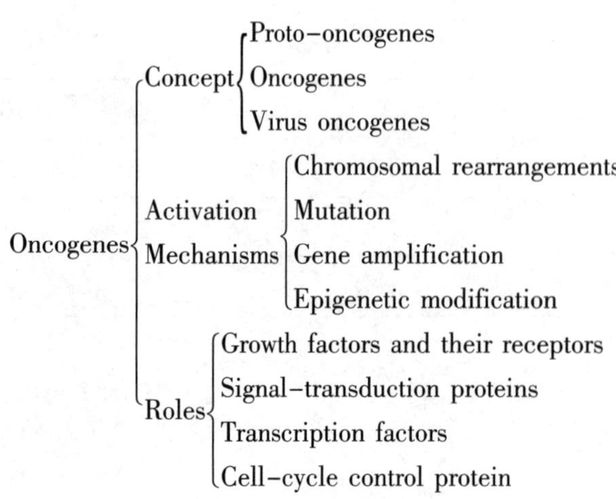

A. Concept

1. **Proto-oncogenes** code for essential proteins involved in maintenance of cell growth, division and differentiation. Proto-oncogenes are functional normal genes in human genome.

2. **Oncogenes** are proto-oncogenes that have been mutated and consequently cause normal cells to grow out of control and become cancer cells.

3. **Virus oncogenes**

Some viruses (tumor virus, most of them are retroviruses) contain oncogenes (**virus oncogenes**, *v-onc*) in their genomes. Virus oncogenes are originally obtained from a cellular host chromosome.

The first recognizedoncogene, *v-src*, was identified in Rous sarcoma virus, a cancer-causing retrovirus. Retroviral oncogenes arose by transduction of cellular proto-oncogenes into the viral genome and subsequent mutation.

4. **Important oncogenes** include *SRC*, *RAS* and *MYC* family etc.

(1) *SRC* family: *SRC*, *LCK*

(2) *RAS* family: *K-RAS*, *H-RAS*, *N-RAS*

(3) *MYC* family: *C-MYC*, *N-MYC*, *L-MYC*

B. Activation Mechanisms of oncogenes

Activation of oncogenes by chromosomal rearrangements, mutations, and gene amplification confers a proliferation advantage or increased survival of cells carrying such alterations.

1. **Chromosomal rearrangements**

(1) Chromosome inversions and translocations are common abnormalities in cancer cells.

(2) Example

● Human Burkitt lymphoma: MYC of chromosome 8 moves to chromosome 14 to up-regulate *MYC* expression.

● CML: translocations between parts of the long arms of chromosome 9 (*ABL* gene) and chromosome 22 (*BCR* gene), resulting the formation of the so-called "Philadephia chromosome (Phchromosomerhe)" and *BCR-ABL* fusion gene.

2. **Mutations**

(1) When an oncogene is mutated and thus activated, the structure of the encoded protein is altered in a way that enhances its transforming activity.

(2) Example: *RAS* oncogenes

● Function: encode proteins with guanosine-nucleotide-binding activity and intrinsic guanosine triphosphatase (GTPase) activity.

● Mutation: in codon 12, 13, or 61, the *RAS* genes encode a protein that remains in the active state and continuously transduces signals by linking tyrosine kinases to downstream serine and threonine kinases.

3. **Gene amplification**

(1) Gene amplification refers to a genomic change that results in an increased dosage of the gene(s) affected.

(2) Amplification represents one of the major molecular pathways through which the oncogenic potential of proto-oncogenes is activated during tumorigenesis.

(3) Example: the amplification of the dihydrofolate reductase gene (*DHFR*) in methotrexate-resistant acute lymphoblastic leukemia (ALL).

4. **Epigenetic modification**

(1) Genomic hypomethylation in promoters and within gene bodies: the putative oncogene, *ELMO3*, is overexpressed in non-small cell lung cancer in combination with hypomethylation of its promoter and these cancer-specific events are associated with the formation of metastases.

(2) microRNA genes: microRNA genes encode for a single RNA strand of about 21 to 23 nucleotides, which regulate gene expression by specifically targeting certain mRNAs in order to prevent them from coding for a specific protein. The first microRNA that has been proven to act as oncogene in human cancer was *miR*-17/92 polycistronic cluster known as *OncomiR*-1, which comprises six microRNAs: *miR*-17, *miR*-18*a*, *miR*-19*a*, *miR*-20*a*, *miR*-19*b*-1, *and miR*-92*a*-1.

C. Roles of oncogenes in the development and progression of cancer and the pathways involved

Among the proteins encoded by proto-oncogenes are positive-acting growth factors and their receptors, signal-transduction proteins, transcription factors, and cell-cycle control proteins.

1. **Growth factor and growth factor receptors**: Growth factors and their transmembrane receptor tyrosine kinases, such as EGF/EGFR, play important roles in cell proliferation, survival, migration and differentiation. ErbB2, belonging to the EGFR family of receptor tyrosine kinases (RTKs), plays an important role in human malignancies via affecting transcription factors and controlling cell cycle.

2. ***MYC***: The *MYC* transcription factor is one of the most important somatically mutated oncogenes in human cancer. The MYC oncoprotein can confer a selective advantage on cancer cells by promoting proliferation, cell survival, differentiation blockade, genetic instability, and angiogenesis, all of which may indirectly contribute to metastasis. Multi-pathways including EMT has been found to be involved in the MYC-mediated alterations of cell behaviors.

Ⅱ. Tumor suppressor genes

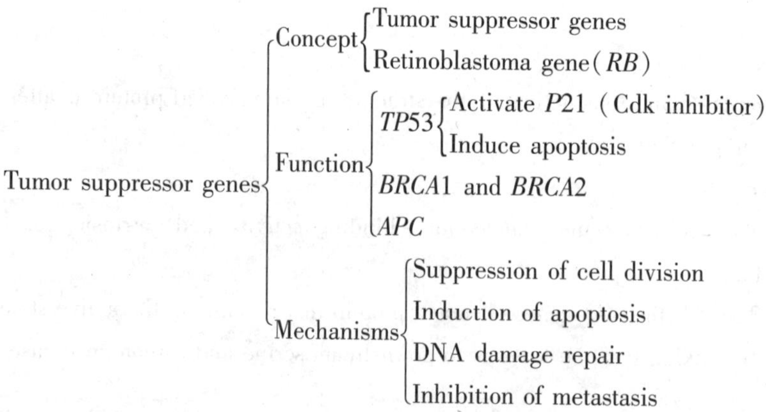

A. Concept

1. **Tumor suppressor genes** can be defined as genes which encode proteins that normally inhibit the formation of tumors.

2. **Retinoblastoma gene(*RB*)**: the first tumor-suppressor gene

(1) The largest percentage (30%) of retinoblastomas contains large scale deletions. Splicing errors, point mutations and small deletions in the promoter region have also been observed in some retinoblastomas.

(2) The major function of pRB is in the regulation of cell cycle progression. Its ability to regulate the **cell cycle** correlates to the state of **phosphorylation** of pRB.

B. Functions of tumor suppressor gene products

Tumor–suppressor genes encode proteins that slow or inhibit progression through a specific stage of thecell cycle, checkpoint–control proteins that arrest the cell cycle if DNA is damaged or chromosomes are abnormal, receptors for secreted hormones that function to inhibit cell proliferation, proteins that promote apoptosis, and DNA repair enzymes.

1. *TP*53

(1) *Rare germ–line mutation of TP*53 leads to the–Fraumeni familial cancer syndrome. On the other hand, somatic mutations of *TP*53 have been foundin the majority of sporadic cancers.

(2) *TP*53 restrains tumor formation by two different mechanisms.

● Activates the *P*21 (Cdk inhibitor) in response to DNA damage and stress. Loss of *TP*53 in cells prevents the *P*21 gene from being transcribed, leading to the increased activity of the multiple Cdks normally turned off by *p*21 and resulting in increased cell proliferation.

● Induce apoptosis.

2. ***BRCA*1 and *BRCA*2**

(1) *BRCA*1 and *BRCA*2 are tumor suppressor genes associated with breast and ovarian cancer, along with several other cancers. Loss of *BRCA*1 or *BRCA*2 leads to the accumulation of other genetic defects, which can then lead to cancer formation.

(2) In addition to their roles in DNA repair, *BRCA*1 and *BRCA*2 have been implicated in a variety of cellular processes, including DNA synthesis, regulation of gene transcription (similar to *TP*53, one target of *BRCA*1 transcriptional activation is the CDK inhibitor *p*21), cell cycle checkpoint control, centrosome duplication and ubiquitination.

3. ***APC gene***

(1) In addition to cancers arising in the setting of FAP, the majority (70% ~ 80%) of non–familial colorectal carcinomas and sporadic adenomas also show homozygous loss of the *APC* gene, thus firmly implicating *APC* loss in the pathogenesis of colonic tumors.

(2) An important function of the APC protein is to cause degradation of β–catenin, thus maintaining low levels in the cytoplasm. Inactivation of the *APC* gene, and consequent loss of the APC protein, increases the cellular level's β–catenin, which, in turn, translocates to the nucleus and promotes cell cycle.

C. Mechanisms involved in tumorsuppressor genes function

1. **Suppression of cell division**

(1) This is **the main mechanism** for most tumor suppressor genes.

(2) *RB*, *APC*, *alternate reading frame* (*ARF*), *RIZ*1, *P*15, *P*16, *P*18, *P*19, *P*21, *P*27, and *TP*53.

2. **Induction of apoptosis**: *TP*53, *APC*, cluster of differentiation 95 (CD95), bridging integrator 1 (*Bin*1) and phosphatase and tensin homolog (*PTEN*).

3. **DNA damage repair**

(1) mutS homolog 2 (*MSH*2), mutL homolog 1 (*MLH*1), Ataxia–telangiectasia–mutated gene product

(*ATM*), breast cancer protein (*BRCA*), Nijmegen breakage syndrome 1 (*NBS*1), Fanconi-Anemia related tumor suppressor (*FA*), and *TP*53.

(2) They are able to fix DNA damages, including mismatch and vast damage to one of the DNA double strands.

4. **Inhibition of metastasis**: Metastin, breast cancer metastasis suppressor 1 (*BRMS*1), tissue inhibitor of metalloproteinase (*TIMP*), cofactor required for specificity protein 1 activation (*CRSP*), and KAL1/CD82.

Ⅲ. Growth factors and their receptors

A. Concept

1. A growth factor is a naturally occurring substance capable of stimulating cellular growth, proliferation, healing, and cellular differentiation. Usually it is a protein or a steroid hormone. Growth factors are important for regulating a variety of cellular processes.

2. Growth factor receptors are transmembrane proteins which bind to specific growth factors and transmit the instructions conveyed by the factors to intracellular space.

B. Activation of growth factor signalling

1. All growth factor receptors are membrane-bound enzyme-linked receptors. Like other membrane receptors, they contain three domains: an extracellular ligand (growth factor) binding domain, a transmembrane domain, and a cytoplasmic domain that acts as an enzyme or forms a complex with another protein that acts as an enzyme. The majority of growth factor receptors are **receptor tyrosine kinases**. Growth factor binding leads to phosphorylation of tyrosine residues on a number of intracellular signaling molecules, and these molecules transmit the signal to the inside of the cell.

2. However, not all growth factor receptors are tyrosine kinase receptors. TGF-β activates receptor **serine-threonine kinases** that phosphorylate the SMAD protein transcription factor, resulting in downstream changes in gene transcription.

[*Practices*]

[A1 type]

[1] Which of the following statements best describes a tumor suppressor gene? ______

A. A gain-of-function mutation leads to uncontrolled proliferation.

B. A loss-of-function mutation leads to uncontrolled proliferation.

C. When it is expressed, the gene suppresses viral genes from being expressed.

D. When it is expressed, the gene specifically blocks the G1/S checkpoint.

E. When it is expressed, the gene induces tumor formation.

[2] Which of the following is the mechanism through which Ras becomes an oncogenic protein?

______

A. Ras remains bound to GAP.

B. Ras can no longer bind cAMP.

C. Ras has lost its GTPase activity.

D. Ras can no longer bind GTP.

E. Ras can no longer be phosphorylated by MAP kinase.

[3] Which of the following statements best describes a characteristic of oncogenes? ____________

A. All retroviruses contain at least one oncogene.

B. Retroviral oncogenes were originally obtained from a cellular host chromosome.

C. Proto-oncogenes are genes, found in retroviruses, which have the potential to transform normal cells when expressed inappropriately.

D. The oncogenes that lead to human disease are different from those that lead to tumors in animals.

E. Oncogenes are mutated versions of normal viral gene products.

[4] When *TP53* increases in response to DNA damage, which of the following events occurs? ____________

A. *TP53* induces transcription of *cdk4*.

B. *TP53* induces transcription of cyclin D.

C. *TP53* binds E2F to activate transcription.

D. *TP53* induces transcription of p21.

E. *TP53* directly phosphorylates the transcription factor *jun*.

[5] Constitutively activated Ras has become insensitive to which of the following elements of the growth factor signaling pathway? ____________

A. Raf-1　B. MEK　C. MAP kinase

D. Ras-GAP　E. Elk-1

[6] Patients with retinoblastoma suffer from a high incidence of tumors arising from clonal outgrowth of some retinal precursor cells due to mutation of the tumor suppressor gene *RB*. Analysis of cells from these tumors indicates that both copies of the *RB* gene are mutated or lost, whereas the surrounding retinal cells have at least one functional *RB* allele. Which of the following terms best describes the genetic phenomenon that leads totumor development in retinoblastoma patients? ____________

A. Loss of imprinting　B. Deregulated expression　C. Incomplete penetrance

D. Gain of function　E. Loss of heterozygosity

[7] Osteosarcoma has recently been diagnosed in a 12-year-old girl. Family history indicates that her paternal aunt died of breast cancer at age 29 after having survived treatment for adrenocortical carcinoma. An uncle died of a brain tumor at age 38 and the patient's father, age 35, has leukemia. An analysis of this patient's DNA would most likely reveal a mutation in which of the following genes? ____________

A. *RB*　B. *RAS*　C. *TP53*

D. *c-ABL*　E. *PKC*

[X type]

[8] Which genes listed below are tumor suppressor genes? ____________

A. *TP53*　B. *BRCA1*　C. *NBS1*

D. *ERBB2*　E. *APC*

[9] For oncogenes, which statement is correct? ____________

A. Inhibit cell proliferation.

B. Induce malignant cell transformation *in vitro*.

C. Induce cell apoptosis.

D. Induce tumor formation *in vivo*.

E. Include virus oncogene and cell oncogene.

[10] Which mechanism is involved in the change for proto-oncogene to oncogene? ____________

A. Down-regulation of tumor suppressor genes

B. Mutation
C. Chromosome translocations
D. Gene amplification
E. Protein modifications

Answers

1	2	3	4	5	6	7	8	9	10
B	C	B	D	D	E	C	ABCE	BDE	BCD

(*Xie Jianjun*)

Chapter 24

Gene Diagnosis and Gene therapy

Examination Syllabus

1. Gene diagnosis: concept, methods and application strategies.
2. Gene therapy: concept, strategies and procedure.

Ⅰ. Gene diagnosis

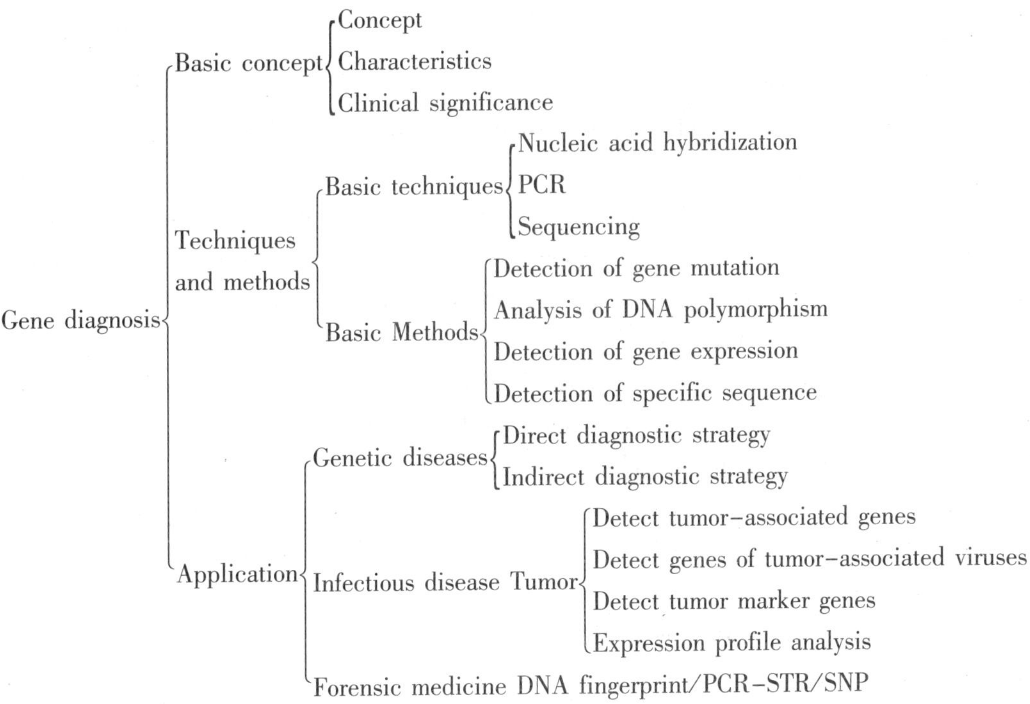

A. Basic concept and principle of gene diagnosis

1. Concept

Gene diagnosis refers to diagnose the specific diseases by detecting the changes of gene structure or expression level with the methods of modern molecular biology and molecular genetics.

(1) Targets (materials): DNA, RNA { DNA diagnosis / RNA diagnosis }

(2) Targets for **Molecular diagnosis**: DNA, RNA, protein

2. Characteristics

(1) Detect gene directly, etiology diagnosis.

(2) High specificity and high sensitivity.

(3) Speed, simplicity.

3. Clinical significance

(1) Confirm diagnosis.

(2) Early diagnosis.

(3) Staging and typing of disease.

(4) Monitor treatment effect.

(5) Prognostic analysis.

B. Techniques and methods used for gene diagnosis

1. Basic techniques

(1) **Nucleic acid hybridization**: direct probe testing

● Southern blot: DNA analysis.

● Northern blot: RNA analysis.

● Dot blot, slot blot: Allele-specific oligonucleotide blot hybridization (ASO).

● In situ hybridization: FISH.

● DNA microarray (gene chip, DNA chip) technique: integrates hundreds of thousands of probes such as DNA, oligonucleotides on the surface of substrates in some way, to generate a two-dimensional array which hybridizes to the labeled samples. After scanning and analyzing, the signals provide information on the identity of samples. The essence of gene chip is **large-scale integrated solid-phase hybridization.**

(2) **PCR (polymerase chain reaction)**: amplification of target gene

● RT-PCR, nested PCR, real time PCR, Quantitative PCR (qPCR).

● PCR/ASO, PCR/SSCP (single-strand conformation polymorphism).

● PCR/analysis of enzymatic cleavage pattern, PCR/RFLP (restriction length fragment polymorphism analysis): cleave DNA molecules using restriction endonuclease, analyze the patterns of restriction sites, RE fragment length and number.

(3) **Gene sequencing**: the most direct and accurate technique, **gold standard.**

Different techniques based on different targets:

● DNA based molecular techniques: Southern blot, FISH, PCR, sequencing, DNA microarray, restriction analysis, RFLP, SSCP.

● RNA based molecular techniques: Northern blot, RT-PCR, realtime qPCR, DNA microarray.

● Protein based molecular techniques: Western blot, immuno-histochemistry, protein microarray.

2. Basic Methods

(1) **Detection of mutations or changes in DNA copy number**

●Mutation detection techniques: DNA Sequencing, PCR, ASO, SSCP, Heteroduplex Analysis (HA), DNA chip, Mass spectrometry, Oligonucleotide-Ligation Assay (OLA), Protein Truncation Test (PTT)

●Examples:

Detect known point mutations: PCR/ ASO; DNA chip

Detect unknown mutations: PCR/SSCP/ Sequencing, DNA chip

(2) **Analysis of DNA polymorphism**: based on genetic marker

①RFLP: restriction analysis; RFLP linkage analysis

②Polymorphism analysis of DNA repetitive sequence

●VNTR (variable number of tandem repeat): 6 ~ 70bp, minisatellite DNA

●STR (short tandem repeat): 1 ~ 4bp, microsatellite DNA, PCR-STR

③SNP (single nucleotide polymorphism)

(3) **Detection of gene expression**

●Quantitative analysis of mRNA: dot blot, RT-PCR, real-time qPCR

●Analysis of mRNA size: Northern blot

●DNA microarray: a single hybridization can detect the expression of thousands of genes in parallel.

(4) **Detection of specific sequences of pathogenic gene**: nucleic acid hybridization (using a virus probe), PCR

C. Application of gene diagnosis

1. Gene diagnosis of genetic diseases

(1) **Direct diagnostic strategy**

●Various genetic mutations that lead to inherited diseases can be directly detected by using the method of detection of gene mutation with various molecular biology techniques (for example, using appropriate probes from defective genes, or PCR). The premise of this strategy is that defective gene must have been cloned and the normal sequence and structure of the gene have been elucidated

●Example: Sickle cell anemia: Southern blot; PCR/ASO

(2) **Indirect diagnostic strategy**

Adopt the method of polymorphism linkage analysis. Look for the chromosome with a genetic defect linking with polymorphism label (RFLP etc.)

2. Gene diagnosis can provide not only definitive pathogen diagnosis for most infectious diseases, but also the diagnosis of carriers and potential infections, pathogen classification and typing etc.

①Nucleic acid hybridization: design a probe for hybridization of specific nucleic acid sequences of pathogens.

②PCR: amplify the conserved sequences of pathogen genes.

③DNA microarray.

3. Gene diagnosis of tumor

(1) Detect tumor-associated genes' mutations and abnormal expression: oncogene, tumor suppressor gene, tumor metastasis gene, tumor metastasis suppressor gene and other gene. The methods of detection of gene mutations and expression could be used.

(2) Detect the genes of tumor-associated viruses

●EB virus: associated with nasopharyngeal cancer and Burkitt lymphoma.

●HPV (human papilloma virus): associated with cervical cancer.

- HBV, HCV: liver cancer.
- HTLV-1related to adult T cell leukemia and lymphoma.

The methods of detection of exogenous genes could be used.

(3) Detect tumor marker genes or mRNA

(4) Expression profile analysis of tumor-associated genes

4. Application of gene diagnosis in forensic medicine

(1) DNA fingerprinting

- Southern blot with oligonucleotide probes targeted to VNTR
- Obtain hybridization bands of different lengths, moreover, the number and size of the hybridization bands are individual-specific, just like a human fingerprint. This hybridization pattern is called DNA fingerprint.

(2) PCR-STR

(3) SNP

Ⅱ. Gene therapy

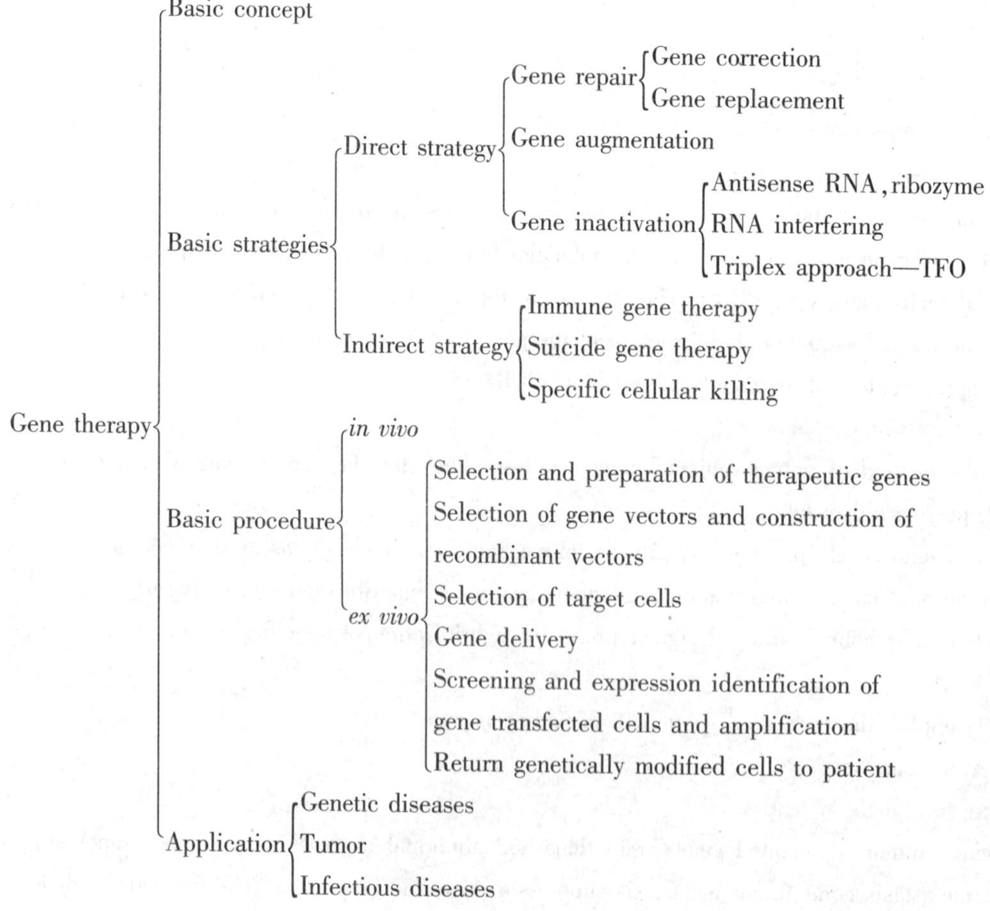

A. Basic concept of gene therapy

Gene therapy is a biomedical intervention based on modification of genetic material in living cells. In other words, gene therapy is the process of transferring certain genetic material into human target cells with molecular biology technology to achieve the therapeutic effect.

B. Basic strategies for gene therapy

1. **Direct strategy**: targeted to defective genes

(1) **Gene correction**: correct the mutated base and keep the normal bases.

(2) **Gene replacement**: integrate normal genes into the genome of target cells to replace defective genes *in situ* by homologous recombination or gene targeting techniques.

Accurate *in situ* repair of defective genes without involving changes ofthe genome is the **ideal therapy strategy**.

New techniques: gene editing techniques: CRISPR (Clustered regularly interspaced short palindromic repeats) /Cas, ZFN (zinc-finger nuclease), TALEN (transcription activator-like effector nucleases).

(3) **Gene augmentation**: gene complementation

Introduce normal functional genes by the way of non-site-specific integration (do not remove abnormal gene), and normal products are expressed to compensate for the function of defective genes or to enhance the original function.

This is the **main strategy** used in clinic at present.

(4) **Gene inactivation**: gene silencing or gene interference

Introduce specific sequences to block the abnormal expression of certain genes.

● Antisense RNA, ribozyme: complement with mRNA——mRNA inactivation or direct cleavage——translation level

● RNA interfering: dsRNA →siRNA (small interfering RNA)→mRNA degradation

● Triplex approach: triplex-forming oligonucleotide (TFO) or oligodeoxyribonucleotide (ODN)——form a triple helix with the DNA double helix molecule of the target gene——inhibit transcription ——DNA level

2. **Indirect strategy**

Introduce therapeutic genes that are not directly related to defective genes.

(1) **Immune gene therapy**

Introduce the gene into disease cells or immune system cells, killing of disease cells because of enhanced immune response.

● Cytokine gene: IL-2, IFN, TNF-α、G-CSF

● MHC I gene, DNA vaccine

● CAR-T

(2) **Suicide gene therapy**

Some genes from virus or bacteria are introduced and expressed to produce enzymes that catalyze the conversion of non-toxic drug precursors into cytotoxic substances. The receptor cells that carry the gene are themselves killed, so the genes are called "suicide genes".

● HSV-TK (thymidine kinase) gene: tk→GCV (ganciclovir) phosphorylation→inhibit DNA polymerase (DNA systhesis) → cell death (including bystander cell)

● E. coli CD (cytosine deaminase) gene: 5′-FC → 5′-FU

(3) **Specific cellular killing**

● The toxin genes such as pseudomonas aeruginosa exotoxin (PE) and diphtheria toxin (DT) gene are recombinedwith TGFα into the fusion gene TGFα-PE or TGFα- DT.

●TGFα can bind to the EGFR (epidermal growth factor receptor) due to its similar structure to EGF (epidermal growth factor). Therefore, this fusion protein can specifically enter and kill tumor cells of bladder, kidney, lung, breast and other tumors which express EGFR highly.

C. Basic procedure of gene therapy

1. Classification of gene therapy

According to the approach of gene delivery, gene therapy can be divided into two types.

(1) ***In vivo***: directly introduce the target gene into the body of the relevant tissue organs, so that it enters the corresponding cells and express.

(2) ***Ex vivo***: introduce the target gene into the target cells *in vitro*, and after screening and proliferation, the cells are returned to the patients and make the gene *in vivo* to effectively express the corresponding products to achieve the purpose of treatment.

2. Basic process of gene therapy

(1) Selection and preparation of therapeutic genes

The primary problem of gene therapy is to select specific target genes that have therapeutic effects on diseases. The target gene can be selected as the normal gene corresponding to the defective gene or the gene unrelated to the defective gene but with therapeutic effect.

(2) Selection of gene vectors and construction of recombinant vectors

It can be divided into **viral vectors** and **non-viral vectors**. RV (retrovirus), AdV (adenovirus), AAV (adenovirus-related virus), and HSV (herpes simplex virus) are currently used as vectors for gene transfer.

The construction of the recombinant vector includes removing the gene sequence of the pathogenic structure of the virus, replacing the blank area with the target gene and inserting the marker gene.

(3) Selection of target cells

The target **somatic cells**: lymphocytes, hematopoietic cells, epithelial cells, keratinocytes, endothelial cells, fibroblasts, liver cells, muscle cells and tumor cells, etc. **Germ cell** gene therapy is considered as unethical.

(4) Gene delivery

Non-viral approach: include physical method and chemical method.

●Physical methods: microinjection, electroporation, direct injection of naked DNA and gene gun method.

●Chemical methods: calcium phosphate precipitation, DEAE glucan method, Plasmid liposome complex, receptor-mediated delivery, etc.

Viral approach: RV, AdV, AAV, and HSV. In the clinical implementation of gene therapy, virus vectors are the main methods.

(5) Screening and expression identification of gene transfected cells and amplification

For *in vitro* cultured cells, gene transfection efficiency is difficult to reach 100%, Therefore, marker genes (such as Neo^R: neomycin phosphotransferase Ⅱ) in the vector should be used to screen transfected cells.

The expression level of the target gene in the selected cells can be further identified by means of Northern blot hybridization or real time PCR.

(6) Return genetically modified cells to patient

The cells modified by the therapeutic gene are delivered back into the body in different ways such as intravenous injection, intramuscular injection, subcutaneous injection, intranasal method, etc.

D. Application of gene therapy

1. Inherited diseases: SCID/ADA gcnc/RV; CF (cystic fibrosis); type B hemophilia
2. Tumor: gene inactivation, gene augmentation, immune gene therapy
3. Infectious disease: gene inactivation

[*Practices*]

[A1 type]

[1] The main advantages of gene diagnosis in infectious diseases are that________

A. etiological diagnosis can be made

B. there is no damage to the body during the diagnosis

C. the method is simple and the cost is low

D. early diagnosis

E. it is helpful in differential diagnosis

[2] Sickle cell anemia is a point mutation in the hemoglobin gene. Which of the following methods is the best one used to detect this mutation?________

A. Isolate DNA from lymphocytes in the blood, examine the exon size by Southern blot.

B. Isolate DNA from lymphocytes in the blood, sequencing the intron of the gene.

C. Isolate DNA from lymphocytes in the blood, amplify the gene by PCR, and then digest it with restriction endonuclease.

D. Isolate DNA from lymphocytes in the blood, amplify the gene by PCR, and then hybridize with ASO probes

E. Isolate DNA from lymphocytes in the blood, examine the exon size by Northern blot.

[3] The molecular basis of DNA fingerprint used for individual identification and paternity test in forensic examination is________

A. polymorphism of DNA

B. polymorphism of RNA

C. polymorphism of amino acids

D. protein polymorphism

E. phenotypic polymorphism

[4] Which of the following is NOT part of the antisense gene therapy:________

A. ribozymes B. antisense RNA C. siRNA

D. TFO E. nuclease

[5] The basic strategy of gene therapy at this stage is________

A. gene replacement

B. in situ genecor rection of the diseased gene

C. gene augmentation or gene interference

D. homologous recombination

E. gene labeling

[6] Gene diagnosis includes________

A. biochemical diagnosis B. serological diagnosis C. prenatal diagnosis

D. pathological diagnosis E. radiographic diagnosis

[7] Which of the following is not common technique for gene diagnosis?________

A. RFLP B. PCR C. Southern blot

D. Western blot E. DNA sequencing

[8] The essence of gene chip is________

A. high density peptide arrays

B. high density monoclonal antibody arrays

C. high density carbohydrate arrays

D. high density oligonucleotide arrays

E. high density protein arrays

[9] Which of the following is not strategy for gene therapy?________

A. Gene augmentation B. Gene replacement C. Immune gene therapy

D. Introduce the antisense nucleic acid E. Gene labeling

[10] Gene therapy using specific antisense nucleic acids to block mutant gene expression is________

A. gene replacement B. gene interference C. gene labeling

D. gene augmentation E. gene correction

[11] Which of the following is not included in the non-viral vector approach to gene therapy?______

A. Receptor-mediated approach

B. Liposome-mediated approach

C. Calcium phosphate co-precipitation method

D. Direct injection

E. Intranasal methodof adenovirus

[A2 type]

[12] A 21-year-old woman was abducted when she went to the local convenience store. Her body was found the next morning in a wooded area behind the store. The autopsy revealed that she had been sexually assaulted and strangled. Crime scene investigators were able to collect a semen sample from vaginal fluid as well as tissue samples from underneath the victim's fingernails. DNA samples were obtained from three suspects besides the victim. A variable number of tandem repeats (VNTR) analysis was performed on the DNA samples from the evidence collected, the victim, and the suspects, and the results were compared. Which of the following techniques is the most appropriately for this analysis?________

A. Allele-specific oligonucleotide probes B. DNA sequencing

C. Northern bolt D. Southern blot E. Western blot

[13] A 10-month-old baby boy presents with steatorrhea, recurrent pulmonary infection, GI upset and foul-smelling stool. Which of the following tests is undertaken to confirm your diagnosis?________

A. Sweat test B. Blood glucose C. CBC

D. RFLP analysis E. Blood fat

[14] A 25-year-old female is sexually active with a new partner and would like to have an HIV test. The current standard for HIV testing is to start with an enzyme-linked immuno sorbent assay (ELISA). HIV antigens are coated on a plate, and the patient's serum is added. If the patient has HIV antibodies, binding of these antibodies to the antigens on the plate will be detected. If the ELISA test is positive, a confirmatory test is done. In this confirmatory test, the patient's serum is added to HIV antigens which are on a nitrocellulose membrane. This confirmatory test is a________

A. Northern blot B. Southern blot C. Western blot

D. gel eletrophoresis E. polymerase chain reaction

[15] DNA fingerprinting is used in the legal world to determine parentage, genealogy, or other genetic relationships, or to implicate suspects in law enforcement investigations involving violent crimes. Fragments produced by various restriction enzymes from a number of different loci can be used to identify individuals with the accuracy of a fingerprint. Therefore, this technique is called "fingerprinting". Which one of the following statements is incorrect about restriction enzymes?________

A. They are site specific endonucleases

B. The restriction sites are palindrome in nature

C. All restriction enzymes are of bacterial origin

D. They always produce complementary single stranded overhangs

E. They are also called "molecular scissors"

[B1 type]

No. 16 ~ 18 share the following suggested answers

A. nucleic acid hybridization B. PCR

C. gene sequencing D. gene library E. gene cloning

[16] The technique of detecting homologous nucleic acid sequences in samples directly by means of probes is________

[17] The technique of detecting sample nucleic acid by amplification of target gene is________

[18] The most direct and accurate gene diagnosis method is________

No. 19 ~ 21 share the following suggested answers

A. gene replacement

B. gene augmentation

C. gene inactivation

D. immune gene therapy

E. suicide gene therapy

[19] The optimal strategy for gene therapy is________

[20] Currently, the most widely used gene therapy strategy is________

[21] The commonly used method of gene therapy in viral disease is________

[X type]

[22] Which of the following are several approaches to gene diagnosis________

A. Detect the mRNA

B. Directly detect mutations in gene structure

C. Detect the changes in protein structure

D. Gene diagnosis is performed according to the preliminary conclusions of clinical diagnosis

E. Detect the morphology changes in protein structure

[23] Non-viral vector approaches to gene therapy include________

A. receptor-mediated approach

B. liposome-mediated approach

C. calcium phosphate co-precipitation method

D. direct injection

E. electroporation

[24] According to the principle, the techniques of gene diagnosis can be divided into the following categories________

A. PCR B. DNA probe C. DNA probe/ PCR

D. DNA sequencing E. DNA probe/ DNA sequencing

[25] Current target cells for gene therapy include________

A. the egg B. the lymphocyte C. bone marrow cells

D. the tumor cells E. sperm cells

Answers

1	2	3	4	5	6	7	8	9	
D	D	A	E	C	C	D	D	E	
10	11	12	13	14	15	16	17	18	
B	E	D	D	C	D	A	B	C	
19	20	21	22	23	24	25			
A	B	C	ABD	ABCDE	ABD	BCD			

Brief Explanations for Practices

2. PCR amplification of DNA fragments can increase the sensitivity of restriction enzyme fragments. The presence of the sickle cell allele was determined by ASO hybridization.

5. Under current technological conditions, in situ correction of diseased genes is only a theoretical idea. It is difficult to replace the diseased gene by homologous recombination, with high risk and low probability. Therefore, normal gene replacement or intervention is the basic strategy of gene therapy at present.

6. The target of gene diagnosis is nucleic acid (DNA or RNA). The object of biochemical diagnosis or serological diagnosis is the cell or compound. The object of pathological diagnosis is tissue section. Prenatal diagnosis is mainly the detection of DNA, belongs to gene diagnosis.

7. Western blot is mainly used to detect proteins, which does not fall into the category of gene diagnosis.

9. Gene labeling do not work as treatments. Marker gene is used to screen the transfected cells.

10. The use of specific antisense nucleic acids to block the abnormal expression of mutant genes and inactivate them belongs to gene intervention.

11. Adenovirus intranasal method USES adenovirus as vector to introduce exogenous genes into host cells, which is not a non-viral vector method.

12. VNTR analysis examines the hypervariable regions of the human genome. These contain sequences that are repeated in tandem a variable number of times and the length is unique for each individual. Because it is DNA fragments that are being analyzed, the Southern blot is the most appropriate technique to use to separate and detect these region. DNA sequencing is too time-consuming to be practical for forensic analyses.

13. A 10-month-old baby boy presents with these syndromes. Genetic disease is highly suspected. RFLP (restriction fragment length polymorphism) analysis detects mutation in the DNA that either introduce or eliminate a recognition site for a restriction enzyme and could confirm the diagnosis.

14. Western blot is used to detect proteins and HIV antibodies are proteins.

15. All statements correctly describe about restriction endonucleases except for the complementary single stranded overhangs (sticky ends) which are not always produced. Blunt ends can also be produced, depending on the nature of the enzyme and the type of cut involved.

22. Detection of protein structure changes is a phenotypic diagnosis. Gene diagnosis must be based on

the preliminary conclusion of clinical diagnosis, not random diagnosis.

24. The techniques of gene diagnosis are mainly based on nucleic acid hybridization, PCR and DNA sequencing.

25. Gene therapy currently bans the use of germ line cells and only allows the use of somatic cells. The target cells that have been used are lymphocytes, hematopoietic cells, epithelial cells, keratinocytes, endothelial cells, fibroblasts, liver cells, muscle cells and tumor cells.

[*Clinical Cases*]

26. Sickle-cell anemia is known to be caused by the substitution of A→T base with the 6th codon of the β-globin gene. The mutation causes restriction enzyme *Mst* Ⅱ recognition sites loss. A patient is suspected to have this disease.

(1) The patient needs to be confirmed by gene diagnosis. Please design four different technique routes of gene diagnosis. Requirement: write down the experimental design idea, the main flow of the experiment, and analysis of expected results of experiments.

(2) Design a technical route for gene therapy of this disease. Requirement: write down the experimental design idea, the main flow of the experiment.

27. Genomic DNAs were extracted from the blood cells of two people. And the restriction enzymes of A, B and C were used for enzyme digestion. Suppose a 10kb length of DNA enzyme digestion map obtained (Figure 24. 1). A mutation was found in sample 2. The loss of B^+ and C^+ enzyme digestion sites was caused. Please design an experimental scheme to perform the RELP analysis and hybridization analysis by using the DNA segment shown in the figure as a probe.

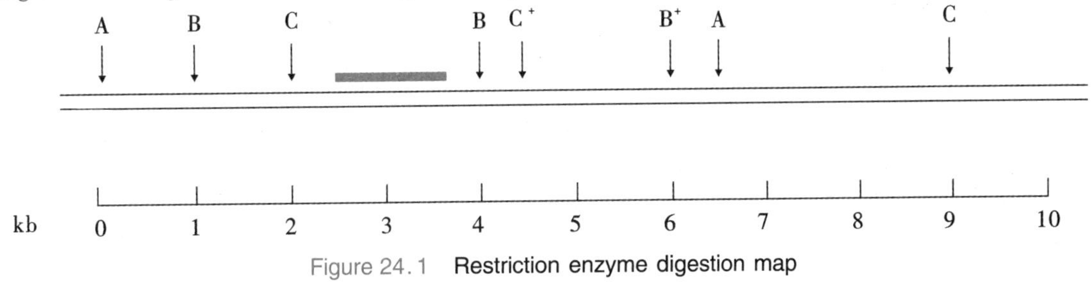

Figure 24. 1 Restriction enzyme digestion map

28. A 3-year-old white boy is brought to the clinic for a chronic productive cough not responding to antibiotics given recently. He has neither fever nor sick contacts. His medical history is significant for abdominal distention, failure to pass stool, and emesis as an infant. He continues to have bulky, foul-smelling stools. No diarrhea is present. He has several relatives with chronic lung and "stomach" problems, and some have even died at a young age. The examination reveals an ill appearing, slender male in moderate distress. The lung exam reveals poor air movement in the base of lungs bilateral and coarse rhonchi throughout both lung fields. A chloride sweat test was performed and was positive, indicating cystic fibrosis (CF).

(1) What is the mechanism of the disease?

(2) How might gel eletrophoresis assist in making diagnosis?

(3) His patient has been diagnosed with cystic fibrosis and has been determined to have the most common mutation, the F_{508} gene. Which of the following is the most cost and time-effective method for testing family members to see who are carriers of the mutation? ____

A. Allele-specific oligonucleotide probe analysis

B. DNA fingerprinting analysis

C. DNA sequencing

D. Restriction length fragment polymorphism analysis

E. PCR

(4) Please design a technical route for gene therapy of this disease.

Answer for clinic case:

27. Restriction endonuclease A was used to perform enzyme digestion of the two samples. Both produce 3.5kb and 6.5kb fragments. The 6.5kb fragment can be hybridized with the probe (Figure 24.2A).

The two samples were cut with enzyme B and the digestion fragment sizes were different. Sample no. 1 got three segments of 3kb, 2kb and 4kb, while sample no. 2 only had two segments of 3kb and 6kb. After hybridization with the probe, only 3kb bands are shown and the two samples are identical (Figure 24.2 B).

After cutting the two samples with the enzyme C, sample no. 1 was cut into 2.5 kb and 4.5kb segments, and only 2.5kb segments could be hybridized with the probe; Sample 2 yields a 7kb fragment that can be hybridized with the probe (Figure 24.2 C).

Therefore, select restriction endonuclease C for enzyme digestion and then hybridize, ideal results can be obtained.

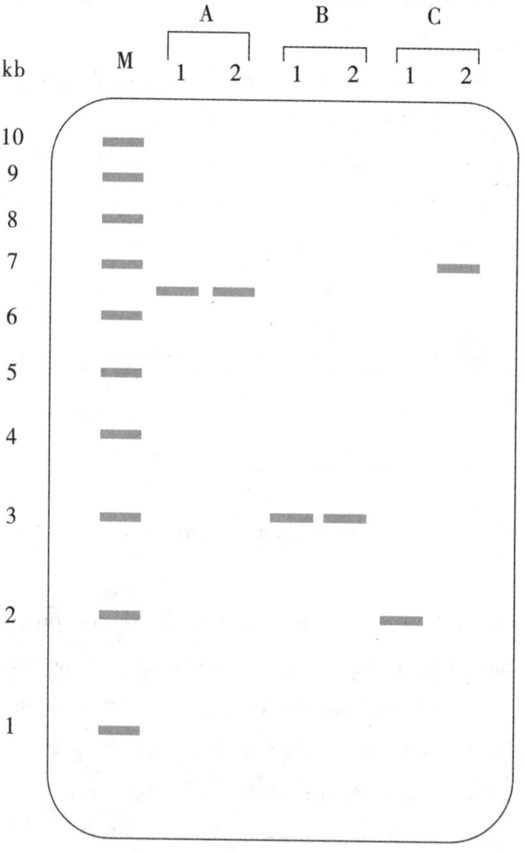

Figure 24.2 Hybridization result

28. Answer: C.

(*Zhu Lina*, *Li Ling*)

Chapter 25

Genetically Engineered Drugs and Vaccines

Examination Syllabus

Genetically engineered drug (protein and peptide drugs) and **genetically engineered vaccines**: major types and examples.

[*Major points*]

Genetically engineered drugs typically include hormones, enzymes, growth and coagulation factors, antibodies as well as vaccines.

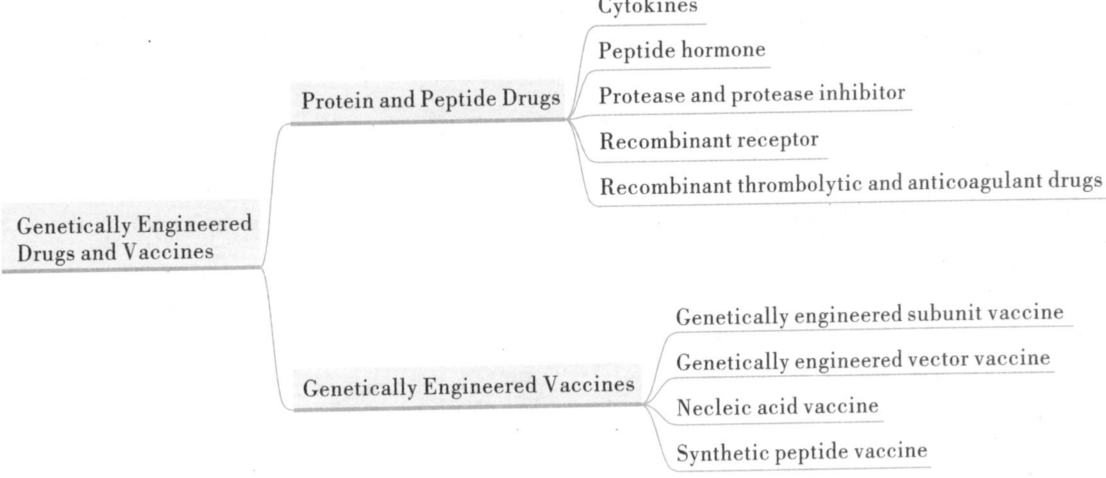

Ⅰ. Protein and Peptide Drugs

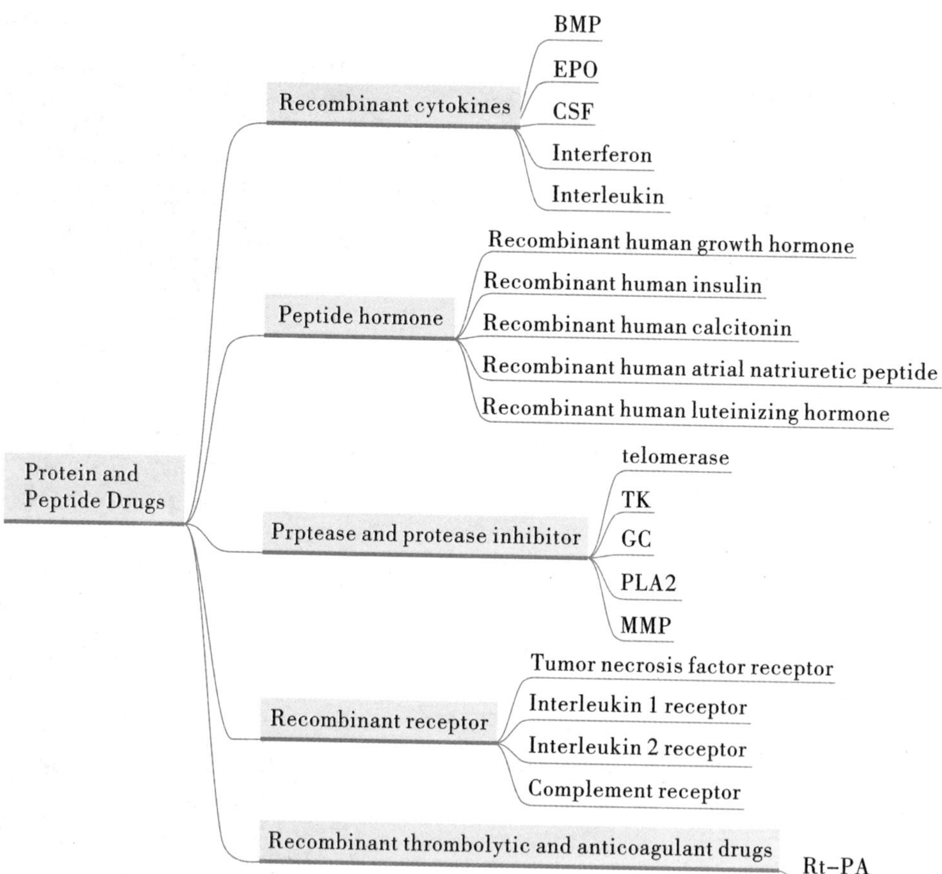

In 1982, the first commercial genetically engineered drugs, human insulin, was approved by the U. S. FDA.

A. Application of genetically engineered drug

Such as in diabetes, cardiovascular disease, viral infection, rheumatoid arthritis, wound repair, anti-tumor, etc.

B. Types of genetically engineered drug

Recombinant cytokine, peptide hormone protease and protease inhibitor, recombinant receptor, recombinant thrombolytic and anticoagulant drugs, etc. (Table 25.1)

Table 25.1 Some examples of protein and peptide drugs produced by recombinant technology

Active molecule	Drug name	disease
Insulin	Humalog, Novo Rapid, Gensulin R, Humulin R, Actrapid HM	for diabetics
Factor Ⅷ	Factor Ⅷ from Bayer	for males suffering from hemophilia A

Continue to Table 25.1

Active molecule	Drug name	disease
Factor Ⅸ	Factor Ⅸ from Bayer	for hemophilia B
Human growth hormone (HGH)	Genotropin, Humatrop, Serostin	for growth hormone deficiency
Erythropoietin(EPO)	Eprex, Epogen	for treating anemia
Interferon	Avonex, Rebif, Betaseron	the hepatitis B virus
α-L-iduronidase (rhIDU; laronidase)	Aldurazyme	mucopolysaccharidosis type I(MPS I; deficiency of α-L-iduronidase) for the treatment of non-neurological symptoms
N-acetylgalactosamine-4-sulfatase(rhASB; galsulfase)	Naglazyme	mucopolysaccharidosis Ⅵ (MPS Ⅵ 4-sulfatase deficiency of N-acetylgalactosamine, Maroteaux-Lamy syndrome)
Granulocyte colony-stimulating factor(G-CSF)	Neupogen	stimulating neutrophil production(e. g., after chemotherapy) and for mobilizing hematopoietic stem cells from the bone marrow into the blood
Tissue plasminogen activator (TPA)	Activase	for dissolving blood clots
Adenosine deaminase (ADA)		for treating some forms of severe combined immunodeficiency(SCID)
Dornase alfa	Pulmozyme	for cystic fibrosis
Glucocerebrosidase	Ceredase	for type 1 Gaucher's disease
Hepatitis B Surface antigen (HBsAg)	Engerix B	to vaccinate against the hepatitis B virus
C1 inhibitor (C1 INH)		used to treat hereditary angioedema
Follicle-stimulating hormone (FSH)	Pergonal	to treat fertility is sues in women, especially women who are anovular and oligoovular

1. Cytokines

Cytokines, produced by a broad range of cells including immune cells, mast cells, endothelial cells, fibroblasts, and various stromal cells, are a broad and loose category of small proteins that are important in cell signaling.

- Bone morphogenetic protein (BMP): used to treat bone-related conditions.
- Erythropoietin (EPO): used to treat anemia.
- Granulocyte colony-stimulating factor (G-CSF): used to treat neutropenia in cancer patients.
- Granulocyte-macrophage colony-stimulating factor (GM-CSF): used to treat neutropenia and fungal infections in cancer patients.
- Interferon alfa: used to treat hepatitis C and multiple sclerosis.
- Interferon-beta: used to treat multiple sclerosis.
- Interleukin 2 (IL-2): used to treat cancer.
- Interleukin 11 (IL-11): used to treat thrombocytopenia in cancer patients.
- Interferon-gamma: used to treat chronic granulomatous disease and osteopetrosis.

2. **Peptide hormone**:

(1) Hormones, produced by glands in multicellular organisms, are signaling molecules that are transported by the circulatory system to target distant organs to regulate physiology and behavior.

(2) **Recombinant peptide hormone** includes recombinant human growth hormone (hGH), recombinant human insulin, recombinant human calcitonin, recombinant human atrial natriuretic peptide (ANP), recombinant human parathyroid hormone (PTH), recombinant human luteinizing hormone (LH), etc.

3. **Genetically engineered proteases**: such as recombinant superoxide dismutase (SOD), telomerase, thymidine kinase (TK), glucocerebri(o)sidase (GC), phospholipase A2 (PLA2), DNA methyltransferase 1 (DNMT1), acetylcholine esterase (AChE), matrix metalloproteinase (MMP), etc.

4. **Recombinant receptor**

(1) Exogenous therapeutic receptors exert therapeutic effect through competitive inhibition of ligand–receptor binding.

(2) **Common recombinant receptors** include tumor necrosis factor receptor, interleukin 1 receptor, interleukin 2 receptor, complement receptor, etc.

5. **Recombinant thrombolytic and anticoagulant drugs**

(1) Thrombosis intimidates human's life and health seriously, and fibrinolytic therapy with the is available.

(2) Recombinant tissue–type plasminogen activator (rt–PA) is the first and the only drug approved by U. S. FDA for clinical thrombolysis so far.

Ⅱ. Genetically Engineered Vaccines

Traditionally, most vaccines were attenuated either by serial passage of the agent in culture or by inactivation of the agent by chemical or physical means. Genetically engineered vaccines are able to circumvent many of these problems by having defined modifications in these agents, giving them specific properties.

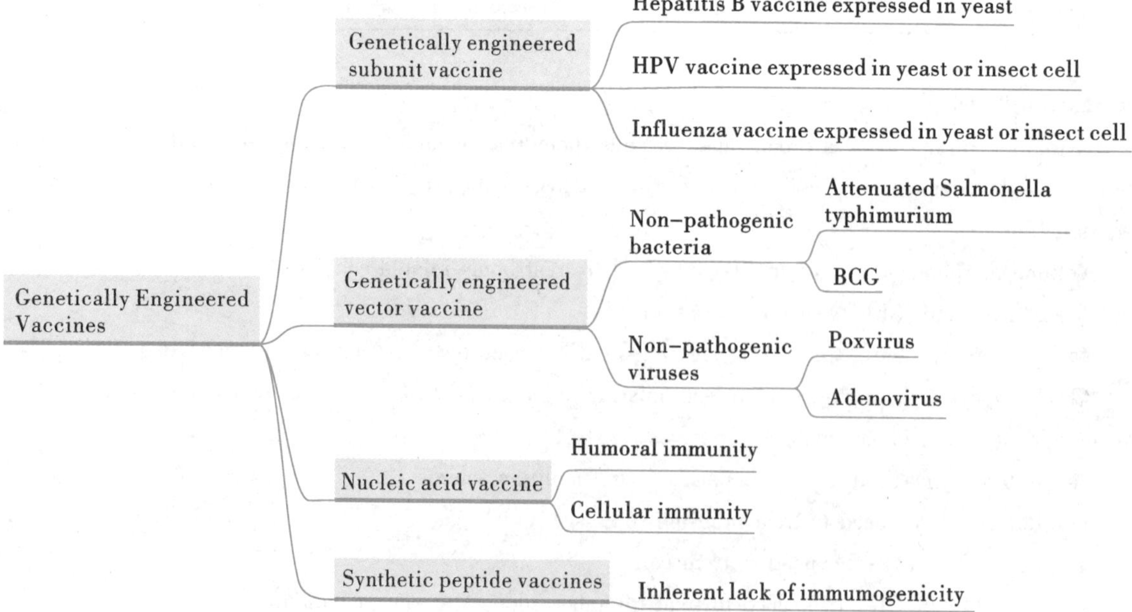

A. Genetically engineered subunit vaccine

1. A subunit vaccine is a vaccine that contains isolated proteins which could induce protective immune response from a virus, but lacks viral nucleic acid.

2. The primary advantage of using a subunit vaccine is that it is very safe, even in people with compromised immune systems.

B. Genetically engineered vector vaccine

Genetically engineered vector vaccine is usually live vaccine and non-pathogenic bacteria or viruses could be used as a vaccine carrier theoretically, such as poxvirus, adenovirus, attenuated Salmonella typhimurium and BCG.

C. Nucleic acid vaccine

1. Nucleic acid vaccination, be heralded as the third generation of vaccine, is a technique for protecting against disease by injection with genetically engineered DNA so cells directly produce an antigen, producing a protective immunological response.

2. Nucleic acid vaccines affect not only humoral immunity but also cellular immunity.

3. Nucleic acid vaccines induce killer cytotoxic T lymphocytes (CTLs), suggesting that an important shift had occurred in non-live vaccine platforms.

D. Synthetic peptide vaccines

1. Synthetic peptide vaccines composed of multiple B cell antigen epitopes and T cell antigen epitopes, which are designed to elicit T cell immunity, are an attractive approach to the prevention or treatment of infectious diseases and malignant disorders.

2. Unfortunately, the advantages that peptide vaccines have to offer are to some extent diminished by their inherent lack of immunogenicity, which so far has been reflected by their not-so-spectacular results in the clinic.

[*Practices*]

[A1 type]

[1] Which of the following does not belong to the genetic engineering drugs?________

A. Interleukin B. Penicillin C. Hepatitis B vaccine

D. Interferon E. Erythropoietin

[2] Which is the first commercial genetic engineering drug approved by the US FDA?________

A. Human insulin B. Recombinant interferon C. BCG

D. Interleukin E. Recombinant human growth hormone

[3] Insulin is generally used for the treatment of which of the following diseases?________

A. Cardiovascular diseases B. Rheumatoid arthritis C. Hepatitis B

D. Diabetes E. Acute myocardial infarction

[4] What kind of disease is the erythropoietin often used to treat? ________

A. Thrombosis B. Hereditary angioedema C. Anemia

D. Myocardial infarction E. Diabetes

[5] What is the role of IFN-α?________

A. Used to treat hepatitis C and multiple sclerosis

B. Used to treat chronic granulomatous disease and osteopetrosis

C. Used to treat neutropenia in cancer patients

D. Used to treat anemia

E. Used to treat myocardial infarction

[6] What kind of disease can adenosine deaminase (ADA) be used to treat?________

A. Hemophilia B. Cystic fibrosis

C. Certain severe combined immunodeficiency disease

D. Mucopolysaccharidosis type I E. Sepsis

[7] Which of the following can be used to treat thrombocytopenia in cancer patients?________

A. IL-11 B. IFN-γ C. EPO

D. TPA E. G-CSF

[8] Which cytokine is mainly produced by kidney cells?________

A. IL-2 B. EPO C. IL-11

D. IFN-β E. Insulin

[9] Which of the following does not belong to the cytokine?________

A. BMP B. G-CSF C. IFN-α

D. hGH E. GM-CSF

[10] Parathyroid hormone is secreted by ________.

A. thyroid B. parathyroid gland C. pituitary

D. hypothalamus E. pancreas

[11] Which of the following are not recombinant peptide hormones?________

A. Recombinant human growth hormone

B. Recombinant human calcitonin

C. Recombinant human atrial natriuretic peptide

D. Recombinant superoxide dismutase

E. Recombinant human parathyroid hormone

[12] Which of the following can be used to treat type I Gaucher's disease?________

A. Thymidine kinase(TK)

B. Matrix metalloproteinase(MMP)

C. Glucocerebrosidase(GC)

D. Acetyl choline esterase (AChE)

E. DNA methyl transferase I

[13] What is the first drug approved by the US FDA for the clinical treatment of thrombosis?________

A. Dornasealfa B. Recombinant tissue-type plasminogen activator

C. Factor VIII D. Glucocerebrosidase E. Acetyl choline esterase

[14] Which of the following statement about the genetic engineering subunit vaccine is incorrect?________

A. Immunogenicity of subunit vaccines can be enhanced by adding adjuvants with immune-enhancing effects.

B. Subunit vaccine is not a complete pathogen

C. Subunit vaccine contains viral nucleic acid

D. Very safe even in people with compromised immune systems

E. Subunit vaccines lack viral nucleic acid

[15] Which of the following does not belong to the vaccine?________

A. Subunit vaccine B. DNA vaccine C. Synthetic peptide

D. Human immunoglobulin preparation E. Hepatitis B vaccine

[16] Which of the following is heralded as the third generation of vaccine?________

A. Subunit vaccine B. Synthetic peptide vaccine C. Nucleic acid vaccine

D. Genetic engineering vector vaccine E. Killed vaccine

[X Type]

[17] Which of the following drugs can be obtained by genetic engineering methods?________

A. Recombinant cell factor B. Peptide hormone C. Protease inhibitor

D. Recombinant thrombolytic agent E. Penicillin

[18] Which of the following cytokines can be used to treat neutropenia in cancer patients?________

A. IL-2 B. G-CSF C. GM-CSF

D. IFN-α E. EPO

[19] Cytokines play an important role in cell signaling and can be produced by which of the following cells?

____

A. Immune cells B. Mast cells C. Endothelial cells

D. Fibroblasts E. Plasma cells

[20] Acute myocardial infarction and cerebral infarction are serious threats to human health, what kind of medicine can be used to treat them?________

A. Recombinant thrombolytic agents B. Interferon C. Anticoagulant drugs

D. Interleukin E. Colony-stimulating factor

[21] What are the methods for obtaining live attenuated vaccines?________

A. Screening attenuated strains from nature

B. Attenuated by the passage of heterologous animals or cells

C. Attenuated by physical and chemical methods

D. Recombination of pathogens using genetic engineering techniques

E. Killing the intact virus or bacteria that are infectious

[22] The advantages of genetically engineered vaccines do not include__________.

A. low production cost

B. low security

C. can be made as multivalent vaccine

D. complete pathogen

E. easy to store and transport

[23] Genetically engineered vaccines include __________.

A. subunit vaccine

B. genetic engineering vector vaccine

C. nucleic acid vaccine

D. synthetic peptide vaccine

E. recombinant cytokine

[24] Which of the following vectors can be used for genetic engineering vector vaccines?________

A. Poxvirus B. Adenovirus C. BCG

D. Attenuated Salmonella typhimurium E. Untreated and highly pathogenic strain

[25] Which of the following statements about nucleic acid vaccines are correct?________

A. Nucleic acid vaccines are divided into DNA vaccines and RNA vaccines.

B. Nucleic acid vaccines affect not only humoral immunity but also cellular immunity.

C. Nucleic acid vaccination can induce the body to produce CTL immune response.

D. Nucleic acid vaccines are absolutely safe.

E. The emergence and development of nucleic acid vaccines is the second revolution in the history of vaccine development.

[26] What are the reasons for the poor effect of synthetic peptide vaccines?________

A. Lack of sufficient immunogenicity

B. It is difficult to synergize B and T cell epitopes.

C. Lack of sufficient stimulation of B cell antigen epitopes

D. Lack of viral nucleic acid

E. It can induce multiple immune responses in the body like protein antigens.

Answers

1	2	3	4	5	6	7	8	9	10
B	A	D	C	A	C	A	B	D	B
11	12	13	14	15	16	17	18	19	20
D	C	B	C	D	C	ABCD	BC	ABCD	AC
21	22	23	24	25	26				
ABCD	BD	ABCD	ABCD	ABC	ABC				

(*Yang Guang*, *Sun Wei*)